AF323292

FLORIDA STATE
UNIVERSITY LIBRARIES

SEP 5 2000

TALLAHASSEE, FLORIDA

BIOMASS BURNING AND ITS INTER-RELATIONSHIPS WITH THE CLIMATE SYSTEM

ADVANCES IN GLOBAL CHANGE RESEARCH

VOLUME 3

Editor-in-Chief

Martin Beniston, *Institute of Geography, University of Fribourg, Perolles, Switzerland*

Editorial Advisory Board

B. Allen-Diaz, *Department ESPM-Ecosystem Sciences, University of California, Berkeley, CA, U.S.A.*

R.S. Bradley, *Department of Geosciences, University of Massachusetts, Amherst, MA, U.S.A.*

W. Cramer, *Department of Global Change and Natural Systems, Potsdam Institute for Climate Impact Research, Potsdam, Germany.*

H.F. Diaz, *NOAA/ERL/CDC, Boulder, CO, U.S.A.*

S. Erkman, *Institute for Communication and Analysis of Science and Technology – ICAST, Geneva, Switzerland.*

M. Lal, *Centre for Atmospheric Sciences, Indian Institute of Technology, New Delhi, India.*

M.M. Verstraete, *Space Applications Institute, EC Joint Research Centre, Ispra (VA), Italy.*

The titles in this series are listed at the end of this volume.

BIOMASS BURNING AND ITS INTER-RELATIONSHIPS WITH THE CLIMATE SYSTEM

Edited by

John L. Innes
*University of British Columbia,
Vancouver, BC, Canada*

Martin Beniston
University of Fribourg, Switzerland

and

Michel M. Verstraete
*Joint Research Center,
Ispra, Italy*

KLUWER ACADEMIC PUBLISHERS
DORDRECHT / BOSTON / LONDON

Library of Congress Cataloging-in-Publication Data

SCI
TD
195
.B56
B55
2000

ISBN 0-7923-6107-5

Published by Kluwer Academic Publishers,
P.O. Box 17, 3300 AA Dordrecht, The Netherlands.

Sold and distributed in North, Central and South America
by Kluwer Academic Publishers,
101 Philip Drive, Norwell, MA 02061, U.S.A.

In all other countries, sold and distributed
by Kluwer Academic Publishers,
P.O. Box 322, 3300 AH Dordrecht, The Netherlands.

Cover: burnt down Alaskan black spruce forest
(photo by Eric Kasischke)

Printed on acid-free paper

All Rights Reserved
© 2000 Kluwer Academic Publishers
No part of the material protected by this copyright notice may be reproduced or
utilized in any form or by any means, electronic or mechanical,
including photocopying, recording or by any information storage and
retrieval system, without written permission from the copyright owner.

Printed in the Netherlands.

Table of contents

List of contributors

MARTIN E. ALEXANDER, Canadian Forest Service, Edmonton, Alberta, Canada.

JOHN C. G. BANKS, Department of Forestry, Australian National University, Canberra, Australia.

BRYAN A. BAUM, Langley Research Center, Hampton, VA USA.

MARTIN BENISTON, Institute of Geography, University of Fribourg, Switzerland.

B. BONSAL, Climate Research Branch, Atmospheric Environment Service, Toronto, Ontario Canada.

LAURA L. BOURGEAU-CHAVEZ, ERIM International, P.O. Box 134008, Ann Arbor, USA.

GUY BRASSEUR, National Center for Atmospheric Research, Boulder, CO, USA.

PIETRO ALESSANDRO BRIVIO, Remote Sensing Dept.-IRRS, National Research Council, Milan, Italy.

DONALD R. CAHOON, Jr., NASA-Langley Research Center, 21 Langley Boulevard, Hampton, USA.

GEOFFREY J. CARY, Department of Forestry, Australian National University, Canberra, Australia.

MARCO CONEDERA, FNP Sottostazione Sud delle Alpi, Bellinzona, Switzerland.

EDWARD DWYER, Global Vegetation Monitoring Unit, Space Applications Institute, Joint Research Centre, European Commission, Ispra, Italy.

SUE A. FERGUSON, Forestry Sciences Laboratory, 4043 Roosevelt Way NE, Seattle, USA.

RICHARD W. FISHER, Air & Watershed Management, USDA, Forest Service, Fort Collins, USA.

DOUGLAS G. FOX, Cooperative Institute for Research on the Atmosphere, Colorado State University, Fort Collins, Colorado, USA.

NANCY H.F. FRENCH, ERIM International, P.O. Box 134008, Ann Arbor, USA.

JOHANN G. GOLDAMMER, Fire Ecology Research Group, University of Freiburg, Germany.

CLAIRE GRANIER, Service d'Aéronomie CNRS, Paris, France. CIRES, University of Colorado, Boulder, CO, USA. NOAA Aeronomy Laboratory, Boulder, CO, USA.

JEAN-MARIE GRÉGOIRE, Global Vegetation Monitoring Unit, Space Applications Institute, Joint Research Centre, European Commission, Ispra, Italy.

D.A. HAUGLUSTAINE, Service d'Aéronomie du CNRS, Université de Paris 6, Paris, France.

JOHN L. INNES, University of British Columbia, Vancouver, Canada.

ERIC S. KASISCHKE, ERIM International, P.O. Box 134008, Ann Arbor, USA.

MURARI LAL, Centre for Atmospheric Sciences, Indian Institute of Technology, New Delhi, India.

CHRIS LARSEN, Department of Geography, State University of New York, University at Buffalo, Buffalo, USA.

JOEL S. LEVINE, Atmospheric Sciences Division - NASA Langley Research Center - Hampton, Virginia, USA.

D.L. MARTELL, Faculty of Forestry, University of Toronto,Toronto, Ontario Canada.

JEAN-FRANÇOIS MÜLLER, Belgian Institute for Space Aeronomy, Brussels, Belgium.

KATHY O'NEILL, Nicholas School of the Environment, Duke University, Durham, USA.

ROGER OTTMAR, Forestry Sciences Laboratory, 4043 Roosevelt Way NE, Seattle, USA.

JOSÉ M.C. PEREIRA, Laboratory for Remote Sensing and Geographical Analysis, Department of Forestry, Instituto Superior de Agronomia, Lisbon, Portugal. Centre for Forest Studies, Department of Forestry, Instituto Superior de Agronomia, Lisbon, Portugal.

ALLEN R. RIEBAU, Wildlife, Fisheries, Watershed and Air Research, USDA, Forest Service, Washington, USA.

DAVID V. SANDBERG, Forestry Sciences Laboratory, 4043 Roosevelt Way NE, Seattle, USA.

A. SHABBAR, Climate Research Branch, Atmospheric Environment Service, Toronto, Ontario Canada.

ROMA SINGH, Centre for Atmospheric Sciences, Indian Institute of Technology, New Delhi, India.

W.R. SKINNER, Climate Research Branch, Atmospheric Environment Service, Toronto, Ontario Canada.

ADÉLIA M. SOUSA, Laboratory for Remote Sensing and Geographical Analysis, Department of Forestry, Instituto Superior de Agronomia, Lisbon, Portugal.

BRIAN J. STOCKS, Canadian Forest Service, Sault Ste. Marie, Ontario, Canada.

DANIELA STROPPIANA, Global Vegetation Monitoring Unit, Space Applications Institute, JRC, Ispra (VA), Italy.

WILLY TINNER, Geobotanisches Institut der Universität Bern, Bern, Switzerland.

MARIA J.P. VASCONCELOS, Centre for Forest Studies, Department of Forestry, Instituto Superior de Agronomia, Lisbon, Portugal. National Centre for Geographical Information (CNIG), Lisbon, Portugal.

MICHEL VERSTRAETE, Joint Research Center, CEC, Ispra, Italy.

MARTIN WILD, Swiss Federal Institute of Technology, Department of Geography, Zurich, Switzerland.

Acknowledgements

We would like to convey our appreciation to all contributing authors for their dedication and effort which has led to this book. We would also like to thank the reviewers for their comments and constructive criticisms.

The financial support of ENAMORS (European network for the development of advanced models to interpret optical remote sensing data) and the Swiss National Science Foundation is gratefully acknowledged. This funding enabled several contributors to attend the Workshop on Biomass Burning and its Inter-relationships with the Climate System, which was held in September 1998 in Wengen (Bernese Alps, Switzerland). The papers in this book are based on material presented at the Wengen-98 Workshop.

We would also like to give special thanks to Ms. Sylvie Bovel-Yerly who finalized the manuscripts in their camera-ready form, in an efficient and timely manner.

Biomass Burning and Climate: An Introduction

JOHN L. INNES
University of British Columbia, Vancouver, Canada

The interactions between biomass burning and climate have been brought into focus by a number of recent events. Firstly, the Framework Convention on Climate Change and, more recently, the Kyoto Protocol, have drawn the attention of policy makers and others to the importance of biomass burning in relation to atmospheric carbon dioxide concentrations. Secondly, the use of prescribed fires has become a major management tool in some countries; with for example the area with fuel treatments (which include prescribed burns and mechanical treatments) having increased on US National Forest System lands from 123,000 ha in 1985 to 677,000 ha in 1998. Thirdly, large numbers of forest fires in Indonesia, Brazil, Australia and elsewhere in 1997 and 1998 received unprecedented media attention. Consequently, it is appropriate that one of the Wengen Workshops on Global Change Research be devoted to the relationships between biomass burning and climate. This volume includes many of the papers presented at the workshop, but is also intended to act as a contribution to the state of knowledge on the inter-relationships between biomass burning and climate change. Previous volumes on biomass burning (e.g. Goldammer 1990, Levine 1991a, Crutzen and Goldammer 1993, Levine 1996a, 1996b, Van Wilgen *et al.* 1997) have stressed various aspects of the biomass–climate issue, and provide a history of the development of our understanding of the many complex relationships that are involved.

Despite the increase in research in biomass burning and climate interactions during the 1990s, the current estimates of biomass consumption by fire remain the same as in 1990. Andreae (1991) estimated that the total biomass consumed by annual burning was on average ca. 8680 teragrams (Tg) (1 teragram = 1 million tons = 10^{12}g) of dry material, with 3690 Tg coming from savanna burning, 2020 Tg from agricultural wastes, 1540 Tg from forests and 1430 Tg from the burning of woodfuels. In total, this results in the release of 3500 Tg of carbon in the form of CO_2 annually. The figure for forests was probably considerably higher in the 1997 and 1998, but no global figures have as yet been published. This compares with an annual production of carbon from fossil fuels and cement production of 6200 ± 600 Tg C (IPCC 1996).

1.   FOREST FIRES AND OTHER FORMS OF BIOMASS BURNING

Biomass burning is a complicated issue as it occurs as a result of both natural and anthropogenic processes. Forest fires, for example, can be started either by lightning or by humans. Increasingly, forest fires are being recognised as a natural feature of many forests, with Oliver and Larson (1996) considering fires to be the most important form of disturbance to forests. However, it is often very difficult to discern which of today's fire regimes are the result of wholly natural processes (see Skinner *et al.*, this volume), and which are the result of changes brought about by humans (see Cary and Banks, this volume; Conadera and Tinner, this volume). In addition, future climate change, whether natural or human-induced, will change fire regimes in many parts of the world (papers by Cary and Banks, Dwyer *et al.*, this volume). Will such changed regimes be considered as natural or anthropogenic? This is an important question in relation to carbon accounting, which currently places great emphasis on "human-induced" changes.

A continuum exists over which the role of fire differs markedly. In some forests, fire is an exceptional occurrence that will have serious, long-term consequences for the forest (Kauffman and Uhl 1990). Most rain forests (both tropical and temperate) fall into this category. Rain forests can burn, as amply demonstrated by the forests of Amazonia, Indonesia and elsewhere (see Levine, this volume, for details of the 1997 fires in Indonesia). The fires are not only in disturbed rain forests, although there is evidence that selectively logged forests have higher flammability levels than undisturbed forests (Uhl and Kauffman 1990, Holdsworth and Uhl 1997), probably because of increases in the volume of woody debris and fine fuels and changes to the forest floor microclimate. In both disturbed and undisturbed forests experiencing several years of low rainfall, fire susceptibility increases (Nepstad *et al.* 1994, 1995). Temperate rainforests can also be burned, with many of the typical species being very fire sensitive. For example, both Sitka spruce (*Picea sitchensis*) and western redcedar (*Thuja plicata*) are considered to have low tolerance to fire (Minore 1979, Agee 1993).

In other forest types, various degrees of fire are required to maintain ecosystem integrity. In many forests, the regular occurrence of relatively low intensity ground fires is necessary for the continued health of the forest. Without such fires, fuel loads increase and when a fire does occur, it rapidly reaches the crown layer, resulting in the destruction of the forest. Such forests are often characteristic of the temperate zone. In Europe, forest fire is not considered to be a widespread natural phenomenon, but evidence from the broadleaf and coniferous temperate forests on North America suggests that fire in such forests may be a relatively frequent disturbance (Harmon 1982, van Lear and Waldrop 1989). Fire in European forests is generally thought of as a hazard, reflecting both the intensive management of the forests and the close juxtaposition of forests to human settlements. Generally in Europe (and North America), the less populated an area is, the more acceptable fire is as an ecological process.

At the opposite extreme to rain forests, some forest tree species, such as jack pine (*Pinus banksiana*) and lodgepole pine (*Pinus contorta*), are adapted to fire, and require it to regenerate adequately. In such forests, stand-destroying fires occur periodically, resulting in a mosaic landscape consisting of even-aged stands at various stages of development. Much of the boreal forest zone is considered to be fire-adapted, with rapid regeneration of the forests following severe fires. Large-scale fires, involving over 100,000 ha, are fairly frequent in the boreal zone, with a return interval of about 100 years (Johnson 1992). One such fire burned in 1950 in northern British Columbia and Alberta, Canada, destroying 1.4 million ha of forest. Another burned from 6 May to 2 June 1987, destroyed 870 000 ha of forest in the Daxinganling Mountains of China (Goldammer and Di 1990). Goldammer and Di (1990) estimate that 1.5 million ha of forest in neighbouring parts of Russia was burnt in the same year. In 1915, an estimated 14 million ha of forest in Siberia was burnt (Shostakovitch 1925). The impacts of these fires depends on their intensity and the tree species being burnt, although it is clear that their impacts on global climate (e.g. Conard and Ivanova 1997) and the global carbon cycle (Kasischke *et al.*, papers in this volume) are significant.

Forest fires are not the only form of natural biomass burning, although the extent to which some of the others are natural or human-induced is unclear. Savanna fires are important over large areas of the semi-humid tropics. Many of these fires originate from lightning strikes, but humans probably set a greater proportion. The areas involved are enormous – Hao *et al.* (1990) estimated that 750 million ha of savannas are burned annually. This is one to two orders of magnitude more than the area of boreal forest that is burned annually. In the absence of deliberate burning, the tree cover in the landscape would probably increase, reducing the susceptibility of the landscapes to fire. Savanna landscapes are highly dynamic, with the balance between trees and grass being determined by plant available moisture, plant available nutrients, fire, herbivory and major anthropogenic events (Stott 1994). Fires in savannas and their associated emissions have been the subject of several large-scale investigations, such as the Fire of Savannas/Dynamique et Chimie Atmosphérique en Forêt Équatoriale (FOS/DECAF) project (e.g. Cachier *et al.* 1991, Lacaux *et al.* 1993), the South Africa Fire–Atmosphere Research Initiative (SAFARI–92) (e.g. Shea *et al.* 1996, Trollope and Trollope 1996) and the NASA Transport and Atmospheric Chemistry Near the Equator—Atlantic (TRACE A) mission (e.g. Fishman *et al.* 1996). Both SAFARI–2 and TRACE A were part of the South Tropical Atlantic Regional Experiment (STARE). These studies have demonstrated the importance of savanna fires in the global carbon budget and the generation of a number of different greenhouse gases.

2.   ANTHROPOGENIC BIOMASS BURNING

Anthropogenic biomass burning is far more important than natural fires, which are responsible for only 10% of total biomass burning (Levine, this volume). Human-initiated fires can take many forms, including the burning of agricultural wastes, domestic fuelwood, biofuels, prescribed burning and destructive burning. Emphasis in this book has concentrated on forest fires (both natural and anthropogenic), with only a single paper devoted to industrial biofuels and domestic fuels. While much attention has focussed on the burning of forests, savannas, grasslands and agricultural land, it is important to realise that of the 3,358 million m^3 of wood used each year throughout the world, 55% is utilised as fuelwood and charcoal (FAO 1999). In some developing countries, biomass accounts for more than 90% of their energy supply (Amous 1998, Denman 1998). Consequently, this book does not claim to give equal emphasis to the different issues involved in climate– biomass burning relationships.

Biofuels are an important element in plans to reduce global carbon dioxide emissions. The creation of plantations will initially represent a sink for atmospheric carbon. However, this will only be a short-term phenomenon. More importantly, the replacement of fossil fuels by renewable sources of energy could result in significant long-term reductions in carbon dioxide emissions (Johansson *et al.* 1993, Lal and Singh, this volume). Biofuels are already important in some countries: more than 17% of the energy use in Finland is derived from biofuels (Pingoud *et al.* 1999) and globally, biomass energy is thought to account for about 15% of the global energy supply (Davidson 1998). However, the value of many biofuels is still limited because of the very low conversion efficiencies, mostly of the order of 15–20%, although there is considerable potential for improvements in these figures (e.g. Shukla 1998). Despite the problems associated with conversion efficiencies, plans exist to increase the proportion of energy generated from biomass in European countries from the current average of 6% to 12% by the year 2010 (Robertson 1998).

Demand for woodfuel in some areas far exceeds the mean annual increment, resulting in progressive deforestation. This has been demonstrated by Horta Nogueira and Trossero (1998), who undertook a detailed analysis of global woodfuel[1] demands. In Table 1, the woodfuel demands of a number of countries are shown.

Table 1. Countries with very high specific woodfuel demands.
From: Nogueira and Trossero (1998).

Country	Population 1000 inhabitants	Specific woodfuel demand m^3 per ha forest
Haiti	7180	439.2
Lesotho	2050	118.2
El Salvador	5768	78.7
Egypt	62931	75.7
Pakistan	140497	29.8
Kenya	28261	25.9
Rwanda	7952	21.6
Yemen	14501	18.3
Burundi	6393	15.2
Bangladesh	120433	14.7

Alone, these figures are difficult to interpret. Evans (1992) provides some information on the productivity of plantation forests in different parts of the Tropics (Table 2). As can be seen, the annual productivity of all but the fastest growing plantations is well below the estimated demands for woodfuels on a per hectare basis in some countries. As a result, substantial afforestation with fast-growing plantations will be required in countries such as Haiti and Lesotho simply to meet woodfuel demands.

With increasing global concern about deforestation, there has been a tendency for the use of selective logging, particularly in tropical forests. This is seen as less damaging that the complete removal of the forest cover. The available information suggests that selective logging will continue and may even be more widespread in future (e.g. Veríssimo *et al.* 1992). There is ample evidence that partially logged forests are more susceptible to fire than undisturbed forests. Fire often spreads into such forests from adjacent land, and may become a regular occurrence, resulting in the progressive deterioration of the forest (Cochrane and Schulze 1999). Consequently, any move towards selective logging as a means of sustainable forest

---

[1] The term woodfuel includes fuelwood (wood that is burnt directly without conversion to any other product), charcoal and black liquor (a by-product of the chemical pulp industry).

management will need to consider the likely impact of fires spreading into the forest from surrounding agricultural areas.

Fire is also deliberately introduced into many forest ecosystems as a management tool. For example, prescribed fire is used in the southeast USA to manage the 54 million ha. loblolly pine (*Pinus taeda*)–shortleaf pine (*Pinus echinata*)–hardwood forest type (Haywood *et al.* 1998). Fire is actually seen as a prerequisite for healthy forests, yet there is the potential for conflict not only with agencies responsible for pollution abatement (see the paper by Fox *et al.*, this book) but also with concerns about the release of carbon dioxide and other greenhouse gases into the atmosphere (see Ferguson *et al*, this volume). Consequently, there are a number of issues associated with prescribed fire that will require attention in the near future.

Table 2. Average growth rates attained in some tropical plantations. From Evans 1992.

Plantation development	Species	Mean annual increment ($m^3\ ha^{-1}\ yr^{-1}$)	Rotation period (years)
Usutu Forest, Swaziland	*Pinus patula*	19	15–17
Viphya Pulpwood Project, Malawi	*Pinus patula*	18	16
Fiji Pine Commission	*Pinus caribaea*	15–20	17–20
Jari Florestal, Brazil	*Pinus caribaea*	20	16
UAIC[1], Congo	*Eucalyptus* hybrids	35	7
Aracruz Florestal, Brazil	*Eucalyptus grandis*	55	7
Shiselweni Forestry, Swaziland	*Eucalyptus grandis*	18	9
PICOP[2], Philippines	*Paraserianthes falcataria*	28	10
Sabah Softwoods, Malaysia	*Paraserianthes falcataria*	30	7–10
Jari Florestal, Brazil	*Gmelina arborea*	15–25	10
CNGT[3], Papua New Guinea	*Araucaria* spp.	20	40
Seaqaqa plantations, Fiji	*Swietenia macrophylla*	14	30

[1]Paper Industries Corporation of the Philippines

[2]Commonwealth New Guinea Timbers

[3]Unite d'Afforestation Industrielle du Congo

3. IMPACTS ON THE CLIMATE SYSTEM

Considerable uncertainty surrounds the calculation of emissions and emission factors. Global calculations of the amounts of greenhouse gases emitted during biomass burning must necessarily make a number of assumptions about the quantities of fuel burnt and the emissions from these fuels. However, this is an extremely complex issue, in which there is the possibility for considerable error (see Ferguson *et al.*, this volume). For example, in prescribed burning, it is recognised that fire behaviour is controlled by a range of factors, including the type, amount and size distribution of the fuel, its moisture content, the microclimate of the site and the techniques used in the burning (e.g. head fires versus backing fires (Clinton *et al.* 1998).

Climate scenarios are being developed using General Circulation Models. These are based on a number of assumptions, and Wild (this volume) illustrates some of the uncertainty that may be introduced as a result of a failure to take into account the impacts of absorbing aerosols released during biomass burning. Recognition of the role of these aerosols could help to produce a new generation of climatic scenarios.

4. REGULATION OF CARBON EMISSIONS

Burning biomass produces smoke and a number of other pollutants. Consequently, problems may arise over atmospheric pollution. Biomass burnt domestically can cause a number of health problems (Smith 1987, 1996, Gupta *et al.* 1998) because of the release of carbon monoxide, methane, nitrogen oxides, benzene, formaldehyde, benzo(a)pyrene, aromatics and respirable particulate matter. These health problems are important given that 75% of the world's population is dependent on domestic biofuels (Robertson 1998). Biomass burning can also result in significant atmospheric pollution, some of which may be transported very long distances, as demonstrated by the SAFARI–2 and TRACE A experiments. The US Clean Air Act contains specific clauses related to fine particulates and other products generated by biomass fires and, as explained in the chapter by Fox *et al.*, there is potential for considerable conflict between land managers and air quality regulators.

Land use change and forestry issues were highly controversial issues in the negotiations leading to the Kyoto Protocol. Even now, many uncertainties remain and these have only just begun to be addressed. For example, Article 3.3 states that human-induced land-use change and forestry activities, limited to afforestation, reforestation and deforestation, shall be used to meet the commitments. Afforestation, reforestation and deforestation are still inadequately defined, and the text appears to exclude prescribed burning (which does not result in loss of forests) while including deliberate fires that result in deforestation. Even this is unclear. If a wild fire is left to burn, as is currently the policy in some parts of the boreal zone, does this represent human-induced deforestation? The Protocol contains the possibility of clarifying these issues, and also adding other elements related to forestry and land use change in Article 3.4, which states that Parties "...shall, at its first session or as soon as practicable thereafter, decide upon modalities, rules and guidelines as to how, and which, additional human-induced activities [besides afforestation, reforestation, and deforestation] related to changes in greenhouse gas emissions and removals by sinks in the agricultural soil and land-use change and forestry categories shall be added to, or subtracted from, the assigned amount for the Parties included in Annex I."

One of the major issues associated with the inclusion of forestry in the Kyoto Protocol is the verification of sinks and sources (Schlamadinger and Marland 1998, Lim *et al.* 1999). Remote sensing provides a useful way in which large-area estimates of burned areas can be obtained. The problems and successes associated with the use of remote sensing are described in this volume in papers by Cahoon *et al.*, Pereira *et al.*, and Stroppiana *et al.*. Other techniques also exist (see Larsen, this volume), and offer considerable potential. However, there are many issues to be resolved before many of these methods can be applied to issues such as carbon accounting.

5. CONFLICTS BETWEEN BIOMASS EMISSIONS AND OTHER RESOURCE MANAGEMENT ISSUES

An interesting phenomenon, the biological effects of which have not been studied, is the occurrence of elevated tropospheric ozone occurring as a result of precursors generated by biomass burning. Elevated ozone concentrations have been documented over Brazil (Kirchhoff 1996, Kirchhoff *et al.* 1996), the South Atlantic Ocean (Fishman *et al.* 1996, Gregory *et al.* 1996) and extend into west Africa (Granier *et al.*, this volume; Hauglustaine *et al.*, this volume). However, to date, there have been no studies of the effects of this ozone on plant communities, despite intensive interest in the biological impacts of ozone in temperate areas.

An increase in the use of woodfuels suggests that demand for forest products will increase. At the same time, there is growing pressure to ensure that the forests of the world are sustainably managed. In some cases, it may be difficult to achieve both these aims simultaneously (Lal and Singh, this volume). For example, there is growing evidence that biomass use is becoming unsustainable in some parts of India (Dwivedi and Kaul 1997, Ramana and Joshi 1997, Shukla 1998). Such potential conflicts pose a major challenge for resource managers attempting to balance local demands for forest products with the need to ensure that the national and international demands for sustainable forest management are being met.

6. CONCLUSIONS

Considerable progress has been made during the 1990s. In 1991, Andrasko *et al.* (1991) identified a number of different options for reducing the atmospheric impacts of biomass burning. They suggested that there were two possible strategies:

1. Reducing the frequency, area and amount of biomass burned in natural and intentional fires. Three options were available to achieve this:
   - Minimising conversion of forest and savanna to other land uses, especially agriculture and pasture, through the use of fire
   - Reducing and improving the efficiency of fire as a land management tool on forest, grassland and agricultural lands.
   - Substituting sustainable land management practices and systems with minimal use of fire for fire-dependent or fire-proclivity systems.
2. Increasing the fraction of primary energy supplied by biomass derived from natural systems, and the efficiency of its utilisation. They identified two possible strategies to do this:
   - Replacing unsustainably harvested biomass with sustainably managed (plantation) biomass as a fuel stock
   - Increasing energy extraction efficiencies for biomass conversion.

Hao *et al.* (1991) also identified a number of specific strategies that might be followed for specific land use types:

1. Croplands
   - Incorporate crop residues into soils instead of burning
   - Burn residues as fuel in household energy systems
   - Replace annual or seasonal crops with tree crops
2. Grassland strategies
   - Increase grassland management to reduce fire frequency and area
   - Substitute game ranching for domestic livestock systems
   - Introduce silvipastoral systems to provide fodder
3. Forest strategies
   - Intensify forest management to reduce wildfires
   - Lengthen rotation times and improve productivity of shifting agriculture
   - Increase productivity of existing agricultural lands
   - Incorporate charcoal into the soil after burning
   - Clear-fell forest before or instead of burning.

Since 1991, we have had the Framework Convention on Climate Change and the Kyoto Protocol. Reductions in carbon dioxide emissions have become a serious policy issue, with major implications for the future management of natural resources throughout the world. In addition, we have had a number of other major policy initiatives, such as the Convention on Biological Diversity and Agenda 21. Taken together, these initiatives provide a framework within which resource management has to be based. The have encouraged resource managers to consider the broader context of their actions. Burning a forest in Brazil or Indonesia not only has local consequences: its effects extend globally. The contributions to this book identify some of the mechanisms behind these consequences and provide details of the difficulties involved in establishing the empirical relationships that are necessary for modelling studies of future climate development.

7.   REFERENCES

Agee, J.K. 1993. *Fire ecology of Pacific Northwest forests*. Island Press, Wahington, DC.

Amous, S. 1998. Biomass data issues and challenges in northern and western Africa. In: Organisation for Economic Co-operation and Development (ed.) Biomass energy: data, analysis and trends. OECD, Paris, pp. 61–65.

Andrasko, K.J., Ahuja, D.R., Winnett, S.M., and Tirpak, D.A. (1991). Policy options for managing biomass burning to mitigate global climate change. In: Levine, J.S. (ed.) *Global biomass burning. Atmospheric, climatic, and biospheric implications*. Mit Press, Cambridge, 445–456.

Andreae, M.O. 1991. Biomass burning: Its history, use, and distribution and its impact on environmental quality and global climate. In: Levine, J.S. (ed.) *Global biomass burning. Atmospheric, climatic, and biospheric implications*. MIT Press, Cambridge, pp. 3–21.

Cachier, H., Ducret, J., Brémond, M.-P., Yoboué, V., Lacaux, J.-P., Gaudichet, A. and Baudet, J. 1991. Biomass burning in a savanna region of the Ivory Coast. In: Levine, J.S. (ed.) *Global biomass burning. Atmospheric, climatic, and biospheric implications*. Mit Press, Cambridge, pp. 174–180.

Clinton, B.D., Vose, J.M., Swank, W.T., Berg, E.C. and Loftis, D.L. 1998. *Fuel consumption and fire characteristics during understory burning in a mixed white pine–hardwood stand in the Southern Appalachians*. US Department of Agriculture Forest Service, Research Paper SRS-12. Southern Research Station, Asheville, N.C.

Cochrane, M.A. and Schulze, M.D. 1999. Fire as a recurrent event in tropical forests of the eastern Amazon: Effects on forest structure, biomass, and species composition. *Biotropica* 31, 2–16.

Conard, S.G. and Ivanova, G.A. 1997. Wildfire in Russian boreal forests – Potential impacts of fire regime characteristics on emissions and global carbon balance estimates. *Environmental Pollution* 98, 305–313.

Crutzen, P.J. and Goldammer, J.G. (eds.) 1993. *Fire in the environment: The ecological, atmospheric, and climatic importance of vegetation fires*. John Wiley, New York.

Davidson, O.R. 1998. Conclusions of last year's workshop: a starting point. In: Organisation for Economic Co-operation and Development (ed.) Biomass energy: data, analysis and trends. OECD, Paris, pp. 13–16.

Denman, J. 1998. IEA biomass energy data: system, methodology and initial results. In: Organisation for Economic Co-operation and Development (ed.) Biomass energy: data, analysis and trends. OECD, Paris, pp. 19–37.

Dwivedi, B.N. and Kaul, O.N. 1997. Forest as biomass energy source in India. In: Ramana, P.V. and Srinivas, S.N. (eds.) Biomass energy systems. British Council and Tata Energy Research Institute, New Delhi, pp. 3–18.

FAO (1999). *State of the World's Forests 1999*. Food and Agriculture Organization of the United Nations, Rome. 154 pp.

Fishman, J., Hoell, J.M., Bendura, R.D., McNeal, R.J. and Kirchhoff, V.W.J.H. 1996. NASA GTE TRACE A Experiment (September–October 1992): Overview. *Journal of Geophysical Research* 101, D19, 23865–23879.

Goldammer, J.G. (ed.) *Fire in the tropical biota: Ecosystem processes and global challenges.* Springer Verlag, New York.

Goldammer JG, Di X (1990) Fire and forest development in the Daxinganling Montane-Boreal coniferous forest, Heilongjiang, Northeast China - a preliminary model. In: Goldammer JG, *Jenkins* MJ (eds) Fire in Ecosystem Dynamics. Mediterranean and northern perspectives. SPB Academic Publishing, The Hague, pp 175-184

Gregory, G.L., Fuelberg, H.E., Longmore, S.P., Anderson, B.E., Collins, J.E. and Blake, D.R. 1996. Chemical characteristics of tropospheric air over the tropical South Atlantic Ocean: Relationship to trajectory history. *Journal of Geophysical Research* 101, D19, 23957–23972.

Gupta, S., Saksena, S., Shankar, V.R., and Josni, V. 1998. Emission factors and thermal efficiencies of cooking biofuels from five countries. *Biomass and Bioenergy* 14, 547–560.

Hao, W.M., Scharffe, D., Lobert, J.M., and Crutzen P.J. 1990. Biomass burning: An important source of atmospheric CO, CO_2, and hydrocarbons. Paper presented at the Chapman Conference on Global Biomass Burning: Atmospheric, Climatic, and Biospheric implications. Williamsburg, Va., 19–23 March.

Harmon, M.E. 1982. Fire history of the westernmost portion of great Smoky Mountains National Park. *Bulletin of the Torrey Botanical Club* 109, 74–79.

Haywood, J.D., Martin, A., Pearson, H.A. and Grelen, H.E. 1998. *Seasonal biennial burning and woody plant control influence native vegetation in loblolly pine stands.* US Department of Agriculture, Forest Service, Research Paper SRS-14. Southern Research Station, Asheville, NC.

Holdsworth, A.R. and Uhl, C. 1997. Fire in eastern Amazonian logged rain forest and the potential for fire reduction. *Ecological Applications* 7, 713–725.

Horta Nogueira, L. and Trossero, M.A. 1998. Introducing WEIS: the FAO wood energy information system. In: Organisation for Economic Co-operation and Development (ed.) *Biomass energy: data, analysis and trends*. OECD, Paris, pp. 115–139.

IPCC 1996. *Climate change 1995: The science of climate change*. Contribution of WGI to the Second Assessment Report of the Intergovernmental Pane on Climate Change. Edited by Houghton, J.T., Meiro Filho, L.G., Callander, B.A., Harris, N., Kattenberg, A. and Maskell, K. Cambridge University press, Cambridge.

Johansson, T.B., Kelly, H., Reddy, A.K.N. and Williams, R.H. (eds.) 1993. *Renewable energy sources for fuels and electricity*. Island Press, Washington DC.

Johnson EA (1992) Fire and vegetation dynamics. Studies from the North American boreal forest. Cambridge University Press, Cambridge

Kauffman, J.B. and Uhl, C. 1990. Interactions of anthropogenic activities, fire, and rain forests in the Amazon Basin. In: Goldammer, J.G. (ed.) *Fire in the tropical biota*. Springer Verlag, New York, pp. 117–134.

Kirchhoff, V.W.J.H. 1996. Increasing concentrations of CO and O3: rising deforestation rates and increasing tropospheric carbon monoxide and ozone in Amazonia. *Environmental Science and Pollution Research* 3, 210–212.

Kirchhoff, V.W.J.H., Alves, J.R. and da Silva, F.R. 1996. Observations of ozone concentrations in the Brazilian cerrado during the TRACE A field expedition. *Journal of Geophysical Research* 101, D19, 24029–24042.

Lascaux, J.-P., Cachier, H. and Delmas, R. 1993. Biomass burning in Africa: An overview of its impact on atmospheric chemistry. In: Crutzen, P.J. and Goldammer, J.G. (eds.) *Fire in the environment: The ecological, atmospheric, and climatic importance of vegetation fires.* John Wiley, New York, pp. 159–191.

Levine, J.S. (ed.) 1991a *Global biomass burning. Atmospheric, climatic, and biospheric implications.* MIT Press, Cambridge

Levine, J.S. 1991b. Global biomass burning: Atmospheric, climatic and biospheric implications. In: Levine, J.S. (ed.) *Global biomass burning. Atmospheric, climatic, and biospheric implications.* MIT Press, Cambridge, xxv–xxx.

Levine, J.S. 1996a. *Biomass burning and global change. Volume 1. Remote sensing, modeling and inventory development, and biomass burning in Africa.* MIT Press, Cambridge.

Levine, J.S. 1996b. *Biomass burning and global change. Volume 2. Biomass burning in South America, Southeast Asia, and temperate and boreal ecosystems, and the oil fires of Kuwait.* MIT Press, Cambridge.

Lim, B., Brown, S. and Schlamadinger, B. 1999. Carbon accounting for forest harvesting and wood products: review and evaluation of different approaches. *Environmental Science and Policy* 2, 207–216.

Minore, D. 1979. *Comparative autecological characteristics of northwestern tree species: a literature review.* US Department of Agriculture Forest Service, General Technical Report PNW-87.

Nepstad, D.C., Jipp, P., Moutinho, P., Negreiros, G. and Vieira, S. 1995. Forest recovery following pasture abandonment in Amazonia: Canopy seasonality, fire resistance and ants. In: Rapport, D., Gander, C. and Calow, P. (eds.) *Evaluating and monitoring the health of large-scale ecosystems.* NATO ASI Series, Springer Verlag, New York, pp. 339–349.

Nepstad, D.C., de Carvalho, C.R., Davidson, E., Jipp, P., Lefebvre, P., Negreiros, G.H., da Silva E.D., Stone, T., Trumbore, S. and Vieira, S. 1994. The role of deep roots in water and carbon cycles of Amazonian forests and pastures. *Nature* 372, 666–669.

Oliver CD, Larson BC (1996) *Forest stand dynamics. Update edition.* John Wiley & Sons, New York.

Pingoud, K., Lehtilä, A. and Savolainen, I. 1999. Bioenergy and the forest industry in Finland after the adoption of the Kyoto protocol. *Environmental Science and Policy* 2, 153–163.

Ramana, P.V. and Joshi, V. 1997. Use of biomass fuels in India – trends and issues. In: Ramana, P.V. (ed.) *Rural and renewable energy: Perspectives from developing countries.* Tata Energy Research Institute, New Delhi, 155–167.

Schlamadinger, B. and Marland, G. 1998. The Kyoto Protocol: provisions and unresolved issues relevant to land-use change and forestry. *Environmental Science and Policy* 1, 313–327.

Shea, R.W., Shea, B.W., Kauffman, J.B., Ward, D.E., Haskins, C.I. and Scholes, M.C. 1996. Fuel biomass and combustion factors associated with fires in savanna ecosystems of South Africa and Zambia. *Journal of Geophysical Research* 101, D19, 23551–23568.

Shostakovitch, V.B. (1925) Forest conflagrations in Siberia. With special reference to the fire of 1915. *Journal of Forestry* 23, 365-371.

Shukla, P.R. 1998. Implications of global and local environment policies on biomass energy demand: a long-term analysis for India. In: Organisation for Economic Co-operation and Development (ed.) *Biomass energy: data, analysis and trends*. OECD, Paris, pp. 267–292.

Smith, K.R. 1987. *Biofuels, air pollution and health: a global review*. Plenum Publishing Corporation, New Delhi.

Smith, K.R. 1996. Indoor air pollution in India. *National Medical Journal of India* 9, 103–104.

Stott, P. 1994. Savanna landscapes and global environmental change. In N. Roberts, ed., *The Changing Global Environment.*, pp. 287–303. Oxford: Basil Blackwell.

Trollope, W.S.W. and Trollope, L.A. 1996. SAFARI–92 characterization of biomass and fire behavior in the small experimental burns in the Kruger National Park. *Journal of Geophysical Research* 101, D19, 23531–23539.

Uhl, C. and Kauffman, J.B. 1990. Deforestation, fire susceptibility, and potential tree responses to fire in eastern Amazonia. *Ecology* 71, 437–449.

Van Lear, D.H. and Waldrop, T.A. 1989. *History, use and effects of fire in the Appalachians*. US Department of Agriculture Forest Service, General Technical Report SE–54, p. 20.

Van Wilgen, B.W., Andreae, M.O., Goldammer, J.G. and Lindesay, J.A. (eds.) 1997. *Fire in southern African savannas: Ecological and atmospheric perspectives*. Witwatersrand University Press, Johannesburg.

Veríssimo, A., Barreto, P., Mattos, M., Tarifa, R. and Uhl, C. 1992. Logging impacts and prospects for sustainable forest management in an old Amazonian frontier: the case of Paragominas. *Forest Ecology and Management* 55, 169–199.

Global Biomass Burning: A Case Study of the Gaseous and Particulate Emissions Released to the Atmosphere During the 1997 Fires in Kalimantan and Sumatra, Indonesia

JOEL S. LEVINE
Atmospheric Sciences Division - NASA Langley Research Center - Hampton, Virginia, USA

Abstract: The roles of biomass burning as a global phenomenon and as a contributor to the global budgets of atmospheric gases and particulates are reviewed. To assess the environmental and health impacts of forest fires, knowledge of the gaseous and particulate emissions produced in the fire and released into the atmosphere is required. Extensive and widespread tropical forest and peat fires swept throughout Kalimantan and Sumatra, Indonesia, in 1997. The fires resulted from burning for land clearing and landuse change. However, the severe drought conditions resulting from El Niño caused small land-clearing fires to become large uncontrolled wildfires. It has been estimated that a total of 45,600 km^2 burned between August and December 1997. The gaseous and particulate emissions resulting from these fires are estimated. The emissions of CO_2, CO, CH_4, NO_x, and particulates from the 1997 Kalimantan and Sumatra fires exceeded the emissions of these species from the Kuwait oil fires of 1991.

1.   BIOMASS BURNING: A GLOBAL PHENOMENON

Biomass burning, the burning of living and dead vegetation for land-clearing and landuse change, has been identified as a significant source of gases and particulates to the regional and global atmosphere (Crutzen et al., 1979; Seiler and Crutzen, 1980; Crutzen and Andreae, 1990; Levine et al., 1995). The bulk of the world's biomass burning occurs in the tropics – in the tropical forests of South America and Southeast Asia and in the savannas of

Africa and South America. The majority of biomass burning (perhaps as much as 90%) is believed to be human-initiated, with natural fires triggered by atmospheric lightning only accounting for about 10% of all fires (Andreae, 1991). Biomass burning is truly a multi-disciplinary subject which includes the following areas: fire ecology, fire measurements and modelling, fire combustion, remote sensing of fires, gaseous and particulate emissions from fires, the atmospheric transport of these emissions and the chemical and climatic impacts of these emissions. Over the last few years, a series of books have documented much of our current understanding of biomass burning. These volumes include: Goldammer (1990), Levine (1991), Crutzen and Goldammer (1993), Goldammer and Furyaev (1996), Levine (1996a), Levine (1996b), and van Wilgen et al. (1997).

Biomass burning is a source of gases and particulates to the atmosphere. Laboratory biomass burning experiments conducted by Lobert *et al.* (1991) have identified the carbon (Table 1) and nitrogen (Table 2) compounds released to the atmosphere by burning.

Table 1. Carbon and gases produced during biomass burning (Lobert *et al.* 1991)

Compound	Mean emission factor relative to the fuel C (%)
Carbon dioxide (CO_2)	82.58
Carbon monoxide (CO)	5.73
Methane (CH_4)	0.424
Ethane (CH_3CH_3)	0.061
Ethene ($CH_2=CH_2$)	0.123
Ethine (CH=CH)	0.056
Propane (C_3H_8)	0.019
Propene (C_3H_6)	0.066
n--butane (C_4H_{10})	0.005
2-butene (cis) (C_4H_8)	0.004
2-butene (trans) (C_4H_8)	0.005
i-butene, i-butene ($C_4H_8 + C_4H_8$)	0.033
1,3-butadiene(C_4H_6)	0.021
n-pentane (C_3H_{12})	0.007
Isoprene (C_5H_8)	0.008
Benzene (C_6H_6)	0.064
Toluene (C_7H_8)	0.037
m-, p-xylene (C_8H_{10})	0.011
o-xylene (C_8H_{10})	0.006
Methyl chloride (CH_3Cl)	0.010
NMHC (As C) (C_2 to C_8)	1.18
Ash (As C)	5.00
Total Sum C	94.92 (including ash)

The major gases produced during the biomass burning process include carbon dioxide (CO_2), carbon monoxide (CO), methane (CH_4), oxides of nitrogen (NO_x = nitric oxide (NO) + nitrogen dioxide (NO_2)), and ammonia

(NH_3). CO_2 and CH_4 are greenhouse gases, which trap Earth-emitted infrared radiation and lead to global warming. CO, CH_4, and NO_x lead to the photochemical production of ozone (O_3) in the troposphere. In the troposphere, O_3 is an irritant and harmful pollutant, and in some cases, is toxic to living systems. NO leads to the chemical production of nitric acid (HNO_3) in the troposphere. HNO_3 is the fastest growing component of acidic precipitation. NH_3 is the only basic gaseous species that neutralises the acidic nature of the troposphere. Particulates, small (usually about 10 micrometers or smaller) solid particles, such as smoke or soot particles, are also produced during the burning process and released into the atmosphere. These solid particulates absorb and scatter incoming sunlight and hence impact the local, regional, and global climate. In addition, these particulates (specifically particulates 2.5 micrometers or smaller) can lead to various human respiratory and general health problems when inhaled. The gases and particulates produced during biomass burning lead to the formation of "smog." The word "smog" was coined as a combination of smoke and fog and is now used to describe any smoky or hazy pollution in the atmosphere.

Table 2. Nitrogen gases produced during biomass burning (Lobert *et al.* 1991)

Compound	Mean emission factor relative to the fuel N (%)
Nitrogen oxides (NO_x)	13.55
Ammonia(NH_3)	4.15
Hydrogen cyanide (HCN)	2.64
Acetonitrile (CH_3CN)	1.00
Cyanogen (NCCN) (As N)	0.023
Acrylonitrile (CH_2CHCN)	0.135
Propionitrile (CH_3CH_2CN)	0.071
Nitrous oxide (N_2O)	0.072
Methylamine (CH_3NH_2)	0.047
Dimetylamine (($CH_3)_2NH$)	0.030
Ethylamine ($CH_3CH_2NH_2$)	0.005
Trimethylamine (($CH_3)N$)	0.02
2-methyl-1-butylamine ($C_5H_{11}NH_2$)	0.04
n-pentylamine (n-$C_5H_{11}NH_2$)	0.137
Nitrates (70% HNO_3)	1.10
Ash (As N)	9.94
Total sum N (As N)	33.66 (Including ash)
Molecular nitrogen (N_2)	21.60
Higher HC and particles	20

Areas of biomass burning are varied and include tropical savannas, tropical, temperate and boreal forests and agricultural lands after the harvest. The burning of fuelwood for domestic use is another source of biomass burning. Global estimates of the annual amounts of biomass burning from these sources are given in Table 3 (Andreae, 1991). In Table 3, the unit of biomass burned is Tg dm/yr (1 Tg = 10^{12} g = 10^6 metric tons; dm = dry matter (biomass matter)). Biomass matter is about 45% by weight composed of carbon. Table 3 also gives estimates of the carbon released ($TgC\ yr^{-1}$) by the burning of this biomass. Combining estimates of the total amount of biomass matter burned/yr (Table 3) with measurements of the gaseous and particulate emissions from biomass burning (Tables 1 and 2) permits estimates of the global production and release into the atmosphere of gases and particulates from burning. Estimates of the global contribution of biomass burning are summarised in Table 4 (Andreae, 1991). The data in Tables 3 and 4 clearly indicate that biomass burning is a global process of major importance in the global budgets of atmospheric gases and particulates.

Table 3. Global estimates of annual amounts of biomass burning and of the resulting release of carbon to the atmosphere (Andreae, 1991)

Source	Biomass burned (Tg dm yr^{-1})	Carbon released (TgC yr^{-1})
Savanna	3690	1660
Agricultural waste	2020	910
Fuel wood	1430	640
Tropical forests	1260	570
Temperate/boreal forests	280	130
World Totals	8680	3910

Table 4. Comparison of global emissions from biomass burning with emissions from all sources (including biomass burning) (Andreae, 1991)

Species	Biomass Burning (Tg element[g] yr^{-1})	All Sources (Tg element[g] yr^{-1})	% attributable to biomass burning
CO_2 (Gross)[a]	3500	8700	40
CO_2 (Net)[b]	1800	7000	26
CO	350	1100	32
CH_4	38	380	10
NMHC[c]	24	100	24
N_2O	0.8	13	6
NO_x	8.5	40	21
NH_3	5.3	44	12
Sulphur	2.8	150	2
COS	0.09	1.4	6
CH_3Cl	0.51	2.3	22
H_2	19	75	25
Tropospheric O_3	420	1100	38
TPM[d]	104	1530	7
POC[e]	69	180	39
EC[f]	19	<22	>86

[a]Biomass burning plus fossil fuel burning.
[b]Deforestation plus fossil fuel burning.
[c]Nonmethane hydrocarbons (excluding isoprene and terpenes).
[d]Total particulate matter (Tg yr^{-1}).
[e]Particulate organic matter (including elemental carbon).
[f]Elemental (black-soot) carbon.
[g]Tg element yr^{-1} where C, N, S Cl are the elements.

2.   A CASE STUDY: THE CALCULATION OF GASEOUS AND PARTICULATE EMISSIONS FROM THE FOREST AND PEAT FIRES IN KALIMANTAN AND SUMATRA, INDONESIA, IN 1997

During the 1997–1998 period, there were a series of extensive and widespread fires in Southeast Asia, South America, Africa, Mexico, Russia, and Florida. The fires in Southeast Asia were particularly extensive and widespread. Some of the consequences of these fires include:
- more than 200 million people exposed to high levels of air pollution and particulates produced during the fires;
- more than 20 million smoke-related health problems;
- fire-related damage in excess of US$ 4 billion;
- the crash of a commercial airliner (Garuda Airlines Airbus 300–B4 on September 26, 1997) in Sumatra due to very poor visibility on landing, with 234 passengers killed; and
- the collision of two ships at sea due to poor visibility in the Strait of Malacca, off the coast of Malaysia, on September 27, 1997, with 29 crew members killed.

International concern about the environmental and health impacts of these fires is evident from the series of United Nations workshops and reports dealing with the Southeast Asia fires, including: The World Meteorological Organization (WMO) Workshop on Regional Transboundary Smoke and Haze in Southeast Asia, Singapore, June 2–5, 1998, The World Health Organization (WHO) Health Guidelines for Forest Fires Episodic Events, Lima, Peru, October 6–9, 1998, and the United Nations Environmental Program (UNEP) *Report on Wildland Fires and the Environment: A Global Synthesis*, published in February, 1999 (Levine *et al.*, 1999).

2.1   Calculation of gaseous and particulate emissions from vegetation and peat fires

The gaseous and particulate emissions produced during the fire and released into the atmosphere must be calculated if both the environmental and health impacts of these fires are to be assessed (Levine, 1999). The calculation of gaseous emissions from vegetation and peat fires can be calculated using a form of an expression from Seiler and Crutzen (1980) for each burning ecosystem/terrain:

$$M = A * B * E \tag{1}$$

where M is total mass of vegetation or peat consumed by burning (tons), A is area burned (km^2), B is biomass loading (tons km^{-2}) and E is burning efficiency (dimensionless) The total mass of carbon ($M(C)$) released to the atmosphere during burning is related to M by the following expression:

$$M(C) = C * M \text{ (Tons of carbon)} \tag{2}$$

C is the mass percentage of carbon in the biomass. For tropical vegetation, $C = 0.45$ (Andreae, 1991); for peat, $C = 0.50$ (Yokelson *et al.*, 1996). The mass of CO_2 ($M(CO_2)$) released during the fire is related to $M(C)$ by the following expression:

$$M(CO_2) = CE * M(C) \tag{3}$$

The combustion efficiency (CE) is the fraction of carbon emitted as CO_2 relative to the total carbon compounds released during the fire. For tropical vegetation fires, $CE = 0.90$ (Andreae, 1991); for peat fires, $CE = 0.77$ (Yokelson *et al.*, 1997).

Once the mass of CO_2 produced by burning is known, the mass of any other species, X_i ($M(X_i)$), produced by burning and released to the atmosphere can be calculated with knowledge of the CO_2-normalized species emission ratio ($ER(X_i)$). The emission ratio is the ratio of the production of species X_i to the production of CO_2 in the fire. The mass of species, X_i, is related to the mass of CO_2 by the following expression:

$$M(X_i) = ER(X_i) * M(CO_2) \text{ (Tons of element } X_i) \tag{4}$$

For the calculations presented in this paper, $X_i = CO, CH_4, NO_x, NH_3$, and O_3. It is important to re-emphasise that O_3 is not a direct product of biomass burning. However, O_3 is produced via photochemical reactions of CO, CH_4, and NO_x, all of which are produced directly by biomass burning. Hence, the mass of O_3 resulting from biomass burning may be calculated by considering the O_3 precursor gases produced by biomass burning.

Values for the CO_2-normalized species emission ratios (ER) for burning tropical vegetation and peat are given in Table 5. The uncertainty in the value of the species emission ratio is typically 30% (Andreae, 1991). These emission measurements were obtained for burning tropical forests in South America, not Southeast Asia. However, studies indicate that the emission ratios from tropical forests in South America should be comparable to those in Southeast Asia (for example, see Andreae, 1991, and Brown and Gaston, 1996). The peat fire emission ratios for gases in Table 5 are based on the measurements of Yokelson *et al.* (1997). These emission measurements were obtained for burning peat from Minnesota and Alaska. Because of the absence in the literature of emission ratio measurements for burning Indonesian peat, values from Yokelson *et al.* (1997) were used. Yokelson *et al.* (1997) did not obtain emission ratios for either NO_x or O_3 from burning peat. To estimate the emission ratio for NO_x from peat fires, the expression for the production ratio of $NH_3/NO_x = 14$ x (1 - MCE), was used where MCE is the modified combustion efficiency (Yokelson *et al.*, 1996). For peat fires, MCE = 0.8 and the ratio of $NH_3/NO_x = 2.80$. For $NH_3 = 1.28\%$ (Yokelson *et al.*, 1997), the value of NO_x is 0.46%. As already noted, CO, CH_4, and NO_x are the chemical precursors in the photochemical production of O_3. For peat fires, the emission ratios for CO, CH_4, and NO_x are 18.15%, 1.04%, and 0.46% (Yokelson *et al.*, 1997). NO_x is the reaction-limiting species in the photochemical production of O_3. To estimate the emission ratio of O_3 for peat fires, we have assumed that the ratio of emission ratio of O_3 to NO_x in forest fires is comparable to that ratio in peat fires. For these assumptions, the emission ratio of O_3 from peat fires is found to be 1.04%.

To calculate the total particulate matter (TPM) released from tropical forest fires and peat fires, we use the following expression (Ward, 1990):

$$TPM = M * P \text{ (Tons of carbon)} \tag{5}$$

where P is the conversion of biomass matter or peat matter to particulate matter during burning. For the burning of tropical vegetation, P = 20 tons of TPM per kiloton of biomass consumed by fire; for peat burning, we assume P = 35 tons of TPM per kiloton of organic soil or peat consumed by fire (Ward, 1990).

Table 5. Emission ratios for tropical forest fires and peat fires

Species	Tropical forest fires	Reference	Peat fires	Reference
CO_2	90.00 %	Andreae (1991)	77.05 %	Yokelson *et al.* (1997)
CO	8.5 %	Andreae *et al.* (1988)	18.15 %	Yokelson *et al.* (1997)
CH_4	0.32 %	Blake *et al.* (1996)	1.04 %	Yokelson *et al.* (1997)
NO_x	0.21 %	Andreae *et al.* (1988)	0.46 %	Derived from Yokelson *et al.* (1997); see text
NH_3	0.09%	Andreae *et al.* (1998)	1.28 %	Yokelson *et al.* (1997)
O_3	0.48 %	Andreae *et al.* (1988)	1.04 %	Derived from Yokelson *et al.* (1997); see text
TPM[1]	20 tons/kiloton	Ward (1990)	35 tons/kiloton	Ward (1990)

[1] Total particulate matter emission ratios are in units of tons/kiloton (tons of total particulate matter/kiloton of biomass or peat material consumed by fire).

2.2   Total area burned and biomass consumed

Perhaps the major uncertainties in the calculation of gaseous and particulate emissions resulting from fires involve poor or incomplete information about four fire parameters:
- the area burned (A);
- the ecosystem or terrain that burned, i.e., forests, grasslands, agricultural lands, peat lands, etc.;
- the biomass loading (B), i.e., the amount of biomass per unit area of the ecosystem prior to burning; and
- the fire efficiency (C), i.e., the amount of biomass in the burned ecosystem that was actually consumed by burning.

The area burned can be determined through the use of satellite measurements. Some operational satellite systems to estimate area burned, as well as monitor active fires, are listed in Table 6.

Table 6. Some operational satellite fire monitoring systems

The following summary gives the current operational satellite systems used in fire monitoring.

Instrument	Details
NOAA (National Oceanic and Atmospheric Administration)/AVHRR (Advanced Very High Resolution Radiometer)	Global 1–km imaging systems. Monitors active fires and burned area
DMSP (Defense Meteorological Satellite Program)/OLS (Operational Linescan System)	Global night-time low light sensor. Monitors active fires.
GOES (Geostationary Operational Environmental Satellite)Imager	Continental high temporal frequency, coarse spatial resolution geostationary imaging. Monitor active fires and smoke.
ERS (European Remote Sensing Satellite)/ATSR (Along-Track Scanning Radiometer) (European Space Agency)	Global 1–km imaging. Monitors active fires and burned areas.
ERS (European Remote Sensing Satellite /JERS (Japanese Earth Resources Satellite) SAR (Synthetic Aperture Radar) (European Space Agency /NASDA (National Space Development Agency of Japan)	Global microwave high resolution system. Monitors burned area.
LANDSAT (Land Satellite) TM (Thematic Mapper)/MSS (Multispectral Scanner System)	Local, high spatial frequency, low temporal frequency. Monitors burned area.
SPOT (Systeme Pour l'Observation de la Terre (CNES) (Centre National d'Études Spatiales)	Local, high spatial frequency, low temporal frequency. Monitors burned area.

Indonesia ranks third, after Brazil and Zaire, in its area of tropical forest. Of Indonesia's total land area of 1.9 million km^2, current forest cover estimates range from 0.9 to 1.2 million km^2, or 48 to 69% of the total. Forests dominate the landscape of Indonesia (Makarim *et al.*, 1998). Large areas of Indonesian forests burned in 1982 and 1983. In Kalimantan alone, the fires burned from 2.4 to 3.6 million ha of forests (Makarim *et al.*, 1998).

It is interesting to note that there is an uncertainty of 1.2 million ha or an uncertainty of 50% in our knowledge of the burned area of fires that occurred 16 years ago.

To date, two different analyses using two different techniques for the determination of the area burned have been reported (Levine *et al.*, 1998 and Liew *et al.*, 1998). Levine *et al.* (1998) calculated burned area in Sumatra and Kalimantan for 1997 from 51 fire maps of regions of the highest density of fires prepared by the USDA Forest Service. Levine *et al.* (1998) found a burned area of 12, 847 km^2 in Sumatra and 4221.9 km^2 in Kalimantan, for a total burned area of 17,068.93 km^2 in 1997. The calculated area burned for Sumatra and Kalimantan for 1997 and 1998 based on USDA Forest Service maps is summarised in Table 7 (Levine *et al.*, 1998).

Table 7. Area burned in Sumatra, Kalimantan, and Java, Indonesia, October–November, 1997 and February–March 1998 (based on USDA Forest Service fire maps; Levine et al., 1998)

Region	Location	Date	Total area
Sumatra	Jambi	15 Oct 97	199.39 km^2
	Muaraenim	15 Oct 97	898.35 km^2
	Sukarami	15 Oct 97	730.87 km^2
	Jambi	19 Oct 97	169.86 km^2
	Rengat South	19 Oct 97	245.71 km^2
	Sukarami	19 Oct 97	403.86 km^2
	Lahat	21 Oct 97	319.95 km^2
	Sukarami	27 Oct 97	4.40 km^2
	Buluranriding River	30 Oct 97	330.76 km^2
	Talang Gelumbang	30 Oct 97	947.70 km^2
	Talang Kubuan	30 Oct 97	2,281.34 km^2
	Kotabumi	04 Nov 97	693.02 km^2
	North Kotaagung	04 Nov 97	242.82 km^2
	Kotabumi	07 Nov 97	73.32 km^2
	Bulau River	09 Nov 97	258.30 km^2
	East Karangagung	09 Nov 97	733.23 km^2
	Jambi	09 Nov 97	35.00 km^2
	Sukadana	09 Nov 97	169.59 km^2
	Bandar	11 Nov 97	86.21 km^2
	Kasui	11 Nov 97	274.53 km^2
	Kotabumi	11 Nov 97	183.79 km^2
	Sukadana	12 Nov 97	79.00 km^2
	Sukadana	12 Nov 97	82.44 km^2
	Bandar	13 Nov 97	15.29 km^2
	Kotabumi	13 Nov 97	115.73 km^2
	Surabaja	13 Nov 97	404.90 km^2
	Bandar	14 Nov 97	90.93 km^2
	Kasui	14 Nov 97	165.15 km^2
	Sukadana	14 Nov 97	37.03 km^2
	Bandar	15 Nov 97	31.45 km^2
	Sukadana	15 Nov 97	93.19 km^2
	Kasui	16 Nov 97	35.80 km^2

	Kotabumi	16 Nov 97	605.02 km^2
	Surabaja	16 Nov 97	47.65 km^2
	Bandar	17 Nov 97	62.63 km^2
	Surabaja	17 Nov 97	54.81 km^2
	Bujut	19 Nov 97	124.02 km^2
	Kotabumi	19 Nov 97	229.04 km^2
	Bujut	20 Nov 97	47.69 km^2
	Sukadana	22 Nov 97	381.61 km^2
	Kotabumi	23 Nov 97	11.34 km^2
	Surabaja	23 Nov 97	360.47 km^2
	Kasui	28 Nov 97	113.29 km^2
	Bandar	30 Nov 97	25.28 km^2
	Surabaja	30 Nov 97	351.27 km^2
	Kampar	06 Mar 98	42.60 km^2
	Kampar	10 Mar 98	44.22 km^2
		Subtotal	*12,933.85 km^2*
Kalimantan	Batuwinang	27 Oct 97	557.28 km^2
	Matua	27 Oct 97	1,140.87 km^2
	Satui	27 Oct 97	852.12 km^2
	Bapuju	28 Oct 97	805.28 km^2
	West Batuwinang	29 Oct 97	791.05 km^2
	Batuwinang	25 Nov 97	75.30 km^2
	Nanang	07 Feb 98	209.27 km^2
	Susang	07 Feb 98	75.42 km^2
	Santan	13 Feb 98	328.26 km^2
	Melintang	19 Feb 98	967.49 km^2
	Susang	19 Feb 98	19.22 km^2
	Mahakam	20 Feb 98	422.57 km^2
	Penawai	20 Feb 98	208.22 km^2
	Sidulang	20 Feb 98	95.07 km^2
	Saliki	21 Feb 98	142.88 km^2
	Sambodja	21 Feb 98	99.97 km^2
	Penawai	22 Feb 98	152.16 km^2
	Mengangau	23 Feb 98	209.45 km^2
	Susang	23 Feb 98	19.30 km^2
	Gitan	24 Feb 98	306.44 km^2
	Djambu	25 Feb 98	66.04 km^2
	Bontang	27 Feb 98	325.93 km^2
	Sedulang	27 Feb 98	562.44 km^2
	Guntung	28 Feb 98	973.46 km^2
	Sangatta	03 Mar 98	670.15 km^2
	Sideman	04 Mar 98	208.41 km^2
	Rapak	13 Mar 98	2,050.68 km^2
	Subtotal		*12,334.73 km^2*
Java	Malang	06 Oct 97	76.19 km^2
	Madium	22 Oct 97	277.77 km^2
	Blitar	23 Oct 97	9.00 km^2
	Subtotal		*362.96 km^2*
	Total		25,631.54 km^2

Liew *et al.* (1998) analysed 766 SPOT "quicklook" images with almost complete coverage of Kalimantan and Sumatra from August to December 1997. Liew *et al.* (1998) estimate the burned area in Kalimantan to be 30,600 km^2 and the burned area in Sumatra to be 15,000 km^2, for a total burned area of 45,600 km^2. The Liew *et al.* (1998) area estimate is a factor of 2.7 greater than the area estimate of Levine *et al.* (1998). This is not surprising as the Liew *et al.* (1998) estimate is based on almost complete coverage of Kalimantan and Sumatra, while the Levine *et al.* (1998) estimate is based on the USDA Forest Service maps prepared for only the very highest density fire regions in Kalimantan and Sumatra. For the calculations reported in this paper, we have used the Liew *et al.* (1998) estimate for total burned area in Kalimantan and Sumatra of 45,600 km^2. (This is equivalent to the combined areas of the states of Rhode Island, Delaware, Connecticut, and New Jersey in the USA). The estimate of Liew *et al.* (1998) represents only a lower limit estimate of the area burned in Southeast Asia in 1997, as the SPOT data only covered Kalimantan and Sumatra and did not include fires on the other Indonesian islands of Irian Jaya, Sulawesi, Java, Sumbawa, Komodo, Flores, Sumba, Timor, and Wetar or the fires in the neighbouring countries of Malaysia and Brunei.

What is the nature of the ecosystem/terrain that burned in Kalimantan and Sumatra? In October 1997, NOAA satellite monitoring produced the following distribution of fire hot spots in Indonesia (UNDAC, 1998): agricultural and plantation areas: 45.95%; bush and peat soil areas: 24.27%; productive forests: 15.49%; timber estate areas: 8.51%; protected areas: 4.58%; and transmigration sites: 1.20%. (The three forest/timber areas add up to a total of 28.58% of the area burned). While the distribution of fire hot spots is not an actual index for area burned, the NOAA satellite-derived hot spot distribution is quite similar to the ecosystem/terrain distribution of burned area deduced by Liew *et al.* (1998) based on SPOT images of the actual burned areas: agricultural and plantation areas: 50%; forests and bushes: 30%; and peat swamp forests: 20%. As the estimates of burned ecosystem/terrain of Liew *et al.* (1998) are based on actual SPOT images of the burned area, their estimates were adopted in our calculations.

What is the biomass loading for the three terrain classifications identified by Liew *et al.* (1998)? Values for biomass loading or fuel load for various tropical ecosystems are summarised in Table 8. The biomass loading for tropical forests in Southeast Asia ranges from 5,000 to 55,000 tons km^{-2}, with a mean value of 23,000 tons km^{-2} (Brown and Gaston, 1996). However, in our calculations we have used a value of 10,000 tons km^{-2} to be conservative. The biomass loading for agricultural and plantation areas (mainly rubber trees and oil palms) of 5,000 tons km^{-2}, is also a conservative value (Liew *et al.*, 1998). Nichol (1997) has investigated the peat deposits of

Kalimantan and Sumatra and used a biomass loading value of 97,500 tons km^{-2} (Supardi *et al.*, 1993) for the dry peat deposits 1.5 m thick as representative of the Indonesian peat in her study. Brunig (1977) gives a similar value for peat biomass loading. The combustion efficiency for forests is estimated at 0.20 and for peat is estimated at 0.50 (Levine and Cofer, 1999). Based on the discussions presented in this section, the values for burned area, biomass loading, and combustion efficiency used in the calculations are summarised in Table 9.

Table 8. Biomass load range and burning efficiency in tropical ecosystems
(from Scholes 1996, except as noted)

Vegetation type	Biomass load range (tons km^{-2})	Burning efficiency
Peat[1]	97500	0.50
Tropical rainforests[2]	5000–55000	0.20
Evergreen forests	5000–10000	0.30
Plantations	500–10000	0.40
Dry forests	3000–7000	0.40
Fynbos	2000–4500	0.50
Wetlands	340–1000	0.70
Fertile grasslands	150–550	0.96
Forest/savanna mosaic	150–500	0.45
Infertile savannas	150–500	0.95
Fertile savannas	150–500	0.95
Infertile grasslands	150–350	0.96
Shrublands	50–200	0.95

[1]Brunig (1977), Supardi *et al.* (1993)

[2]Brown and Gaston (1996)

Table 9. Parameters used in calculations

1. Total area burned in Kalimantan and Sumatra, Indonesia in 1997: 45,600 km^2

2. Distribution of Burned Areas, Biomass Loading and Combustion Efficiency

Landuse	Percentage burned	Biomass loading	Combustion efficiency
Agricultural and plantation areas	50%	5,000 tons km^{-2}	0.20
Forests and bushes	30%	10,000 tons km^{-2}	0.20
Peat swamp forests	20%	97,500 tons km^{-2}	0.50

2.3   Results of calculations: gaseous and particulate emissions

The calculated gaseous and particulate emissions are summarised in Table 10. For each of the seven species listed, the emissions due to agricultural/plantation burning (A), forest burning (F), and peat burning (P) are given. The total (T) of all three components (A+F+P) is also given. The "best estimate" total emissions are: CO_2: 191.485 million metric tons of C (MtC); CO: 32.794 MtC; CH_4: 1.845 MtC; NO_x: 5.898 MtN; NH_3: 2.585 MtN; O_3: 7.100 MtO_3; and total particulate matter: 16.154 MtC.

Scholes *et al.* (1996) calculated the biomass consumed by burning in 11 different ecosystems in another tropical ecosystem, southern Africa, using equations (1) – (3). Scholes *et al.* (1996) performed a detailed statistical analysis of the errors associated with the calculated values of biomass consumed by fire using a statistical procedure of Nelson (1992), which assumes that all error terms (e) are independent. In the error analysis of Scholes *et al.* (1996), the total error (E_{total}), which corresponds to the 3-sigma (99%) confidence level, is estimated using the following expression:

$$e_{total} = (e^2_{burned\ area} + e^2_{fuel\ load} + e^2_{combustion\ completeness})^{1/2} \qquad (6)$$

Table 10. Gaseous and particulate emissions from the fires in Kalimantan and Sumatra in 1997 (for total burned area = 45,600 km^2)

For each species, the best estimate emission value is on first line and the range of emission values in parenthesis under best guess (See text for discussion of emission estimate range and uncertainty calculations). Units of emissions: Million metric tons (Mt) of C for CO_2, CO, and CH_4; Mt of N for NO_x and NH_3; Mt of O_3 for O_3; MtC of particulates; 1 Mt = 10^{12} grams = 1 Teragram, Tg.

Species	Agricultural/ plantation fire emissions	Forest fire emissions	Peat fire emissions	Total fire emissions
CO_2	9.234 (4.617 – 13.851)	11.080 (5.54 – 16.62)	171.170 (85.585 – 256.755)	191.485 (95.742 – 287.226)
CO	0.785 (0.392 – 1.177)	0.942 (0.471 – 1.413)	31.067 (15.533 – 46.600)	32.794 (16.397 – 49.191)
CH_4	0.030 (0.015 – 0.045)	0.035 (0.017 – 0.052)	1.780 (0.89 – 2.67)	1.845 (0.922 – 2.767)
NO_x	0.023 (0.011 – 0.034)	0.027 (0.013 – 0.040)	0.921 (0.460–1.381)	0.971 (0.485–1.456)
NH_3	0.010 (0.005 – 0.015)	0.012 (0.006 – 0.018)	2.563 (1.281 – 3.844)	2.585 (1.292 – 3.877)
O_3	0.177 (0.088 – 0.265)	0.213 (0.106 –0.319)	6.710 (3.35 – 10.06)	7.100 (3.55 – 10.65)
TPM	0.460 (0.23 – 0.69)	0.547 (0.273 – 0.820)	15.561 (7.780 – 23.341)	16.568 (8.284 – 24.852)

Scholes *et al.* (1996) assumed the following uncertainties for each calculation parameter: $e_{burned\ area}$ = 30%, $e_{fuel\ load}$ = 30% and $e_{combustion\ completeness}$ = 25%. In the error analysis of the calculations presented in this paper, the error associated with uncertainties in the emission ratio ($e_{emission\ ratio}$) (30%) has also been included. The uncertainties in the calculation parameters in the detailed error of analysis of Scholes *et al.* (1996) were adopted in this study, with the exception of the uncertainty in the burned area of 30%. The burned area determination used in Scholes *et al.* (1996) was based on satellite measurements of active fires that were converted to burned area, which is not a simple one-to-one transformation and introduces errors. The burned area determination used in the present paper was more straight-forward as it was based on direct satellite photography of burned areas using SPOT images (Liew *et al.*, 1998). Hence, there is less uncertainty in this burned area determination and an uncertainty of 10% was assumed. (Liew *et al.* (1998) did not give a burned area uncertainty in their paper). Using equation (6), the calculated uncertainty in the emission calculations presented in this paper is 50.2%. The uncertainty range for each species emission is shown in parentheses under the "best estimate" value in Table 2.

However, it is important to re-emphasise that these emission calculations represent lower limit values as the calculations are only based on burning in Kalimantan and Sumatra in 1997. The calculations do not include burning in Java, Sulawesi, Irian Jaya, Sumbawa, Komodo, Flores, Sumba, Timor, and Wetar in Indonesia or in neighbouring Malaysia and Brunei.

It is interesting to compare the gaseous and particulate emissions from the 1997 Kalimantan and Sumatra fires with those from the Kuwait oil fires of 1991, described as a major environmental catastrophe. Laursen *et al.* (1992) have calculated the emissions of CO_2, CO, CH_4, NO_x, and particulates from the Kuwait oil fires in units of metric tons per day. The Laursen *et al.* (1992) calculations are summarised in Table 10. To compare these calculations with the calculations presented in this paper for Kalimantan and Sumatra (Table 9), we have normalized our calculations by the total number of days of burning. The SPOT images (Liew *et al.*, 1998) covered a period of five months (August–December 1997) or about 150 days. For comparison with the Kuwait fire emissions, we divided our calculated emissions by 150 days. These values are summarised in Table 11. The gaseous and particulate emissions from the fires in Kalimantan and Sumatra significantly exceeded the emissions from the Kuwait oil fires. The 1997 fires in Kalimantan and Sumatra were a significant source of gaseous and particulate emissions to the local, regional, and global atmosphere.

Table 11. Comparison of gaseous and particulate emissions: The Indonesian fires and the Kuwait oil fires. (Units of emissions: Metric tons per day of C for CO_2, CO, and CH_4; metric tons per day of N for NO_x; metric tons per day for particulates; 1 million metric tons = 10^{12} g = 1 Teragram, Tg)

Species	*Indonesian fires*	*Kuwait oil fires*[1]
CO_2	*1.28 x 10^6*	*5.0 x 10^5*
CO	*2.19 x 10^5*	*4.4 x 10^3*
CH4	*1.23 x 10^4*	*1.5 x 10^3*
NO_x	*6.19 x 10^3*	*2.0 x 10^2*
Particulates	*1.08 x 10^5*	*1.2 x 10^4*

[1] Laursen *et al.* (1992)

3. ACKNOWLEDGEMENTS

This research was funded by the Global Change Program of the U.S. Environmental Protection Agency under Interagency Agreement DW 80936540–01–1 and the Director's Discretionary Fund of NASA Langley Research Center.

4. REFERENCES

Andreae, M. O. 1991, Biomass burning: Its history, use, and distribution and its impact on environmental quality and global climate. Global Biomass Burning: Atmospheric, Climatic, and Biospheric Implications (J. S. Levine, Ed.). The MIT Press, Cambridge, Massachusetts, 3–21.

Andreae, M. O., E. V. Browell, M. Garstang, G. L. Gregory, R. C. Harriss, G. F. Hill, D. J. Jacob, M. C. Pereira, G. W. Sachse, A. W. Setzer, P. L. Silva Dias, R. W. Talbot, A. L. Torres and S.C. Wofsy 1998, Biomass burning emission and associated haze layers over Amazonia. Journal of Geophysical Research, 93, 1509–1527.

Blake, N. J., D. R. Blake, B. C. Sive, T. -Y. Chen, F. S. Rowland, J. E. Collins, G. W. Sachse and B. E. Anderson 1996, Biomass burning emissions and vertical distribution of atmospheric methyl halides and other reduced carbon gases in the South Atlantic Region. Journal of Geophysical Research, 101, 24151–24164.

Brown, S. and G. Gaston 1996, Estimates of biomass density for tropical forests. Biomass Burning and Global Change, Volume 1 (J. S. Levine, Ed.). The MIT Press, Cambridge, Massachusetts, 133–139.

Brunig, E. F., 1977. The tropical rainforest–A wasted asset or an essential biospheric resource? *Ambio*, 6, 187–191.

Crutzen, P. J., and M. O. Andreae 1990, Biomass burning in the tropics: Impact on atmospheric chemistry and biogeochemical cycles. Science 250, 1679–1678.

Crutzen, P. J., and J. G. Goldammer (Editors) 1993, Fire in the Environment: The Ecological, Atmospheric, and Climatic Importance of Vegetation Fires. John Wiley and Sons, Chichester, England, 400 pages.

Crutzen, P. J., L. E. Heidt, J. P. Krasnec, W. H. Pollock and W. Seiler 1979, Biomass burning as a source of atmospheric gases CO, H_2, N_2O, NO, CH_3Cl, COS. Nature, 282, 253–256.

Goldammer, J. G. (Editor) 1990, Fire in the tropical biota: ecosystem processes and global challenges. Springer-Verlag, Berlin, Germany, 497 pages.

Goldammer, J. G., and V. V. Furyaev (Editors) 1996, Fire in Ecosystems of Boreal Eurasia Kluwer Academic Publishers, Dordrecht, The Netherlands, 528 pages.

Laursen, K. K., R. J. Ferek, and P. V. Hobbs 1992, Emission factors for particulates, elemental carbon, and trace gases from the Kuwait oil fires. Journal of Geophysical Research, 97, 14491–14497.

Levine, J. S. 1999, The 1997 fires in Kalimantan and Sumatra, Indonesia: Gaseous and particulate emissions. Geophysical Research Letters, 26, 815–818.

Levine, J. S. (Editor) 1991, Global Biomass Burning: Atmospheric, Climatic, and Biospheric Implications. The MIT Press, Cambridge, Massachusetts, 569 pages.

Levine, J. S. (Editor) 1996a, Biomass Burning and Global Change: Remote Sensing, Modeling and Inventory Development, and Biomass Burning in Africa. The MIT Press, Cambridge, Massachusetts, 581 pages.

Levine, J. S. (Editor) 1996b, Biomass Burning and Global Change: Biomass Burning in South America, Southeast Asia, and Temperate and Boreal Ecosystems, and the Oil Fires of Kuwait. The MIT Press, Cambridge, Massachusetts, 377 pages.

Levine, J. S., T. Bobbe, N. Ray, R. G. Witt, and A. Singh 1999, Wildland Fires and the Environment: A Global Synthesis. Environment Information and Assessment Technical Report 99–1, The United Nations Environmental Program, Nairobi, Kenya, 46 pages.

Levine, J. S. and W. R. Cofer 1999 Boreal forest fire emissions and the chemistry of the atmosphere. Fire, Climate Change and Carbon Cycling in the North American Boreal Forests (E. S. Kasischke and B. J. Stocks, Eds.), Ecological Studies Series, Springer-Verlag, New York, (In press).

Levine, J. S., W. R. Cofer, D. R. Cahoon, and E. L. Winstead 1995, Biomass burning: A driver for global change. Environmental Science and Technology, 29, 120A–125A.

Levine, J. S., T. D. Edwards, T. E. McReynolds, and C. W. Dull 1998 Gaseous and particulate emissions from the fires in Kalimantan and Sumatra, Indonesia. Report of WMO Workshop on Regional Transboundary Smoke and Haze in South-East Asia, World Meteorological Organization, Geneva, Switzerland.

Liew, S. C., O. K. Lim, L. K. Kwoh, and H. Lim 1998, A study of the 1997 fires in South East Asia using SPOT quicklook mosaics. Paper presented at the 1998 International Geoscience and Remote Sensing Symposium, July 6–10, Seattle, Washington, 3 pages.

Lobert, J. M., D. H. Scharffe, W.-M. Hao, T. A. Kuhlbusch, R. Seuwen, P. Warneck, and P. J. Crutzen 1991,. Experimental evaluation of biomass burning emissions: Nitrogen and carbon containing compounds. Global Biomass Burning: Atmospheric, Climatic, and Biospheric Implications (J. S. Levine, Ed.). The MIT Press, Cambridge, Massachusetts, 289–304.

Makarim, N., Y. A. Arbai, A. Deddy, and M. Brady, 1998: Assessment of the 1997 Land and Forest Fires in Indonesia: National Coordination. International Forest Fire News, United Nations Economic Commission for Europe and the Food and Agriculture Organization of the United Nations, Geneva, Switzerland, No. 18, January 1998, 4–12.

Nelson, L.S. 1992. Technical aids: Propagation of error. Journal of Qualitative Control, 24, 232–235.

Nichol, J. 1997. Bioclimatic Impacts of the 1994 smoke haze event in Southeast Asia. Atmospheric Environment, 44, 1209–1219.

Scholes, R. G., J. Kendall, and C. O. Justice 1996. The quantity of biomass burned in southern Africa. Journal of Geophysical Research, 101, 23667–23676.

Seiler, W. and P.J. Crutzen 1980, Estimates of gross and net fluxes of carbon between the biosphere and the atmosphere from biomass burning. Climatic Change, 2, 207–247.

Supardi, A. D. Subekty and S. G. Neuzil 1993, General geology and peat resources of the Siak Kanan and Bengkalis Island peat deposits, Sumatra, Indonesia. Modern and Ancient Coal Forming Environments (J. C. Cobb and C. B. Cecil, Eds.). Geological Society of America Special Paper, Volume 86, 45–61.

UNDAC (United Nations Disaster Assessment and Coordination Team) 1998. Mission on Forest Fires, Indonesia, September–November 1997. International Forest Fire News, United Nations Economic Commission for Europe and the Food and Agriculture Organization of the United Nations, Geneva, Switzerland, No. 18, January 1998, 13–26.

van Wilgen, B. W., M. O. Andreae, J. G. Goldammer, and J. A. Lindesay (Editors) 1997, Fire in Southern African Savannas: Ecological and Atmospheric Perspectives. Witwatersrand University Press, Johannesburg, South Africa, 256 pages.

Ward, D. E. 1990, Factors influencing the emissions of gases and particulate matter from biomass burning. Fire in the Tropical Biota: Ecosystem Processes and Global Challenges (J. G. Goldammer, Ed.). Springer-Verlag, Berlin, Ecological Studies Volume 84, 418–436.

Yokelson, R. J., D. W. T. Griffith, and D. E. Ward 1996, Open-path Fourier transform infrared studies of large-scale laboratory biomass fires. Journal of Geophysical Research, 101, 21067–21080.
Yokelson, R. J., R. Susott, D. E. Ward, J. Reardon and D. W. T. Griffith 1997, Emissions from smouldering combustion from biomass measured by open-path Fourier transform infrared spectroscopy. Journal of Geophysical Research, 102, 18865–18877.

Modelling the Effect of Landuse Changes on Global Biomass Emissions

SUE A. FERGUSON, DAVID V. SANDBERG and ROGER OTTMAR
Forestry Sciences Laboratory, 4043 Roosevelt Way NE, Seattle, USA

Abstract: The rate and magnitude of emissions from prescribed burns and wildfires in wildland areas throughout the world are related to biomass consumption, which is controlled by total biomass, fuel moisture, fuel distribution (fuel size and arrangement), and ignition pattern. Consequently, landuse practices, which can affect many of these components, play a crucial role in determining the rate and magnitude of smoke production from biomass burning. The variability of landuse and its relation to the magnitude and rate of smoke production, however, usually are not considered when estimating biomass emissions. For example, much prescribed wildland burning in the United States has changed from high-intensity slash burning associated with land clearing activities, in which 20 hectare fires typically emit more than 10,000 grams/second of particles within an hour or two, to low-intensity understory burning related to health management where 120 hectare fires emit less than 2,000 grams/second of particles for several hours to days. Total emissions may be similar but the duration of emissions and associated heat release rates are significantly different, causing vastly different impacts on visibility, human health, and climatic forcing. Despite changes in landuse and fire, many regional and global estimates of biomass emissions in the United States continue to assume that most emissions result from land-clearing type slash burns.

Meanwhile, in South America estimates of biomass emissions typically assume dry fuels, yet most burning occurs within a few months of harvesting. The large logs remain wet, reducing emissions by more than 50%, which is unaccounted for in global emission estimates. Also, while land-clearing remains vigorous in the tropics, a change toward using fire for health management already has begun.

In this paper, an emission production model is used to show the differences in emission magnitudes and rates for prescribed fires in rain forests of Washington State and the Brazilian Amazon, and in dry forests of Oregon State and the Brazilian cerrado. In addition to emissions of particles and carbon gases, the model estimates heat release rates that affect plume buoyancy. These values are used to evaluate impacts on human health, visibility, and components of climate forcing.

1.   INTRODUCTION

Biomass fires are a significant cause of regional air pollution and an important global source of carbon to the atmosphere. Because the magnitude of emissions from biomass fires is so large, many attempts have been made to assess their effects on the global carbon budget. Efforts are underway to reduce emissions that contribute to climate change and affect health and welfare. The effectiveness of these efforts can be evaluated by assessing their influence on the number and size of fires that occur, the consumption of biomass by those fires, and emission characteristics. An Emission Production Model, EPM, is used here as a way to refine large-scale estimates of fire emissions, track incremental changes in emissions over time, and to evaluate the effect of landuse and fire-management practices.

Emissions from biomass burning contribute roughly 6% of the particulate emissions from all global sources (Andreae, 1991), but the regional contribution to the global budget is changing with time because of changing landuse practices. There has been significant recent emphasis on tropical areas because the rate of tropical deforestation clearly dominates current global emissions from fires (Laursen and Radke, 1996). During the agricultural revolution in the late 19th century, however, northern latitude emissions were globally significant (Holdsworth *et al.*, 1996). Also, recent experiments suggest that current emissions from North American and boreal fires may be much greater than some recent estimates (Hegg *et al.*, 1990; Cofer *et al.*, 1996). Predicting future emissions from fires will depend significantly on how well changes in landuse and climate are predicted, how well we model the effects those changes have on emissions, and how consistently we are able to model changes across the biomes of the world.

Changes in climate and landuse are incremental, so the methods we use to assess the impact of change must be dynamic enough to capture transient responses. Global estimates of fire emissions, however, typically are based on static models that make simple calculations about biomass consumption, and assume an average pre-burn loading and constant fraction of consumption in each biome. Seasonality, climate variability, and differing burning characteristics often are simplified or ignored. Diurnal timing and buoyancy are not generally considered.

The Emission Production Model (EPM: Sandberg and Peterson 1984) can address much of the variability currently lacking in global biomass emission estimates. It was initially designed over a decade ago to estimate emissions and heat-release rates from individual prescribed fires, so that management options to reduce emissions and their local impact on air quality could be evaluated. Development since then has incorporated more robust algorithms that address a wide range of burning styles, vegetation types, and fuel conditions (Sandberg and Ferguson, in preparation). Use of the model is expanding from its initial local application to having greater utility in assessing large-scale regional and global emissions as well as other fire effects.

2.   CHARACTERISTICS OF FIRES

Wildland fires are increasing in many parts of the world due to increased human pressure and an apparent increase in the severity of climatic conditions leading to large catastrophic fires. Although attribution is difficult, the observed warming of the planet during the past two decades has been coincident with an increase in fires in ecosystems ranging from the tropics to the boreal forests during the last two decades (e.g., Agee, 1993; Prins and Menzel, 1994; Larsen, 1996). Partly as a result of these natural disasters, policy makers and scientists are gaining appreciation for the values at risk from fire as well as the ecological importance of fire. The use of prescribed fire to sustain ecosystems, prevent catastrophes, and manage natural resources is on the increase in many countries of the world. Also, of course, the use of fire to clear forests for conversion to agriculture, especially in the tropics, continues at an alarming rate.

We use the term "fires" to be inclusive of "prescribed biomass fires" (i.e., those that are intentionally used to accomplish resource and landuse management objectives) and "wildland fires" (i.e., all fires that are unintentional). The term "wildland fuelbeds" excludes agricultural fuelbeds, but includes all of the live and dead biomass between the mineral soil and the top of the dominant vegetation canopy of ecosystems. Fires vary widely in their intensity (heat release per unit time) and severity (heat release per unit area) because of differences in the physical characteristics of wildland fuelbeds, the condition of the fuel elements (especially fuel moisture), the current weather, and the nature of ignition.

Fire intensity, which in part controls combustion efficiency and plume rise, varies over several orders of magnitude according to natural and managed variability in fuel condition, weather conditions, and ignition pattern. At one end of the spectrum are prescribed fires used for landuse conversion or wildfires during periods of optimum conditions, with heat release rates sufficient to loft plumes high into the troposphere. At the other end of the spectrum are smouldering ground fires or fires in very light fuels, with heat release rates so low that plumes rarely exceed surface boundary layer heights. Intensity is likely to vary dramatically with diurnal winds and humidity if fires burn more than several hours. Freely spreading fires also vary in intensity from minute-to-minute as wind and other burn conditions change.

Differences in fire severity cause ranges in fuel consumption from about 1,000 to 500,000 kg per hectare. Fires used for land clearing in converting forests to agriculture, grazing, or urban development are intentionally high intensity and high severity with as much as 70% to 90% fuel consumption. The highest consumption rates occur in dry fuels that are densely distributed or piled then ignited almost instantaneously. If burned early in a dry season and soon after logging, however, less than 25% of the fuels may be consumed because high fuel moistures reduce combustion. Fires in the understory of forests or woodlands may consume less than 10% of accumulated biomass as fuels are sparsely scattered and ignition is gradual or spotty. In the same biome, however, if fires involve the dominant vegetation of connected canopies, as much as 60% of total above-ground biomass may be consumed. Fire severity can be controlled by mechanically manipulating the fuel bed, and/or scheduling intentional fires and controlling the ignition pattern. Other policy options that can limit biomass consumption include preventing wildfires and prohibiting prescribed fires when high severity is expected.

Fire duration, typically ranging from an hour to several weeks, is another important variable. Emissions and biomass consumption can be minimised by nearly instantaneous ignition, creating a short-duration convection column that collapses soon after flaming stops, followed by very little smouldering. Fires that last for many days promote smouldering combustion, higher emission factors (i.e., mass of emissions per mass of biomass consumed), and serve as an ignition source for wildfires. Generally speaking, management practices that promote fires of shorter duration are favoured for reducing emissions and for limiting the impact of non-buoyant plumes.

Despite the desire to reduce emissions by reducing fire duration, long-duration fires are becoming increasingly common as land managers promote low intensity, meandering fires that remove fine fuels but do not damage large trees. This is causing the diurnal cycle to play an increasingly significant role in emission rates and related impacts. During the night, emission rates usually decrease. At the same time, however, threats to human health increase. During the day, when emission rates usually increase, greater dispersion allows lower surface concentrations but contributions to regional haze and its impact on visibility and radiative flux become pronounced. Figure 1 shows the light-scattering coefficient, which is proportional to particle mass, from a nephelometer that was placed approximately 4km down-valley from a 360 hectare, prescribed understory burn in northeastern Oregon. Ignition began at 1100 Pacific Daylight Time (PDT) on 13 May 1997 and flaming was complete by approximately 1600 PDT the same day. Smoke entered the valley from the smouldering fire as soon as radiative cooling at night diminished lofting.

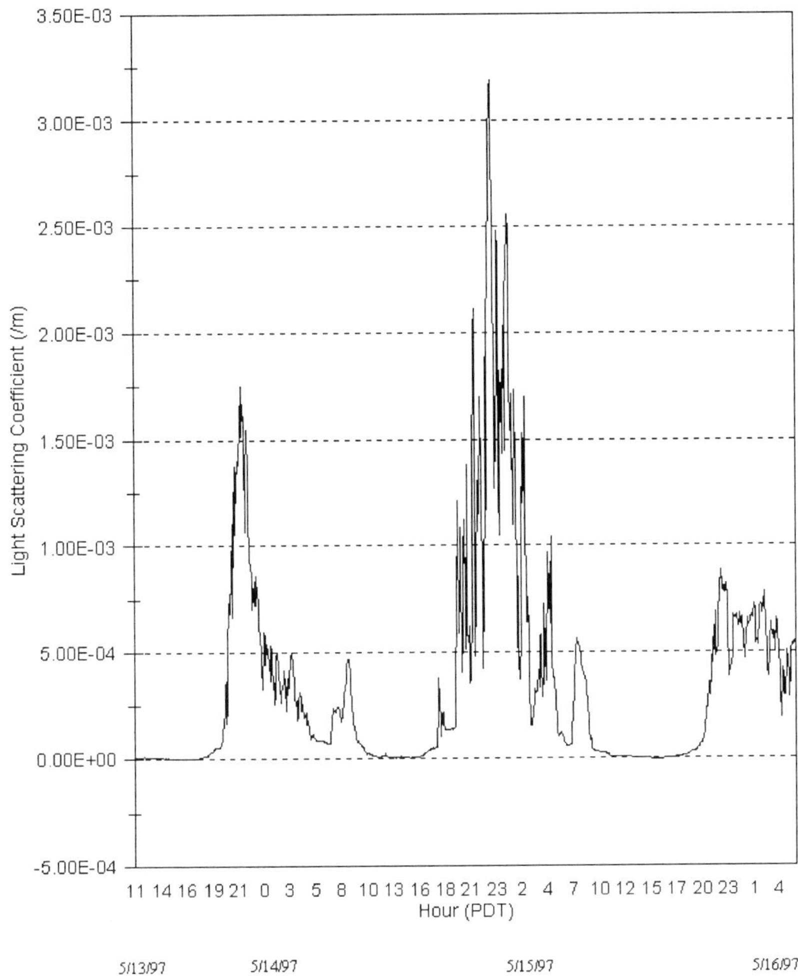

Figure 1. Hourly light-scattering coefficient from measurements taken approximately 4 km down-valley from a 360-hectare smouldering under-story burn in northeastern Oregon. Time is in Pacific Daylight

Emissions continued for several days, barely noticeable during the day when emissions were lofted away from the valley, but each night smoke settled into the valley as smouldering from rotten logs and old stumps continued. Total particulate emissions from this fire were estimated to be nearly 20 x 10^6 kg. Less than one half of this total was emitted during the first few hours of ignition when heat release rates were relatively high and buoyant emissions were dispersed widely by upper-level winds. The remaining smoke, with over 10 x 10^6 kg of particulate matter, was emitted in the next several days after ignition while the fuel smouldered independently. The weakly buoyant emissions lingered close to the ground surface where they were transported by topographically controlled thermal winds. An estimate of impact based on total emissions would have missed the diurnal variability in emission rates and concentrations.

3.   MODELLING EMISSIONS

The Emission Production Model (EPM) provides a time-sequence of heat release rate, biomass consumption, and emission production from any fires in wildland fuelbeds (Sandberg and Peterson 1984). The principal purpose of EPM is to anticipate and manage air quality problems and it is a principal source-strength estimator for a number of smoke dispersion models (Ferguson and Hardy 1994, Breyfogle and Ferguson 1996). EPM has been in widespread use for planning and screening prescribed fires in the USA for fifteen years. The model development is ongoing as science progresses in fuel consumption and heat flux, and as new databases are generated that describe the spatial distribution of biophysical parameters, including vegetation type, fuel characteristics, and ambient weather or climate. Details of the model physics are explained by Sandberg and Ferguson (in preparation). The following describe primary features of the model.

The inputs to EPM are a description of the fuelbed, the rate of fire ignition, and the amount of biomass consumed. EPM[1] is designed to link with a biomass consumption model[1] that provides model estimates of fuel consumption in each combustion stage (flaming and smouldering) for duff, rotten logs and stumps, shrubs, grasses, leafy and needle canopies, and several size classes of sound, dead woody material. Typically, the user inputs ignition intervals for prescribed burns and ignition rates are estimated from the source of ignition (e.g., helicopter dropped incendiaries, hand-held torch, etc.). EPM is being linked with a fire behaviour model (FARSITE: Finney 1995), however, so ignition rates can be automatically calculated from the fire spread. This is especially useful for ignition rates in wildfires. For regional and global applications, ignition rates are estimated from the dominant landuse activity. For example, piles usually are ignited within a few minutes, dispersed harvest residue requires several minutes to about an hour for ignition, while the ignition of understory burns usually continues for

---

[1] Currently algorithms from CONSUME (Ottmar *et al.* 1993) and SMS_INFO (Ottmar 1992) are embedded into the EPM code. Modifications are being made, however, to remove these internal algorithms and replace them with a link to CONSUME v2.0 (in preparation) or other fuel consumption models.

several hours. In all applications, ignition rates are assumed to be constant during the ignition period.

EPM calculates the length of each combustion stage at a point directly from the characteristic dimension of fuel reduction obtained from CONSUME or its substitute. For example, based on laboratory experiments of Anderson (1969) EPM presumes that the flaming stage lasts slightly over 3 minutes for each centimetre of woody-fuel diameter reduction. Once a fuel ceases flaming, the duration of its initial smouldering stage is characterized by an exponential-decay constant, which is derived empirically from the estimates of mass consumed and their characteristic dimensions of reduction (e.g., diameter reduction of wood and height reduction of duff). For example, field experiments have shown that the smouldering rate in short-needled conifer duff (partially decomposed litter) diminishes by a factor of (1-1/e) every 6 minutes for each centimetre of duff consumed. Other fuelbed components can control the rate of initial smouldering as well, so a set of heuristics is included in EPM to choose the optimum method of estimating the exponential decay constant in each fuelbed type.

In addition to a dependent smouldering stage, related to the die-back of flaming combustion, there also is an independent smouldering process that occurs in porous fuels like rotten wood, litter, duff (fermentation and humus layers), peat (organic soils), and moss. Independent smouldering rates diminish as a negative exponential, similar to dependent smouldering but with time constants of hours to days instead of minutes. The rate of independent smouldering consumption is related to moisture content, porosity, and inorganic content. Currently a simple kinetic approximation is used with empirical time constants that are based on only a few observations, but research is underway to sharpen this estimate.

The rate of biomass consumption in each combustion stage is calculated within EPM based on fuel description, ignition pattern, duration of each combustion stage, and the exponential time constants. Emission rates and heat fluxes are calculated every 3 minutes by assuming that the flaming and smouldering stages of combustion are at their peak intensity as soon as the fuelbed becomes fully involved in that combustion stage. This assumption has been verified in numerous field studies, and allows the independent calculation of a rate at any time.

An emission factor is derived from look-up tables (US Environmental Protection Agency 1991) according to fuel type, and mass-weighted at each time step according to the ratio of smouldering-to-flaming consumption. The emission factor multiplied by the rate of biomass consumption yields the total emission rate for each pollutant, including several carbon compounds (carbon monoxide [CO], carbon dioxide [CO_2], and methane [CH_4]), particulate size classes (particles less than 10 micrometers in diameter

[PM$_{10}$] and less than 2.5 micrometers in diameter [PM$_{2.5}$]), and heat release rates (BTU). The model is being updated to include output of large particles and gases that affect regional haze and ozone production, such as sulphur and nitrogen oxides.

The current version of EPM assumes that all of the generated heat is transferred to vertical flux and entrained into a single convective column. It has long been known, however, only 40–80% of the heat is convected, the rest is radiated away from glowing embers and flames or conducted to the ground (e.g., McCarter and Broido 1965). Also, Byram and Nelson (1974) speculate that the convective pattern breaks into separate cells when burning rates are slow relative to the area of burn. Although the assumptions in EPM should overestimate plume rise, tests with a simple dispersion model indicate that modelled plume rise is nearly equal or slightly lower than observed over relatively intense slash burns in the northwestern United States (Hardy *et al.* 1993). Clearly, more work on this aspect of the model is needed.

The diagram in Figure 2 summarises the data and computational flow of EPM. Inputs of fuel characteristics are measured fuel volumes or estimated loadings from photo inventories or tables of fuel characteristic classes. Fuel moisture also can be measured immediately preceding a fire or estimated from fire weather or climate indices. These components provide input to fuel consumption algorithms. Currently, the only type of consumption model that distinguishes between combustion phases is CONSUME (Ottmar *et al.* 1993) but links to future models and other algorithms is possible. EPM then uses the ignition pattern and theoretical equations of combustion processes to calculate rates of combustion from the mass of each fuel element consumed during flaming and smouldering. The ignition pattern is user defined, provided by fire spread models (e.g., FARSITE), or estimated from landuse patterns. Emission factors from a variety of published sources (e.g., Ward *et al.* 1989) are used to determine emission rates for each pollutant.

Figure 3 shows the characteristics of a typical biomass burn measured over a 8.5 hectare fire (Ferguson and Hardy 1994). Shortly after ignition a flaming phase continues for several minutes during which time maximum release rates of heat, particles, and gases occur. After flaming, sound woody material begins smouldering with smouldering emissions decreasing exponentially[2]. In places with deep layers of duff or rotten wood, an independent smouldering phase can continue for several hours to days. This sequence occurs in all types of fires, from stand-replacement and forest clearing to slash reduction and ecosystem restoration. The magnitude of fuel consumption, and consequently emissions, depends on fuel loading, fuel

---

[2]   A sharp rise in emission rate seen at about hour 1900 is thought to be due to a practice of lowering the sample packages over the smoldering fire as soon as flaming ceases. Other variability is due to ambient wind and the complex fuel array.

condition, and ignition rate. The duration of each phase primarily is a function of the size distribution of fuels and fuel moisture.

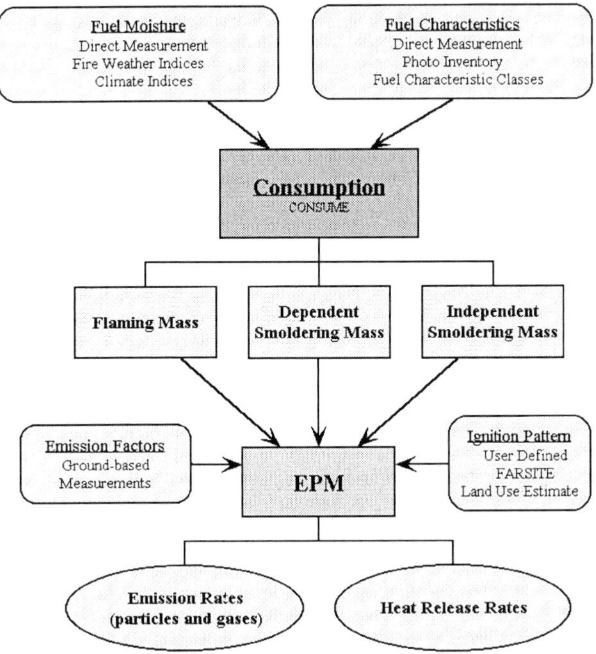

Figure 2. Input requirements and computational flow diagram for EPM

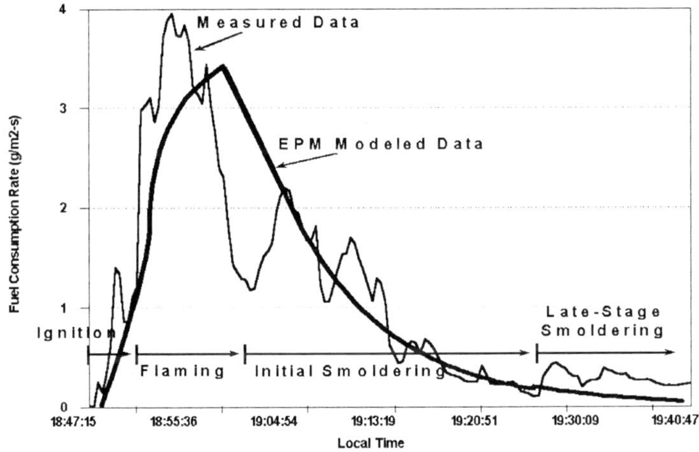

Figure 3. Fuel consumption rate over 0.5 hectare of a typical biomass burn. The thin line shows measured data from sampling packages elevated over the point of ignition. The thick line shows modelled data from EPM.

In addition to measured values of fuel consumption, Figure 3 shows modelled emission rates from EPM. The magnitude and duration of flaming and initial smouldering phases are successfully captured by the model. Because rotten wood was not inventoried in this case, EPM did not predict an independent smouldering phase.

4.   EFFECTS OF LANDUSE

To show how differences in landuse may affect biomass emissions, several test cases were derived for EPM. The cases consider typical fuel loading, fuel condition, and ambient weather in broadcast slash, stand replacement, and understory burns. Fuel loading data were selected from measurements in Amazonian primary tropical forest (Kauffman *et al.* 1994), western Washington Douglas fir rain forest (Hobbs *et al.* 1996), eastern Oregon ponderosa pine (Ottmar *et al.*, in press), and cerrado *sensu stricta-denso* near Brasilia (Kauffman *et al.* 1994), a savanna forest. It should be noted that while EPM is being modified to accept data on rotten stumps and logs, live standing and herbaceous fuels, and grass and mosses, no data on these fuel types were available for the selected case studies. Table 1 summarises the available pre-burn load of fuel elements in each example.

Table 1. Pre-burn fuel loadings (Mg/ha) for each size class of woody fuels and duff depth. Data for case examples from Kauffman *et al.* (1994), Hobbs *et al.* (1996), and Ottmar *et al.* (in press)

| | woody | fuels | | | | | Duff Depth (cm) |
	0 to 0.6 cm	0.6 to 2.5 cm	2.5 to 7.6 cm	7.6 to 22.8 cm	22.8 to 50.8 cm	50.8+ cm	
Amazon – slash	17.6	16.9	38.7	70.6	145.0	0.0	1.3
Cerrado – stand	12.1	0.7	1.0	26.1	1.2	0.0	0.8
Oregon Pine – stand	0.2	3.9	10.2	5.1	7.0	11.4	2.0
Oregon Pine – slash	1.2	5.6	11.6	10.4	15.0	0.0	2.0
Oregon Pine – understory	0.2	3.9	10.2	5.1	7.0	11.4	2.0
Washington Fir – dry slash	6.3	10.4	17.4	52.9	118.7	73.2	7.1

Under dry, well-cured, conditions in the Amazon, Washington, and Oregon forest, fuel moistures were assumed to be 25% for 1,000-hour fuels (7.6 – 15.2 cm diameter wood) and 10% for 10-hour fuels (0.6 – 2.5 cm diameter wood). Burning slash in the Amazon, however, can occur before total drying, within three months after harvest. In this case fuel moistures are close to 70% in 1,000-hour fuels and 15% in 10-hour fuels. In the cerrado, which tends to be drier than elsewhere, fuel moistures were assumed to be 10% for 1,000-hour and 5% for 10-hour fuels. Ignition of 20 hectare fires was assumed to require one hour, except in the ponderosa pine understory burn where it is more usual to burn large areas more slowly so 120 hectares were assumed to ignite in six hours. All cases were on flat terrain.

Figure 4 shows results of the EPM test-runs for three types of tropical burns. In the Amazon primary forests, many trees are harvested during land clearing. This removes large logs and limits smouldering. Usually burning takes place within three months of logging and before the wood has completely dried. This wet-slash type of burn emits nearly 7,000 grams per second (g s^{-1}) of particles during its flaming phase, which is followed by a short smouldering phase. When fuels are completely dry, usually one full year after harvesting, nearly 11,000 g s^{-1} are emitted in a flaming fire and there is almost no smouldering component. In the cerrado, where small trees and grasses are burned prior to any working of the landscape, so-called stand-replacement fires emit about 4,500 g s^{-1} during the flaming stage and a short smouldering phase follows.

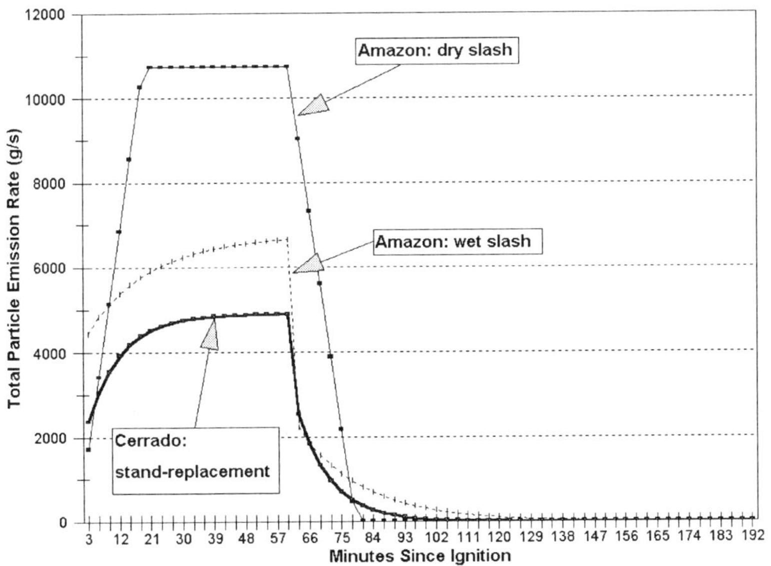

Figure 4. Modelled emissions for typical fuel loading, fuel condition, ambient weather, and ignition pattern in 20-hectare fires for broadcast slash burning and stand-replacement in two forest types of Brazil.

In the United States, emission rates from four types of fires are shown in Figure 5. Slash burning in the temperate rain forest of the Olympic Peninsula in Washington State is comparable to the burning of dry slash in the Amazon rain forest, with over 11,000 g s^{-1} emitted during the peak flaming stage. A very long smouldering period continues because of deep duff layers that accumulate in old-growth Douglas fir forests and the large old stumps and rotten logs left over from a bountiful harvest. Significantly less smoke is emitted from fires in the interior Pacific Northwest. In the ponderosa pine forests of northeastern Oregon, the smoke from burning harvest residue and stand-replacement fires both emit nearly the same rate of particles, about 3,500 g s^{-1} during the flaming stage. This rate is significantly lower than emission rates in rain forest slash and somewhat lower than savanna emission rates because the fuel in pine forests are much more sparsely distributed. Understory fires, which are typical of frequent small fires in fire-adapted ecosystems, and the style of prescribed fire used to restore a natural fire regime, show very small rates of particle emissions (less than 400 g s^{-1} from the 120 hectare plot) over a long period, reflecting the slow and deliberate ignition rate.

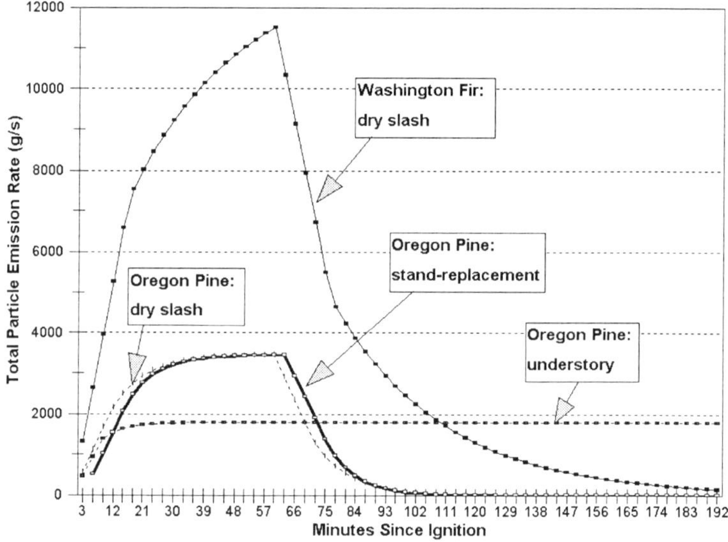

Figure 5. Modelled emissions for typical fuel loading, fuel condition, ambient weather, and ignition pattern in 20-hectare fires for broadcast slash burning and stand-replacement, and for a 120-hectare understory burn in two forest types of the Pacific northwestern United States.

In addition to gas and particles, EPM calculates heat release rates. Therefore, it is possible to estimate the effect on plume rise for each fire. This is an important step in determining the contribution of biomass emissions to the climate system as well as its impact on human health and visibility. Buoyant plumes can carry emissions high into the atmosphere where long residence times are possible, furthering the potential for chemical changes that affect radiation and create ozone. Less buoyant plumes cause emissions to remain relatively close to the ground, compounding human health issues and causing visibility problems that degrade transportation safety and scenic vistas. In estimating the global impacts of biomass burning, Liousse *et al.* (1996) showed that injection height of aerosols significantly affects predicted concentrations.

To show the variability that may be possible among different landuse examples, the very simple approximation proposed by Manins (1985) is used to estimate plume rise in a stably stratified atmosphere,

$$Z = 1434 \; x (P)^{1/4},$$

where Z = plume rise in meters and P = maximum power in gigaWatts (GW). Manins compared this formula with observed plume heights and approximate power estimates ranging from 20 to 60,000 GW from natural fires, volcanic eruptions, and wartime firestorms. The resulting mean square error was estimated at 30% (Harrison and Hardy 1992). Table 2 summarises the maximum power modelled by EPM and the estimated plume rise from Manin's formula.

Tableau 2. Modelled estimates of maximum power from EPM and estimated plume rise from Manins' formula (1985) for each case example

Fire Type	Peak Power (GW)	Plume Rise (meters)
Amazon – dry slash	17.34	2926
Amazon – wet slash	9.37	2509
Cerrado – stand	4.56	2096
Oregon Pine – stand	4.17	2049
Oregon Pine – slash	5.09	2154
Oregon Pine – understory	0.61	1267
Washington Fir – dry slash	15.91	2864

From this exercise it is clear that burning dry slash in both the tropical rainforests of the Amazon and temperate rainforests of western Washington create the greatest heat and therefore the highest plumes, reaching close to 3000 meters. Often cumulus and cumulonimbus clouds develop over these large fires and further turbulence and entrainment enhances their buoyancy. Modelled peak power is significantly less for Oregon pine understory burning but plumes from these fires can exhibit reasonable buoyancy shortly after ignition, which usually coincides with maximum power output. The smouldering phase of a fire outputs a fraction of the maximum power and by the time the independent smouldering stage is reached, power is close to 0.1 GW or less, causing plumes to remain well below 800 meters above ground level.

5.   USING EPM FOR GLOBAL ESTIMATES OF BIOMASS EMISSIONS

The locally derived model, EPM, can be used to estimate global biomass emissions and gain a more complete understanding of the impact of wildland fires. For example, Mack *et al.* (1996) estimated global biomass emissions by developing a global fire model that estimates burning efficiencies from fine dead fuel moisture and by introducing some ability to model the effects of climate and landuse changes over time. Components that significantly affect the rate and magnitude of emissions, however, were not considered. These include ignition patterns, large dead-fuel moisture, and unnatural fuel distributions (e.g., piles and harvest slash), which are explicitly resolved in EPM. Therefore, linking EPM with global models could provide more accurate assessments of the impacts of biomass burning on carbon budgets and the global climate.

In addition to modelling, global emissions can be estimated by remote sensing of smoke. Aircraft or satellite based sensors, however, must make certain assumptions about the character of biomass burns to properly interpret sensed values. For example, Kaufman *et al.* (1996) assumed a simple model of mixed energy when interpreting data from the 1.6 μm radiation channel on NASA's ER–2 MODIS Airborne Simulator (MAS) over the SCAR–C Quinault prescribed fire, which was the same fire for which we obtained fuel loadings in our idealised case of burning dry harvest residue in a fir rainforest. The MAS-derived mixed-energy results were compared with another approach that considered each pixel as black body having homogeneous temperature. Little disagreement was found and their conclusion was that the derived flux of radiative energy was not very sensitive to assumptions in the interpretation algorithms.

When comparing MAS-derived results with EPM output for the Quinault prescribed fire, however, significant differences were found. For example, heat release rates in EPM rose steeply during the ignition phase of the fire whereas the derived values began instantaneously at about 75% of maximum. According to ground-based observers, ignition occurred in 10 stages over a period of about an hour and build-up was rapid, like the EPM simulation, but not instantaneous. Also, the maximum heat release rate from EPM was about 35% greater than the derived maximum radiative energy. Because EPM models total heat at the source, which includes radiative energy as well as convective energy, and the MAS derived values only consider radiative energy, the difference in maximum heat release rates is expected. Another aspect of radiation measurements affect calculated fire size, which can be overestimated if horizontal fluxes are not considered in the remote sensing algorithms.

In their work, Kaufman *et al.* (1996) reported that the MAS-derived particle emission rate was about twice as great as the EPM values. This is not surprising because it appears that they used the EPM $PM_{2.5}$ output, which is only a fraction of total particles. If compared against the EPM total particle emission rate, the maximum EPM rates were about 80% of maximum MAS-derived rates. This is well within the error bounds for both model and remote sensing data.

The MAS-derived radiative energy and particle emission rates reflect similar patterns that are distinctly different than EPM patterns of heat release and particle emissions, especially in the ignition phase of the fire. This suggests problems in the timing and proportion of flaming to smouldering combustion that are used in remote sensing algorithms. It is our belief that use of ground-based observations and models like EPM can improve remote sensing capabilities.

6.   SUMMARY

The EPM test cases show that land clearing and burning of residue slash in dense rain forests (whether temperate or tropical) causes the highest rates of biomass particulate emissions. Also, with comparable heat-release rates (over 15 GW from the 20 hectare plots), plumes from these types of fires are likely to reach high levels of the atmosphere and, therefore, be transported over longer distances. Higher fuel moistures, however, can reduce emissions by 30% or more. Slash burning and stand-replacement fires in dry pine and cerrado forests, where fuels typically are more sparsely distributed, emit less smoke and are less likely to loft above typical mixing heights. Long duration, low magnitude emissions are common in understory burns like those intentionally ignited for ecosystem management in ponderosa pine forests of the Pacific Northwest.

In this paper the EPM model was used to show the distinctive differences in biomass emission rates caused by different landuse patterns. We believe that this type of modelling can provide more accurate estimates of biomass emissions by improving remote sensing algorithms and coupling with global models. For example, errors occur when remote sensing algorithms misrepresent heat release, timing of combustion phases, and the proportion of flaming to smouldering. Likewise, models that neglect ignition patterns, large fuel moisture, and unnatural fuel distributions may produce inaccurate results.

Some caution with these results is necessary. While EPM compares well with tower and aircraft measurements of biomass emissions from broadcast slash burns, and qualitative comparisons between EPM model output and emissions from wildfires, stand-replacement fires, and understory burns appear promising, more quantitative verification is needed. Nevertheless, the EPM model is a viable tool for estimating biomass emission characteristics in a variety of burning styles and fuel structures.

Development of EPM is part of a larger strategy by the Fire and Environmental Research Applications (FERA) team of USDA Forest Service Research to provide one information system that supports; a) single-event risk assessments in fuel management and fire management decisions, b) programmatic risk assessments in developing fire management strategies, and c) large-scale assessments of global change and landuse policy options. Along with EPM, we are developing:

1. simple techniques (such as Photo Series) to inventory the characteristics of wildland fuelbeds;
2. Fuel Characteristic Classes to systematically classify fuelbeds;
3. algorithms that predict the moisture content of fuels based on ambient weather;
4. algorithms that predict fuel consumption during the flaming and smouldering;
5. identification of fire severity thresholds (i.e., conditions of non-linearity of fire effects); and
6. mesoscale climate scenarios, based on historic data or modelled future climates.

These tools are being developed primarily to support fuel management decisions on federal lands, as part of the US Joint Fire Sciences Program. However, companion research is being done along a pole-to-pole "Transect of the Americas" to expand their applicability to all common boreal and tropical ecosystems. We are currently revising the model to modernise its user interface, improve its technical performance relative to long-smouldering fires and non-buoyant plumes, and add data defaults and linkages. We propose use of EPM to develop more precise estimates of global biomass emissions and improved understanding of the impact of biomass emissions on global climate.

7.   REFERENCES

Agee, J.K. 1993. Fire ecology of Pacific Northwest forests. Island Press. Washington, D.C. 493 pp.

Anderson, H.E. 1969. Heat transfer and fire spread. United States Department of Agriculture, Forest Service, Research Report INT–69. Ogden, Utah. 20 pp.

Andreae M.O. 1991. *Biomass burning: its history, use, distribution and its impact on environmental quality and global climate.* In Global Biomass Burning: Atmospheric, Climatic, and Biospheric Implications. Cambridge, MA: MIT Press. J.S. Levine, editor.

Breyfogle, Steve; Ferguson, Sue A. 1996. *User assessment of smoke-dispersion models for wildland biomass burning.* Portland, OR: U.S. Department of Agriculture, Forest Service, Pacific Northwest Research Station; PNW–GTR–379. 30.

Byram, G.M. and R.M. Nelson, Jr. 1974. *Buoyancy characteristics of a fire heat source.* Fire Technology. February:68–79.

Cofer III W. R., E. L. Winstead, B. J. Stocks, L. W. Overbay, J. G. Goldammer, D. R. Cahoon, and J. S. Levine. 1996. *Emissions from Boreal Forest Fires: Are the Atmospheric Impacts Underestimated?* Chap. 79 in *Biomass Burning and Global Change, Volume II.* Cambridge, MA: MIT Press. 834–9.

Ferguson, S.A. and Hardy, C.C. 1994. Modeling smoldering emissions from prescribed broadcast burns in the Pacific Northwest. Int. J. Wildland Fire. 4(2): 135–142.

Finney, M. A. 1998. *FARSITE Fire Area Simulator Version 1.0: Users Guide and Technical Documentation.* Systems for Environmental Management, P.O. Box 8868, Missoula MT 59807.

Hardy, C.C., S.A. Ferguson, P. Speers-Hayes, C.B. Doughty, and D.R. Teesdale. 1993. Assessment of PUFF: *A Dispersion Model for Smoke Management.* Final Report to the Pacific Northwest Region USDA Forest Service, Portland, OR.

Harrison, H. and C. C. Hardy. 1992. Plume rise from prescribed fires: model and data. In: 1992 Annual Meeting of the Pacific Northwest International Section of the Air and Waste Management Association, N. Maykut, Tech. Ed. 11–13 November 1992; Bellevue, WA. Air and Waste Management Association, Pittsburgh, PA. [not paged].

Hegg, D.A., L.F. Radke, P.V. Hobbs, R.A. Rasmussen, and P.J. Riggan. 1990. *Emissions of some trace gases from biomass fires.* J. of Geophys. Res. 95(D5):5699–75.

Hobbs, P.V., J.S. Reid, J.A. Herring, J.D. Nance, R.E. Weiss, J.L. Ross, D.A. Hegg, R.D. Ottmar, and C. Liousse. 1996. *Particle and trace-gas measurements in the smoke from prescribed burns of forest products in the Pacific Northwest.* In Biomass Burning and Global Change Volume 2, Biomass Burning in South America, Southeast Asia, and Temperate and Boreal Ecosystems, and the Oil Fires of Kuwait. J.S. Levine, editor. 2:697–715.

Holdsworth, G., K. Higuchi, K. Zielinski, P.A. Mayewski, M. Wahlen, B. Deck, P. Chylek, B. Johnson, and P. Damiano. 1996. *Historical biomass burning: late 19th century Pioneer Agriculture Revolution in northern Hemisphere ice core data and its atmospheric interpretation.* J of Geophysical Res. 101(D18):23317–23334.

Kauffman, J.B., R.W. Shea, R.F. Hughes, D.L. Cummings, E.A. Castro, and R.D. Ottmar. 1994. *Total aboveground biomass, fuel loads, and combustion factors of Brazilian tropical forests and savannas: a data and photographic summary.* unpublished.

Kaufman, Y.J., L.A. Remer, R.D. Ottmar, D.E. Ward, R.R. Li, R. Kleidman, R.S. Fraser, L. Flynn, D. McDougal, and G. Shelton. 1996. Relationship between remotely sensed fire intensity and rate of emission of smoke: SCAR–C experiment. In Biomass Burning and Global Change Volume 2, Biomass Burning in South America, Southeast Asia, and Temperate and Boreal Ecosystems, and the Oil Fires of Kuwait. J.S. Levine, editor. 2:685–696.

Larsen, C.P.S. 1996. Fire and climate dynamics in the boreal forest of northern Alberta, Canada, from AD 1850 to 1989. The Holocene. 6(4):449–456.

Liousse, C., J.E. Penner, C. Chuang, J.J. Walton, H. Eddleman, and H. Cachier. 1996. A global three-dimensional model study of carbonaceous aerosols. J. of Geophys. Res. 101(D14):19411–19432.

Laursen, K.K. and L.F. Radke. 1996. *Biomass burning smoke in the tropics: from sources to sinks.* In Biomass Burning and Global Change. Volume 1: Remote Sensing, Modeling and Inventory Development, and Biomass Burning in Africa. J.S. Levine, editor. 1:193–201.

Mack, F. J., Hoffstadt, G. Esser, J.G. Goldammer. J.S. Levine. 1996. *Modeling the influence of vegetation fires on the global carbon cycle.* In Biomass Burning and Global Change. Volume 1: Remote Sensing, Modeling and Inventory Development, and Biomass Burning in Africa. J.S. Levine, editor. 149–159.

Manins, P.C. 1985. Cloud heights and stratospheric injections from a thermonuclear war. J. Atmos. Sci. 19:1249–1255.

McCarter, R.J. and A. Broido. 1975. *Radiative and convective energy from wood crib fires.* Pyrodynamics. 265–85.

Ottmar, R.D. 1992. *Overview of SMS_INFO.* U.S. Department of Agriculture, Pacific Northwest Research Station, Seattle Forestry Sciences Laboratory. [Unpublished].

Ottmar, R. D., M. F. Burns, A. D. Hanson, and J. N. Hall. 1993. *CONSUME Users Guide.* U.S. Department of Agriculture, Forest Service, Pacific Northwest Research Station, Portland, OR.

Ottmar, R.D., E. Alvarado, P.F. Hessburg, *et al..* In Press. *Historical and current forest and range landscapes in the Interior Columbia River Basin and portions of the Klamath and Great Basins.* Part II: Linking vegetation patterns and potential smoke production and fire behavior. Gen. Tech. Rep. PNW–GTR.

Prins, E.M. and W.P. Menzel. 1994. Trends in South American biomass burning detected with the GOES visible infrared spin scan radiometer atmospheric sounder from 1983 to 1991. J. of Geophys. Res. 99(D8):16719–16735.

Sandberg, D.V.; Peterson, J. 1984. *A source of strength model for prescribed fires in coniferous logging slash.* 12–14 November 1984; Portland, Oregon: Air Pollution Control Association: 10.

Sandberg, D.V.; Ferguson, S.A. In preparation. *A dynamic model of initial effects from wildland biomass fires: EPMv2.0.*

U.S. Environmental Protection Agency. 1991. Supplement A to compilation of air pollutant emission factors, Volume I, AP–42. U.S. Environmental Protection Agency, Office of Air Quality Planning and Standards, Research Triangle Park, N.C.

Ward, D.E., C.C. Hardy, D.V. Sandberg, T.E. Reinhardt. 1989. Part III – emissions characterization. In Sandberg, D.V., D.E. Ward, R.D. Ottmar and others. Mitigation of prescribed fire atmospheric pollution through increased utilization of hardwoods, piled residues, and long-needled conifers. Final report: U.S. DOE, EPA. Available from U.S. Department of Agriculture, Forest Service, Pacific Northwest Research Station, Seattle, WA.

Direct Effects of Fire on the Boreal Forest Carbon Budget

ERIC S. KASISCHKE[1], BRIAN J. STOCKS[2], KATHY O'NEILL[3], NANCY H.F. FRENCH[1] and LAURA L. BOURGEAU-CHAVEZ[1]

[1]*ERIM International, P.O. Box 134008, Ann Arbor, USA*
[2]*Canadian Forest Service, P.O. Box 490, Sault Ste. Marie, Ontario, Canada*
[3]*Nicholas School of the Environment, Duke University, Durham, USA*

Abstract: Past approaches to estimating the amounts of carbon released during fires in boreal forests have depended on two types of data:
1) those collected during prescribed burns; or
2) those collected from a limited number of points in naturally-occurring fires.
Neither of these approaches is felt to produce information that is reliable in terms of describing the conditions that exist in natural fires. In this paper, we present the results of the study of two fires located in the boreal forest of interior Alaska, one during the summer of 1990 and the second during the summer of 1994. In this study, we utilized remote sensing imagery for two purposes. First, pre-fire imagery was used to map the distribution of vegetation/forest cover prior to the fire. A field-sampling campaign was then conducted to estimate the levels of above and below-ground biomass levels associated with each vegetation/forest category. Second, approaches were developed to use satellite imagery from the Landsat Thematic Mapper to estimate three levels of fire severity in the burned forests. Ground data were again collected to estimate the percentage of biomass consumed during fires in each of these classes. By combining the two sets of information, it was possible to estimate the amounts of carbon-based greenhouse gases released during each fire. In the two fires, 25 and 38 t C ha^{-1} were released, values which are two to three times greater than previously estimated for boreal forest ecosystems. These studies show that most (85%) of the carbon released originates from the burning of biomass in the ground layer (e.g., mosses, litter, and organic soil), which contrasts greatly with biomass burning in other biomes, where almost all the biomass consumed is in above-ground vegetation. To estimate the different species of gases released, we developed a fire behaviour model to divide the fires into two classes: flaming and smouldering. We then used published values to estimate the amount of different greenhouses released during these two fire phases. In a relative sense, much lower levels of carbon dioxide and higher levels of methane and carbon

monoxide are released during flaming fires in boreal forests because smouldering fires are responsible for burning ground-layer organic matter.

1.   INTRODUCTION

Over the last few decades, much attention has focused on identifying the terrestrial sources and sinks of atmospheric carbon (Tans *et al.*, 1992; Fan *et al.*, 1998). This issue is driven by the fact that over the past four decades the rise in the atmospheric concentration of carbon dioxide has much been much lower than the amounts of carbon being released into the atmosphere through the use of fossil fuels, the clearing of tropical forests, and the burning of biomass and other materials. The rate of increase in atmospheric carbon is about 3.3 Gt (Gt = gigaton = 10^{15} g) yr^{-1}, much lower than the current levels of anthropogenic additions (currently 7.6 Gt C yr^{-1}). While it is thought that the increased partial pressure of atmospheric CO_2 results in 2.0 Gt C yr^{-1} being absorbed by the world's oceans, the fate of the remaining carbon is still unknown. The consensus within the scientific community is that a carbon sink of about 2 Gt yr^{-1} exists within the terrestrial ecosystems of the northern hemisphere (Schlesinger, 1997; Fan *et al.*, 1998).

Four biomes in the northern hemisphere are possibilities for the terrestrial carbon sink: (1) boreal forests; (2) northern peatlands; (3) tundra, and (4) temperate forests. Table 1 summarises the carbon stored in each of these biomes. All available evidence indicates that most of the carbon present in these biomes is in the soil layer.

Table 1. Summary of total amounts of carbon present in northern
terrestrial biomes of Alaska and Canada, and Eurasia

Biome	*Area* *(x 10^6 ha)*	*Plant Biomass* *Carbon[1] (Gt)*	*Soil Carbon[2]* *(Gt)*
Boreal forest[3,4]	1249	78	227
Peatlands[5]	260	< 0.1	397
Tundra[3]	890	< 0.1	180
Temperate Forest[6]	1040	21	100

[1]Includes plant biomass and plant detritus for the boreal forest

[2]Includes forest soils and peat

[3]Apps *et al.* (1993)

[4]Alexyev and Birdsey (1997)

[5]Zoltai and Martikainen (1996)

[6]Heath *et al.* (1993)

A recent study involving inverse modelling of atmospheric concentrations of CO_2 presents an argument that the main terrestrial sink for atmospheric carbon lies in the temperate forests of North America below 51° N latitude, with the North American and Eurasian boreal forest serving as a weak carbon sink during the period of 1988 to 1992, averaging an uptake of 0.4 Gt C yr $^{-1}$ (Fan *et al.*, 1998). However, it is important to understand the time scales over which different biomes are acting as carbon sources or sinks. For example, the reason that the temperate forest is serving as a carbon sink today stems from the significant regrowth of forest on lands that were cleared over the past three centuries, particularly in the eastern United States and Canada. In this case, human landuse and management practices are playing a direct role in carbon source–sink relationships.

This is not the case in the boreal region where the dominant disturbances are fires whose ignition source in remote areas is lightning. While forest clearing and land conversion has led to some loss of terrestrial carbon in Russia over the past century (Melillo *et al.*, 1988), the scale of these losses is much lower than found in temperate and tropical forests.

In this chapter and its companion, we discuss the role that fire plays in the boreal forest carbon budget. In this chapter, we present a generalised model for quantifying the boreal forest carbon budget and discuss direct emissions of carbon from fires. In a companion chapter in this volume (Kasischke *et al.* 1999a), we discuss the indirect effects that fire has on the carbon budget, as well as how future climate changes are likely to influence the fire regime in the boreal forest, and the implications for changes to the carbon budget.

The present chapter is organised into five sections, including this introduction. Section 2 presents an overview of the boreal forest carbon budget, while Section 3 presents a basic model for estimating carbon release from fires in the boreal forest. Section 3 also discusses how this model was exercised for the North American boreal forest. Section 4 discusses a particularly important aspect of the fire–carbon model, the consumption of ground layer biomass, including organic soil, during fires. Finally, Section 5 discusses the global implications of the results presented in this chapter.

2. THE BOREAL FOREST CARBON BUDGET

Quantifying the rates carbon exchange between the atmosphere and the boreal forest requires measuring or modelling four distinct processes. There is only one process that removes carbon from the atmosphere and add it to the boreal forest – photosynthesis (P_s). In contrast, there are three processes that remove carbon from the boreal forest and add it to the atmosphere: plant respiration (R_p); heterotrophic (soil) respiration (R_h); and oxidation of plant material during fires (O_f). Net biome production (NBP: in gm C m^{-2} yr^{-1}) is simply expressed as

$$NBP = P_s - R_p - R_h - O_f \qquad (1)$$

Two parameters derived from the terms in Eq. (1) are presently being measured by the scientific community: Net Primary Production (NPP) and Net Ecosystem Production (NEP), where

$$NPP = P_s - R_p \qquad (2)$$

and

$$NEP = NPP - R_h = P_s - R_p - R_h \qquad (3)$$

NPP represents the net seasonal growth added to living plant material, while NEP represents the net plant biomass (living and dead) added to a biome since it accounts for decomposition. The reason that the boreal forest contains so much carbon is not because of high rates of NPP, but because of low rates of heterotrophic respiration. These low decomposition rates are caused by the generally low soil temperatures found throughout the year in the boreal region, which in many cases result in the formation of permafrost.

Only recently have scientists been able to estimate NPP and NEP in different vegetated biomes. Methods have been developed to use satellite imagery to monitor seasonal patterns of vegetation greenness and canopy temperature. These parameters are then used as inputs for models estimating NPP over large areas (Liu *et al.*, 1997; Running and Hunt, 1993). Advances in technology and field data collection approaches through tower eddy correlation measurements allow for the estimation of NEP over kilometre-scale patches of uniform vegetation cover with an accuracy of ± 20 gm C m^{-2} yr^{-1} (Wofsy *et al.*, 1993; Goulden *et al.*, 1998). If properly scaled through collection of data in different forest ecosystems of different ages, a continuous series of tower measurements over decadal scales offer the opportunity to not only estimate NEP, but to understand how inter-annual variations in photosynthesis and plant and soil respiration vary as a function of climate.

However, the tower measurements cannot be used to estimate NBP because they do not measure losses from fire. Hence, a network of eddy correlation towers deployed throughout the boreal region will not enable the determination of whether or not the boreal forest is a net source or sink of atmospheric carbon. This requires computing the annual carbon losses through biomass burning during wildfires.

Studies have shown NPP ranges from 300 to 400 g C m^{-2} yr^{-1} for the more productive forests found in the southern part of the North American boreal forest to 10 to 25 g C m^{-2} yr^{-1} for the less productive forests in the northern part (Gower *et al.*, 1997; Liu *et al.*, 1997). NEP has been shown to be 15 to 20 g C m^{-2} yr^{-1} for the least productive (black spruce – *Picea mariana* [Mill.] B.S.P.) forests of this region (Frolking, 1997) and 130 g C m^{-2} yr^{-1} for the more productive (aspen – *Populus* spp.) forests (Black *et al.*, 1996).

3. ESTIMATING CARBON RELEASE DURING FIRES

Seiler and Crutzen (1980) postulated the basic equation used to estimate carbon released during biomass burning

$$O_f = A \; B \; f_c \; \beta \tag{4}$$

where A is the total area burned during a specified period of time (in hectares), typically on an annual basis, B is the biomass density (t ha^{-1}) exposed to fire, f_c is the carbon fraction of this biomass, and β is the fraction of the biomass that is burned or oxidised during the fire. The initial studies of carbon release from biomass burning by Seiler and Crutzen (1980) only considered consumption of above-ground biomass. Biomass burning in the boreal forest is more complex because the deep layers of mosses and organic soils common in this biome can also be consumed during fires. To account for this variation, we modified Eq. (4) to

$$O_f = A \; [(B_a \; f_{ca} \; \beta_a) + (C_g \; \beta_g)] \tag{5}$$

where the a and g subscripts refer the above-ground and ground-layer biomass components (mosses, lichen, litter, and the deep organic soil layers), respectively, and C_g is the average carbon density of the ground-layer components.

The model expressed in Eq. (5) was used to estimate carbon release from fires in the North American boreal forest during the period of 1980 to 1989. The first step in this study was to divide the territory of Canada and Alaska into ecoregions (Figure 1) based upon the ecosystem classifications of Wikens *et al.* (1993) and Gallant *et al.* (1995). Five of these ecoregions comprise the boreal forest of North America: subarctic, Alaska subarctic, boreal east, boreal west, and cordilleran. Using these divisions, the parameters in Eq. (5) were obtained in the following manner.

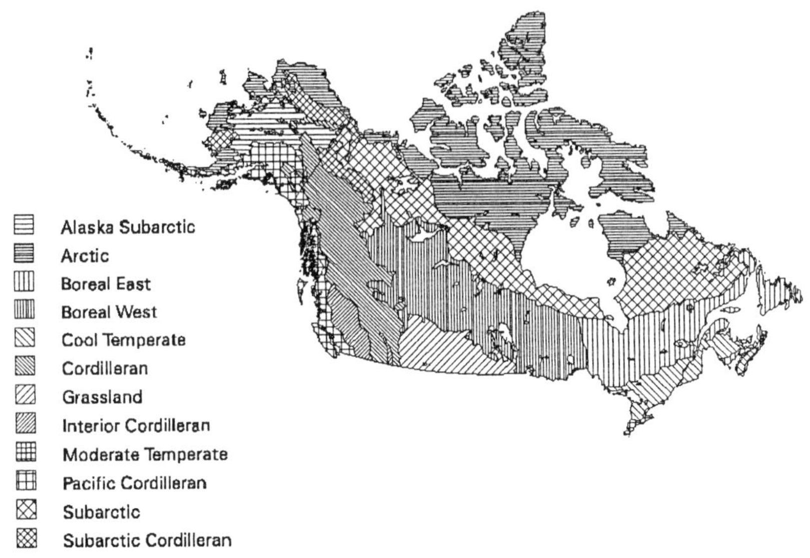

Figure 1. Major forest ecoregions of Canada and Alaska.

Total area burned for the different ecoregions (A) was obtained by reviewing records maintained by different fire and resource management agencies in Canada and Alaska. These records contained an annual summary of fire sizes and locations, as well as maps of fire boundaries based on field records, aerial surveys and analyses of satellite imagery.

In terms of understanding how much forest area is burned, the information contained in the fire records may in fact represent an overestimation. The information contained within fire records typically represents the area within the outer perimeter of a fire, A_p. A certain fraction of the area within the perimeter is actually forest (F), but perimeters can also contain non-forested areas that do not burn, such as wetlands. Thus, to estimate total area burned, we used the following relationship

$$A = A_p F \tag{6}$$

Copies of fire perimeter maps from the 1980 through 1989 fire seasons were digitised and entered into a geographic information system. Merging these maps with overlays of the ecoregion boundaries within a geographic information system enables the estimation of the annual area burned in each ecoregion on an annual basis (the fourth column in Table 2), as well plotting the locations of individual fires.

We assumed that the fraction of forest cover within a given burn was identical to the fraction of area covered by forest in that region (column 2 in Table 2) based on data from the Canadian Forest Service (Penner *et al.*, 1997).

Figure 2 presents a plot of annual area burned for the different ecoregions during the 1980s. The data show that 84% of the area burned in the North American boreal forest occurs within two regions: the boreal west (64%) and the subarctic (20%). They also show that two-thirds of the area burned during the 1980s occurred in just three years: 1980, 1981 and 1989.

Table 2. Parameters used to model carbon release during biomass burning in the North American boreal forest

Ecoregion	Fraction of Forest Cover (F)	Area (10^6 ha)	Annual Area Burned (ha) (A_p)	Carbon Density (t ha^{-1})		Average Fraction Carbon Consumption (minimum – maximum values)	
				Above-ground ($f_{ca}B_a$)	Ground-layer (C_g)	Above-ground (β_a)	Ground-layer (β_g)
Subarctic	0.65	170.02	567,772	10	110	0.25(0.13–0.35)	0.05(0.02–0.25)
Alaska Subarctic	0.75	40.24	131,817	19	56	0.23(0.11–0.33)	0.38(0.19–0.48)
Boreal West	0.90	159.61	1,859,059	17	108	0.26(0.13–0.36)	0.06(0.03–0.26)
Boreal East	0.94	91.00	134,938	27	134	0.22(0.11–0.32)	0.06(0.03–0.26)
Cordilleran	0.67	98.52	190,808	60	57	0.13(0.06–0.23)	0.39(0.19–0.59)
Total	0.78	559.39	2,884,394				

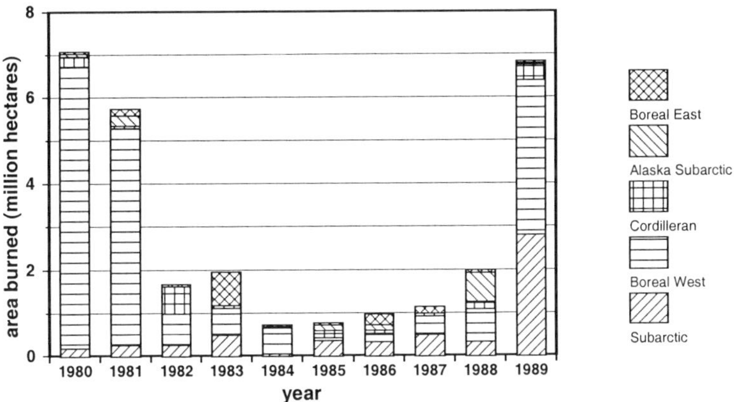

Figure 2. Annual area burned in the different ecoregions of the North American boreal forest for 1980 to 1989

Figure 3 presents total area burned for the entire territory of Canada and Alaska for the period of 1970 through 1998 along with area burned for the five boreal forest ecoregions for the years 1980 to 1989. This plot shows that the majority of fires in Canada and Alaska (97%) in the time period from 1980 to 1989 occurred within the boreal forests of this region. Preliminary analysis of fire record data from the 1990s shows that during the large fire years of 1994, 1995 and 1998, most of the area burned also occurred in the boreal ecoregions (B.J. Stocks, unpublished data).

Studies of the relationship between climate and fire occurrence in the Canadian boreal forest show that large-scale outbreaks only occur during periods when the climate is warmer and drier than the longer-term norms, e.g., when drought conditions make the forests susceptible to fire (Flannigan and Harrington, 1988). During the six most active fire years between 1970 and 1998, a total of 37.2 million ha burned (6.2 million ha yr^{-1}), while during the remaining 23 years, a total of 35.5 million ha burned (1.5 million ha yr^{-1}). Figure 3 also suggests there has been a marked increase in fire activity over the past three decades. The average area burned during the 1970s was 1.5 million ha yr^{-1}; during the 1980s, 3.0 million ha yr^{-1}; and during the 1990s, 3.2 million ha yr^{-1}).

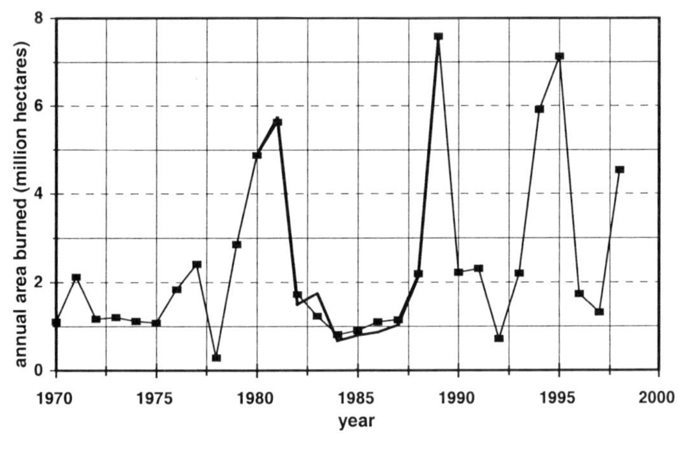

Figure 3. Annual area burned throughout Canada and Alaska for 1970 through 1998.

Above-ground biomass densities (B$_a$) for the different ecoregions were based on summary data for Canada (Kurz and Apps, 1993) and Alaska (Kasischke *et al.*, 1995a,b). The carbon fraction of above-ground vegetation (f$_{ca}$) was assumed to be 0.5. The average above-ground carbon values are presented in column 5 in Table 2. The carbon densities of the ground layer biomass (C$_g$) were again obtained from published data summaries (Kasischke *et al.*, 1999b; Tarnocai, 1998) (column 6 in Table 2).

The average levels of biomass consumption during fires (β$_a$ and β$_g$) in the different ecoregions were obtained from data generated during controlled fires in Canada (Stocks and Kauffman, 1997) and field observations in Alaska (Kasischke *et al.*, 1999b) (columns 7 and 8 in Table 2). These studies have clearly shown that the fraction of biomass consumed during fires varies a great deal, depending on fuel (forest) type and fuel moisture conditions at time of fire. Higher levels of biomass consumption occur during years with warmer, drier conditions. As permafrost hinders soil drainage, the fraction of consumption of ground-layer biomass tends to be higher when fires occur later in the growing season, when the depths to permafrost are greatest. Deeper depths to permafrost result in better drained soil and drier fuel conditions.

A simple model was used to account for variations in fraction of biomass consumed based on the close correlation between both annual area burned and fraction of biomass consumption and inter-annual variations in climate. Larger areas are burned and higher levels of biomass consumption are experienced during warmer, drier years and conversely, during cooler, wetter years, total area burned and fraction of biomass consumed are lower.

For each region, during the lowest fire year it was assumed

$$\beta_{a\text{-low}} = 0.5\ \beta_{a\text{-avg}} \tag{7a}$$

$$\beta_{g\text{-low}} = 0.5\ \beta_{g\text{-avg}} \tag{7b}$$

During the highest fire year in each region, it was assumed that:

$$\beta_{a\text{-high}} = \beta_{a\text{-avg}} + 0.1 \tag{8a}$$

$$\beta_{g\text{-high}} = \beta_{g\text{-avg}} + 0.2 \tag{8b}$$

Figure 4 presents a plot of the fraction of carbon consumed for the boreal west region based on the average values presented in Table 2 and the values calculated using Equations (7) and (8) plotted as a function of the area burned for the three years (minimum, average, and maximum). These plots show a linear relationship between annual area burned and burn fraction. In a similar fashion, linear relationships between annual area burned and fraction of biomass consumed were established for all regions.

Figure 5 summarises average carbon release from fires in the North American boreal forest ecoregions. Two cases are presented. The first case represents the average for all fires during the 1980s, while the second case represents the year with the highest burned area, 1980. During the 1980s, the model predicts that an average of 7.6 g C m^{-2} (of total land area) yr^{-1} was released during fires, with a total carbon release of 0.055 Gt C yr^{-1}. During the severe fire year (1980), an average of 23.2 g C m^{-2} yr^{-1} was released during fires, with a total carbon release of 0.189 Gt C yr^{-1}.

During the severe fire year of 1980, an average of 114.8 g C m^{-2} yr^{-1} was released during fires in the western boreal forest zone (Figure 1), the region of North America where most fires occurred during the 1980's (Table 2). The forest cover of this region is dominated by black spruce and jack pine (*Pinus banksiana*), with some aspen forests. Based on eddy correlation measurements made in these different forest stands (Black *et al.*, 1996; Goulden *et al.*, 1998; Gower *et al.*, 1997), it is estimated that NEP for these forests is of the order of 60 to 80 g C m^{-2} yr^{-1}. This means that during severe fire years, these forests may be serving as a net carbon source for the atmosphere, not a sink.

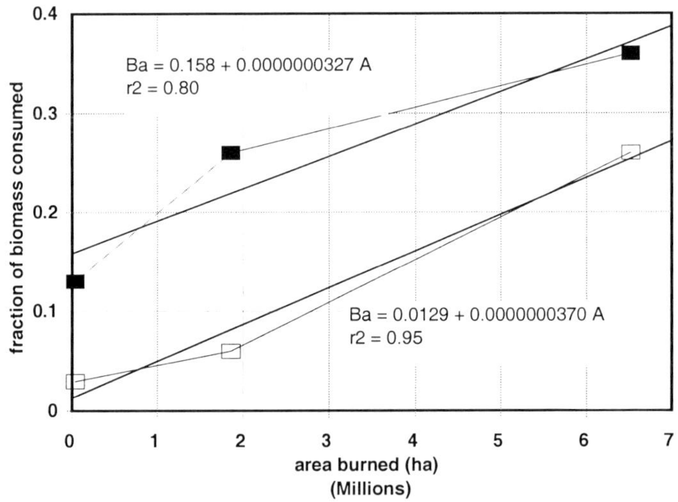

Figure 4. Illustration of the linear relationship between annual area burned and fraction of biomass consumed in the western boreal ecoregion based on the relationships expressed in Eqs. (7) and (8)

Caution needs to be taken when considering the long-term implications of the above conclusion that certain regions of the boreal forest may be serving as a net carbon sink. First, scientists are only just beginning to obtain the long-term measurements in a wide range of forest types that will allow estimation of NEP. At the same time, relatively little effort has been devoted to quantifying the patterns of biomass consumption during fires in different forest types. As illustrated in the next section, there is a high degree of variability in biomass burning in boreal forests. What is clear is that if fire activity continues to increase in the boreal forest, then estimating the amounts of carbon released directly by fire will become an extremely important term in understanding whether or not the boreal forest remains a carbon sink, or becomes an atmospheric carbon source (Kasischke 1996; Kasischke *et al.* 1995b).

Figure 5a. Average carbon release during the 1980's.

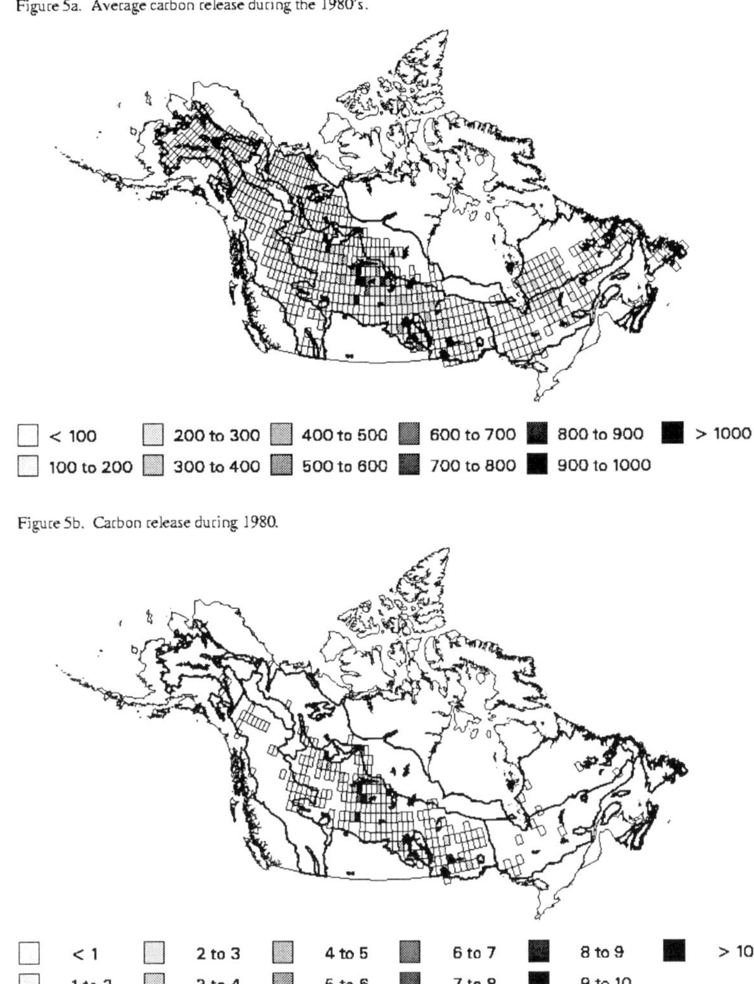

Figure 5b. Carbon release during 1980.

Figure 5. Total carbon release from fires in the North American boreal forest).
(a) Average for the years 1980 to 1989 (units are Kt C = 10^9 g);
(b) estimates for severe fire year of 1980 (units are Mt C = 10^{12} g).

4. PATTERNS OF GROUND-LAYER BIOMASS CONSUMPTION

As discussed above, boreal forests are unique because a significant fraction of the biomass present in the ground layer can and does burn during fires. The components of the ground layer biomass include horizontal layers of litter, lichen, living and dead moss, fibric soil and humic soil. The combined depth of these layers often is 30 to 40 cm or deeper, with the levels of stored carbon reaching 120 to 150 t ha^{-1}.

One of the major unresolved issues in estimating carbon release from boreal forest fires is the degree to which ground layer biomass is burned. Controlled burns in Canada and Russia showed modest levels of ground-layer biomass burning, with carbon releases of 5 to 10 t ha^{-1} (Firescan Science Team, 1996; Stocks and Kauffman, 1997). These controlled burns, however, occurred in forests with relatively shallow organic soil layers or under conditions where the moisture content of this layer of the fuel matrix was high. Therefore, these observations may not be representative of biomass burning patterns throughout the boreal forest, especially during drought years.

A recent study by Kasischke *et al.* (1999b) investigated patterns of ground-layer biomass burning in naturally-occurring fires located in the Alaskan boreal forest. A set of 13 study areas were located in 10 different black spruce stands, a single sample was collected in a white spruce (*Picea glauca*) stand and a single sample in an aspen stand. For each of the study areas, a burned site was located immediately adjacent to an unburned site. The size and density distributions of the overstory trees at each site were selected to be nearly identical; the comparison of data from the burned versus unburned sites enabled the estimation of how much ground-layer biomass was consumed during fires.

The first step in the analysis procedure was to measure the average depth of the various ground layers (live and dead moss, lichen, litter, fibric soil, humic soil) in a burned site and adjacent unburned site. Ground layer depths were measured at 40 different randomly located points along transects in each burned and unburned site. Samples were also collected from different ground biomass layers for laboratory analysis to determine bulk density and percent carbon content. These measurements were used to estimate the carbon density per unit depth for each biomass layer.

Figure 6 presents the average depths for each sample location and Figure 7 presents the carbon levels derived from these data. The average fraction of carbon consumed in the ground layer was 0.423 for the black spruce stands, 0.665 for the white spruce stand, and 0.821 for the aspen stand. The average carbon release for the black spruce stands was 36.1 t ha^{-1}, 20.1 t ha^{-1} for the white spruce stand, and 9.1 t ha^{-1} for the aspen stand. Studies have shown that when the distribution of forest types are accounted for, the average carbon release from fires in the Alaskan boreal forest (including above-ground and ground-layer biomass) ranged from 25 to 35 t ha^{-1} (Kasischke *et al.*, 1999b; Michalek *et al.*, 1999). These data were used to estimate the average fraction of biomass consumed for the Alaskan subarctic ecoregion in Table 2.

a. unburned soil depths

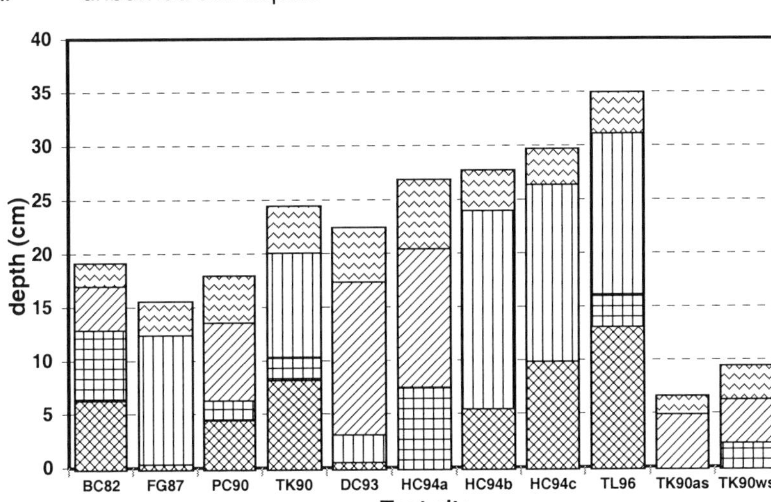

b. burned soil depths

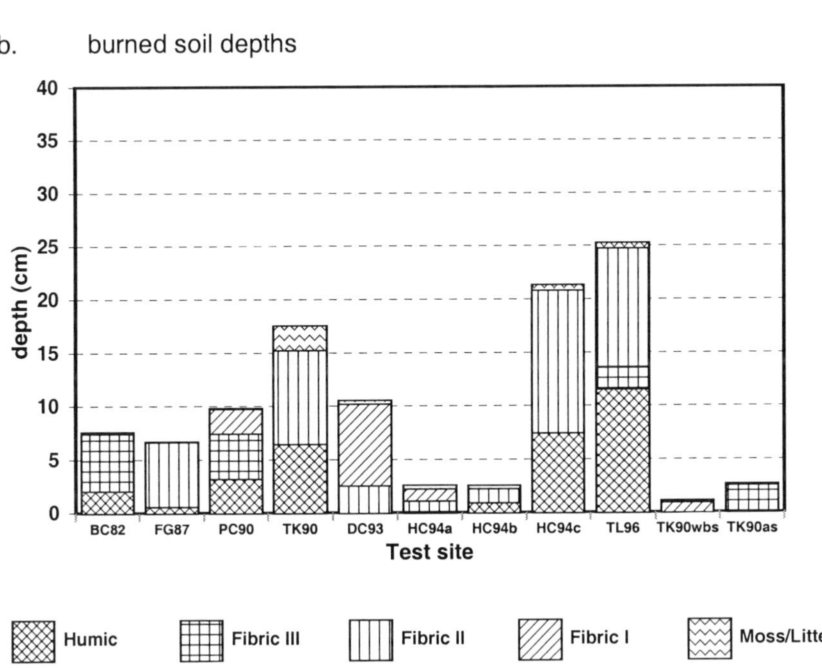

Figure 6. Depths of ground layer biomass present in (a) unburned and (b) burned forest stands located in interior Alaska. (Note the letter/number combinations on the X-axis are designators for the various test sites. All the sites were black spruce stands with the exception of TK90as [aspen] and [TK90ws]. The letters denote a geographical location, while the numbers represent the year of the burn. For example, BC82 is the designator for a fire that occurred near Bonanza Creek in 1982).

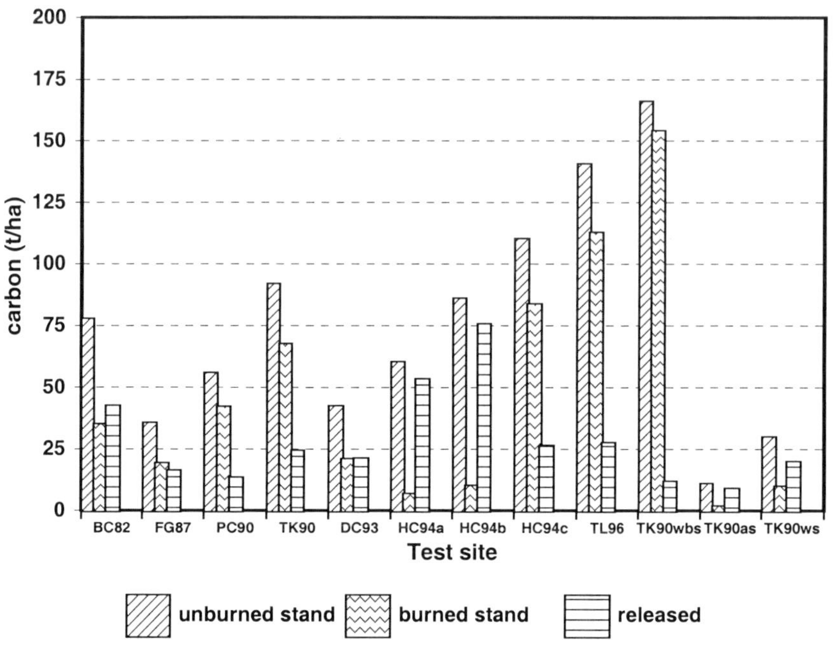

Figure 7. Estimates of carbon release from burning of
ground-layer biomass in Alaskan test stands (see Figure 6).

Finally, the data presented in Figure 7 were analysed with respect to effects of timing of the fire on ground-layer biomass burning. Fires in black spruce stands that burned in the May – July time period experienced 35% consumption of the ground layer biomass, whereas fires that burned in August – September experienced 83% consumption. This result was not surprising as the depth to the top of the permafrost increases later in the growing season, resulting in better soil drainage and lower soil moistures.

In summary, the results from the Alaskan study clearly show that the deeper organic soils common to many boreal forest types burn during fires. Fires that occur late in the growing season when fire conditions are extreme have a high probability of consuming large amounts of ground-layer biomass (> 90% of the carbon released), releasing correspondingly large amounts of carbon (50 – 75 t C ha^{-1} burned) to the atmosphere.

5.   GLOBAL IMPLICATIONS

In their landmark study of carbon releases from fires in different biomes, Seiler and Crutzen (1980) estimated that only 0.023 Gt C yr^{-1} were being released during biomass burning in the boreal forest. It is now generally agreed that this estimate is low because of: (1) underestimation of the annual area burned in the boreal forest and, (2) the approach did not take into account the large amounts of ground-layer biomass consumed by fires.

Although the results presented in this chapter demonstrate that we have the capability to estimate carbon release from the North American boreal forest, similar studies have yet to be carried out in the Russian boreal forest (where two-thirds of the world's boreal forests are located). In addition, there is considerable debate in the boreal forest community as to not only how much fire actually occurs in the Russian boreal forest, but also on the characteristics of these fires.

Figure 8 presents a plot of annual area burned in the North American and Eurasian boreal forests based on official government statistics. Russian fire statistics do not cover their entire boreal forest region, but only that portion where active fire suppression occurs. This protected zone consists of two-thirds of Russia's forests. North American fire scientists have a difficult time believing that even though the area of the Russian forests covered by the data presented in Figure 8 is 50% larger that the North American boreal forest, the average area burned is less than 40% of that reported for North America. The strong suspicion that the Russia fire statistics were under-reported was confirmed by an analysis of AVHRR imagery collected during the summer of 1987. This analysis showed that at least 12 million hectares of boreal forest in Russia burned that year, most of it within the areas that were monitored by Russian foresters (Cahoon *et al.*, 1994). This compares to the official estimate of 1.27 million hectares of fire. More recently, analysis of AVHRR imagery reveals that at least 5.7 million hectares of Russian boreal forest burned in 1998 in the Russian Far East, compared to government estimates of 2.0 million hectares for that region (Kasischke *et al.*, 1999c). Interviews with local forest officials have since revealed the systematic under-reporting of fires in Russia because of financial incentives provided to those regions that achieved a high level of fire suppression (Conard and Ivanova, 1998).

Another point of debate between North American and Russian fire scientists is the types of fires that occur in the Russian boreal forests. In North America, most fires (>90%) are crown fires, while the Russians report that most fires (>90%) in the Eurasian boreal forest are surface fires. Examination of AVHRR imagery clearly shows that the fires in Russia during the severe fire years of 1987 and 1998 are high energy events identical to crown fires observed in the North American boreal forest. Based on this comparison, we believe that during severe fire years, most of the area burned in Russia occurs in crown fires.

Overall, we conclude that the patterns of fire in North America and Russia are quite similar, in both their extent and severity. We feel the results of the North American boreal forest study can be extrapolated to the entire boreal forest. It is estimated that on average, 0.164 Gt C yr^{-1} were released from fires in the boreal forest during the 1980s. During the most severe fire years, we estimate that between 0.50 and 0.60 Gt C yr^{-1} were released during fires.

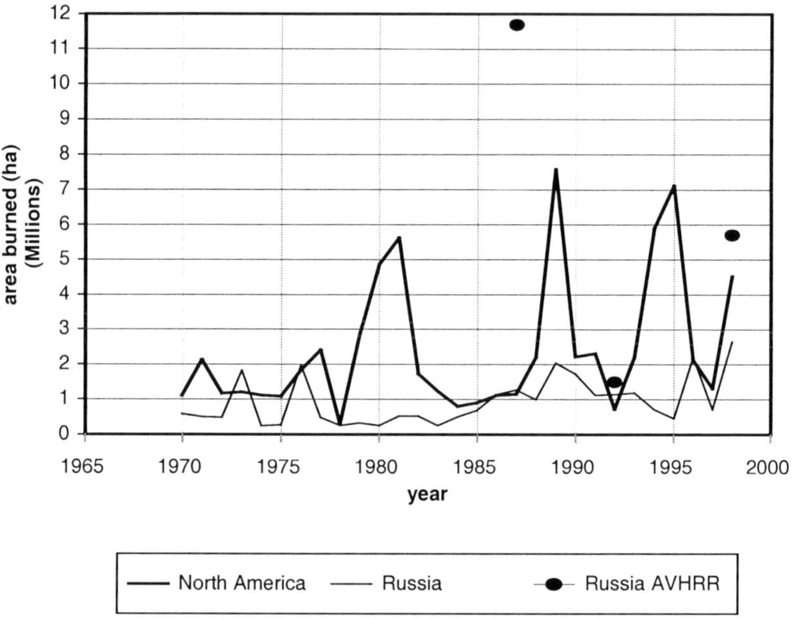

Figure 8. Comparison of annual area burned in North American and Russian boreal forests.

6.   ACKNOWLEDGEMENTS

The research summarised in this chapter was supported by a series of grants from the National Aeronautics and Space Administration and the U.S. Environmental Protection Agency (EPA). It should be noted that the research discussed in this chapter has not been subjected to review by these agencies and therefore does not necessarily reflect the views of these agencies and no official endorsement should be inferred.

7.   REFERENCES

Alexeyev, V.A. and R.A. Birdsey (eds.), 1997. *Carbon Storage in Forests and Peatlands of Russia* (translated from Russian). U.S. Forest Service.

Apps, M.J., W.A. Kurz, R.J. Luxmoore, L.O. Nilsson, R.A. Sedjo, R. Schmidt, L.G. Simpson, and T.S. Vinson 1993. Boreal forests and tundra. *Water, Air, and Soil Pollution 70*:39–53.

Black, T.A., G. den Hartog, H.H. Neumann, P.D. Blanken, P.C. Yang, C. Russel, Z. Nesic, X. Lee, S.G. Chen, and R Staebler, 1996. Annual cycles of water vapour and carbon dioxide fluxes in and above a boreal aspen forest. *Glob. Change. Biol. 2*: 219–239.

Cahoon, D.R., Stocks, B.J., Levine, J.S., Cofer, W.R. III, and Pierson, J.M. 1994. Satellite analysis of the severe 1987 forest fires in northern China and southeastern Siberia. J. Geophys. Res. 99, 18627–18638.

Conard, S.G. and G. A. Ivanova. 1998. Wildfire in Russian boreal forests – potential impacts of fire regime characteristics on emissions and global carbon balance estimates. *Environmental Pollution* 98: 305–313.

Fan, S. M. Gloor, J. Hahlman, S. Pacala, J. Sarmiento, T. Takahashi, and P. Tans, 1998. A large terrestrial carbon sink in North America implied by atmospheric and oceanic carbon dioxide data and models. *Science 282*: 456–458.

FIRESCAN Science Team, 1996. Fire in ecosystems of Boreal Eurasia: the Bor Forest Island fire experiment fire research campaign Asia-north. in *Biomass Burning and Climate Change – Volume 2 – Biomass Burning in South America, Southeast Asia, and Temperate and Boreal Ecosystems, and the Oil Fires of Kuwait*, edited by J.L. Levine, MIT Press, Cambridge, MA, pp. 848–873.

Flannigan, M.D., and J.D. Harrington 1988. A study of the relation of meteorologic variables to monthly provincial area burned by wildfire in Canada. *J. Appl. Meteorol.* 27:4441–4452.

Frolking, S., 1997. Sensitivity of spruce/moss boreal forest net ecosystem productivity to seasonal anomalies in weather. *J. Geophys. Res. 102*, 29,053–29,064.

Gallant, A.L., E.F. Binnian, J.M. Omernik, and M.B. Shasby, 1995. *Ecoregions of Alaska*. U.S. Geological Survey Professional Paper 1567, Washington, D.C., 73 pp.

Goulden, M.L., S.C. Wofsy, J.W. Harden, S.E. Trumbore, P.M. Crill, S.T. Gower, T. Fries, B.C. Daube, S.-M. Fan, D.J. Sutton, A. Bazzaz, and J.W. Munger, 1998. Sensitivity of boreal forest carbon balance to thaw. *Science 279*:214–217.

Gower, S.T., J.G. Vogel, J.M. Norman, C.J. Kucharik, S.J. Steele, and T.K. Snow, 1997. Carbon distribution and aboveground net primary production in aspen, jack pine, and black spruce stands in Saskatchewan and Manitoba, Canada. *J. Geophys. Res. 102:* 29029–29041.

Heath, L.S., P. Kauppi, P. Burschel, H.D. Gregor, R. Guderian, G.H. Kohlmaier, S. Lorenz, D. Overdieck, F. Scholz, H. Thomasius, and M. Weber 1993. Carbon budget of the temperate forest zone. *Water, Air, and Soil Pollution 70*: 55–69.

Kasischke, E.S., 1996. Fire, climate change, and carbon cycling in Alaskan boreal forests. *Biomass Burning and Climate Change – Volume 2 – Biomass Burning in South America, Southeast Asia, and Temperate and Boreal Ecosystems, and the Oil Fires of Kuwait*, edited by J.L. Levine, MIT Press, Cambridge, MA, pp. 827–833.

Kasischke, E.S., N.H.F. French, L.L. Bourgeau-Chavez, and N.L. Christensen, Jr., 1995a. Estimating release of carbon from 1990 and 1991 forest fires in Alaska. *J. Geophys. Res. 100:* 2941–2951.

Kasischke, E.S., N.L. Christensen, Jr., and B.J. Stocks, 1995b. Fire, global warming and the mass balance of carbon in boreal forests. *Ecol. Appl. 5*: 437–451.

Kasischke, E.S., L.L Bourgeau-Chavez, K.P. O'Neill, and N.H.F. French, 1999a. Indirect and long-term effects of fire on the boreal forest carbon budget, (this volume).

Kasischke, E.S, K. P. O'Neill, N.N.H.F. French, and L.L. Bourgeau-Chavez, 1999b. Patterns of biomass burning in Alaskan boreal forests. in *Fire, Climate Change and Carbon*

Cycling in the North American Boreal Forests, Kasischke, E.S., and B.J. Stocks (editors), Ecological Studies Series, Springer-Verlag, New York, (in press).

Kasischke, E.S., K. Bergen, R. Fennimore, F. Sotelo, G. Stephens, A. Janetos, and H.H. Shugart 1999c. Satellite imagery gives a clear picture of Russia's boreal forest fires. *EOS – Transactions of the American Geophysical Union* 80, 141, 147.

Kurz, W.A. and M.J. Apps. 1993. Contribution of northern forests to the global C cycle: Canada as a case study. *Water, Air and Soil Pollution 70*:163–176.

Liu, J., J.M. Chen, J. Cihlar, and W.M. Park 1997. A process-based boreal ecosystem productivity simulator using remote sensing inputs. *Remote Sens. Environ. 62:* 158–175.

Melillo, J.M., J.R. Fruci, R.A. Houghton, B. Moore, III and D.L. Skole, 1988. Land-use change in the Soviet Union between 1850 and 1980: causes of a net release of CO_2 to the atmosphere. *Tellus, 40B:* 116–128.

Michalek, J.L., N.N.H.F. French, E.S. Kasischke, R.D. Johnson, and J.E. Colwell, 1999. Using Landsat TM data to estimate carbon release from burned biomass in an Alaskan spruce forest complex. *Int. J. Remote Sens.* (in press).

Penner, M., Power, K., Muhariwe, C., Teller, R., and Wang, Y. (editors), 1997. *Canada's Forest Biomass Resources: Deriving Estimates from Canada's Forest Inventory.* Canadian Forest Service Information Report BC–X–370, Victoria, BC.

Running, S.W. and E.R. Hunt, Jr., 1993. Generalization of a forest ecosystem process model for other biomes, BIOME–BGC, and an application for global scale models. pp. 141–158 in J.R. Ehleringer and C. Field (eds). *Scaling Processes between Leaf and Landscape Levels.* Academic, San Diego.

Schlesinger, W.H. 199xx. *Biogeochemistry – An Analysis of Global Change, 2nd Edition.* Academic Press, San Diego, 629 pp.

Seiler, W. and P.J. Crutzen, 1980. Estimates of gross and net fluxes of carbon between the biosphere and atmosphere from biomass burning. *Clim. Change, 2:* 207–247.

Stocks, B.J., and Kauffman, J.B, 1997. Biomass consumption and behavior of wildland fires in boreal, temperate, and tropical ecosystems: parameters necessary to interpret historic fire regimes and future fire scenarios. p.169–188 in *Sediment Records of Biomass Burning and Global Change.* J.S. Clark, H. Cachier, J.G. Goldammer and B.J. Stocks (eds), NATO ASI Series, Subseries 1, "Global Environmental Change", Vol. 51, Springer-Verlag, Berlin, Germany.

Tans, P.P., I.Y. Fung and T. Takahashi 1990. Observational constraints on the global atmospheric CO_2 budget. *Science 247*:1431–1438.

Tarnocai, C., 1998. The amount of organic carbon in various soil orders and ecological provinces in Canada. pp. 81–92 in R. Lal, J.M. Kimble, R.F. Follet, and B.A. Stewart (eds), *Soil Processes and the Carbon Cycle,* CRC Press, Boca Raton, Florida.

Wikens, E.B., C.D.A. Rubec, G. Ironside, 1993. Canada terrestrial ecoregions, in *National Atlas of Canada, 5th edition.* (MCR 4164). Canada Centre for Mapping, Energy, Mines and Resources Canada, and State of the Environment Reporting, Environment Canada, Ottawa, Ontario.

Wofsy, S.C., M.L. Goulden, J.W. Munger, S.-M. Fan, P.S. Bakwin, B.C. Daube, S.L. Bassow, and F.A. Bazzaz, 1993. Net exchange of CO_2 in a mid-latitude forest. *Science 260*: 1314–1417.

Zoltai, S.C. and P.J. Martikainen 1996. The role of forested peatlands in the global carbon cycle. pp. 47–58 in M.J. Apps and D.T. Price, editors. *Forest Ecosystems, Forest Management and the Global Carbon Cycle.* NATO Advanced Science Institutes Series, Vol. 140. Springer-Verlag, Heidelberg.

The Impact of Biomass Burning on the Global Budget of Ozone and Ozone Precursors

CLAIRE GRANIER[1,2,3], JEAN-FRANÇOIS MÜLLER[4] and
GUY BRASSEUR[5]
[1]*Service d'Aéronomie CNRS, Paris, France*
[2]*CIRES, University of Colorado, Boulder, CO, USA*
[3]*NOAA Aeronomy Laboratory, Boulder, CO, USA*
[4]*Belgian Institute for Space Aeronomy, Brussels, Belgium*
[5]*National Center for Atmospheric Research, Boulder, CO, USA*

Abstract: Biomass burning contributes significantly to the emissions of atmospheric trace gases such as carbon monoxide, nitrogen oxides, and hydrocarbons. For example, current evaluations estimate that about half of the surface emissions of carbon monoxide result from biomass burning. As these trace species act as ozone precursors, biomass burning plays an important role in the ozone budget of the troposphere. A three-dimensional chemical-transport model of the troposphere has been used to quantify the impact of biomass burning on the distribution of the main chemical tropospheric chemical species. The model results are compared with available observations, and the discussion focusses mostly on carbon monoxide, nitrogen oxides, and ozone global distributions. The contribution of biomass burning in each continent to the global budget of CO is evaluated. This contribution is compared to other emissions of carbon monoxide such as emissions related to industrial activities, soils and oceans, and net production of CO resulting from the oxidation of natural and anthropogenic hydrocarbons. The global and regional production of ozone resulting from these emissions is also discussed. The results of this study show that biomass burning plays a very important role in controlling the distributions of carbon monoxide, not only in the tropics, but also at mid- and high-latitudes. This emphasizes the need for accurate global estimations of the distribution of biomass burning as well as of emission factors for the study of past and future evolution ozone and other key chemical tropospheric species.

1. INTRODUCTION

Biomass burning has both a natural and a human-related origin. Lightning associated with thunderstorms has been responsible for very large fires, especially in boreal forests. Biomass burning has also been used for centuries for various landuse practices, such as deforestation, shifting cultivation, domestic energy production and burning of agricultural residues, especially in tropical areas. These fires affect the atmosphere, as they represent a significant source of chemical species including carbon dioxide, carbon monoxide, methane, non-methane hydrocarbons, nitrous and nitrogen oxides, methyl chloride and methyl bromide, etc. and particles in the lower part of the atmosphere (Levine *et al.*, 1995). These trace gases and aerosols play an important role in the chemistry of the atmosphere and global climate. However, this role is difficult to quantify accurately, as the emissions are themselves difficult to evaluate, due to a large diversity of combustion products, which result from wide ranges in fuel types, fuel chemistry and fire behaviour.

High values of the tropospheric ozone column over western Africa at subtropical latitudes have been observed during the September to November period through the analysis of satellite data (Fishman and Brackett, 1997; Hudson and Thompson, 1998). The contribution of biomass burning to this maximum at the end of the dry season was confirmed by observations made during different field campaigns (Anderson *et al.*, 1993; Thompson *et al.*, 1996; Browell *et al.*, 1996). Measurements performed at different latitudes in the Atlantic Ocean (Weller *et al.*, 1996) have shown that, during summer and spring, low latitudes tropospheric ozone concentrations are higher in the southern hemisphere than in the northern hemisphere. Tropospheric ozone enhancements have also been measured in the Indian Ocean during the September–November period (Baldy *et al.*, 1996), and have been related to biomass burning.

In this paper, we report the results of calculations performed with a three-dimensional chemical-transport model that quantify the importance at the global scale of biomass burning on tropospheric chemical tracers such as CO, nitrogen oxides, non-methane hydrocarbons, ozone and the hydroxyl radical, OH.

2. THREE-DIMENSIONAL MODELLING OF THE TROPOSPHERE USING THE IMAGES MODEL.

2.1 Description of the model

The impact of biomass burning emissions has been quantified by using the most recent version of the three-dimensional chemistry-transport model called IMAGES, and described by Müller and Brasseur (1995), Pham *et al.*

(1995) and Granier *et al.* (1996). The model calculates the distributions of 56 species, with a 5 degree resolution in both latitude and longitude. The vertical domain extends from the surface to 50 hPa, and 25 vertical levels are considered. Large-scale transport of chemical species is computed on the basis of monthly averaged winds and diffusivities derived from the analysis of ECMWF wind fields. Boundary layer processes and the effect of deep convection are parameterised.

The chemical scheme accounts for the most important chemical species in the troposphere, which can affect both the oxidising capacity of the atmosphere, and the photochemical production of ozone. The list of species includes O_x, HO_x, NO_x and SO_x compounds, several hydrocarbons (methane, C_2-C_3 alkanes and alkenes, isoprene, terpenes, acetone, and a surrogate species, called OTHC, for the other non-methane hydrocarbons not explicitly calculated, and their oxidation products). In the current version of the model, the integration time-steps are 30 minutes and 6 hours to simulate respectively the diurnally-varying and diurnally averaged evolution of the chemical compounds. The model results have been extensively evaluated and compared with observations (Müller and Brasseur, 1995; Friedl *et al.*, 1997).

2.2   Surface emissions

Global distributions of surface emissions and of deposition velocities used in the first version of the model were based on the inventory developed by Müller (1992). In the current version of the model, the available emission inventories developed within the IGAC/GEIA (Global Emission Inventory Analysis) are used: they include the fossil fuel emissions of nitrogen and sulphur oxides described by Benkovitz *et al.* (1996), and the biogenic continental emissions of isoprene, terpenes and other non-methane hydrocarbons (Guenther *et al.*, 1995).

Emissions from biomass burning result mainly from forest and savanna fires, agricultural residues and fuelwood burning. Savanna burning is responsible for about 40% of the total biomass burned each year (Andreae, 1991). The amount of biomass burned is evaluated on the basis of the total area burned annually, the above-ground biomass density, the fraction of the biomass burned and the burning efficiency of the above-ground biomass. The total mass of carbon released in the atmosphere is determined assuming that about 45% of the biomass by weight is composed of carbon. In the model, the distribution of CO_2 released by biomass burning is specified on a monthly mean basis following the most recent version of the 5x5 degree resolution inventory developed by Hao and Liu (1994) for tropical biomass and by Müller *et al.* (1992) for temperate and boreal biomass burning. The global estimates of the total amount CO_2 released are given in Table 1, where they are compared with values proposed by Andreae (1991). These two estimates differ by about 25% for the total emissions, and by almost a factor of 2 for the different emission categories, reflecting the limited information available on the amount of biomass burned, as well as the large year-to-year variation in the amount of biomass burned.

Table 1. Global amount of biomass burnt and CO2 emitted as used in the IMAGES model. Comparison with the global estimates of Andreae et al. (1991)

	Total CO_2 emitted from biomass burning (Tg C/yr.) (this study)	Total CO_2 emitted from biomass burning (Tg C/yr.) (Andreae, 1991)
Tropical forests	930	570
boreal and temperate forests	220	130
savannas	1245	1660
fuelwood	425	640
agricultural waste	300	910
Total	3120	3910

The emission ratios, defined as the mass of each trace species produced during the burning process normalized with respect to CO_2, have been measured in the laboratory or during field campaigns. For each species, this emission ratio depends on the type of ecosystem, as shown for example by Hao and Ward (1993): the values used in the model, specified in Table 2, are revised values from Hao and Ward (1993). The amounts of chemical species produced by biomass burning for each type of burning process are then determined by multiplying the total mass of carbon released by the appropriate emission factor.

The total amount emitted at the surface for each chemical species is shown in Table 3. The Table specifies the yearly average emissions from industrial activities, biomass burning and other factors, i.e. generally, oceanic and biogenic sources. The percentage of biomass burning emissions relative to the total emissions is also given in the Table. Biomass burning represents more than 30% of the surface emissions of CO, C_2H_4, C_2H_6 and C_3H_6, between 10 and 20% for NO_x and propane, while it represents less than 10% of the global methane emissions. In order to highlight the regional differences in the emissions from biomass burning, Table 4 provides the emission of CO resulting from industrial activities, biomass burning, and natural processes, as well as the percentage resulting from biomass burning for different areas of the world. Biomass burning represents a yearly average of more than 65% of total emissions in Africa and South America, 18 and 32% in Europe and Northern America, respectively, and around 50% in Asia, where both industrial and biomass burning are important sources of atmospheric gases.

Table 2. Emission ratios using CO2 as a reference for trace gases emissions from biomass burning (in %)

	tropical forests	temp. and boreal forests	savanna	fuelwood	agricultural waste
CO	12.4	14.9	4.2	9.9	8.7
CH_4	1.57	1.05	0.24	2.49	0.44
NO_x	0.21	0.21	0.074	0.65	0.19
C_2H_4	0.36	0.24	0.055	0.57	0.10
C_2H_6	0.12	0.078	0.018	0.18	0.033
C_3H_6	0.11	0.072	0.016	0.17	0.10
C_3H_8	0.044	0.029	0.044	0.70	0.083
OTHC	0.16	0.107	0.024	0.25	0.045

Table 3. Surface emissions of species as used in IMAGES. Total emissions are in Tg C, except for CO (in Tg CO) and NOx (in Tg N)

	Industrial	Biomass burning	Other emissions	Total	% due to biomass burning
NO_x	22	7	5	34	21
CO	471	626	186	1283	49
CH_4	164	42	335	541	8
C_2H_4	5	15	13	33	45
C_2H_6	6	5	2	13	38
C_3H_6	2	6	11	19	32
C_3H_8	10	2	3	15	13
OTHC	83	13	277	373	3
isoprene	0	0	442	442	0
terpenes	0	0	112	112	0

Table 4. Regional surface emissions of CO (in Tg CO/year)

	Industrial emissions	Biomass burning emissions	Other surface emissions	Total	Percentage due to biomass burning emissions
Europe	171	41	12	224	18
Asia	99	157	42	298	53
Africa	24	196	40	260	75
Australia	8	20	8	36	55
North America	145	77	19	241	32
South America	24	135	49	208	65
Oceans			16	16	0
Total	471	626	186	1283	49

3. RESULTS OF THE SIMULATION

Two-year runs of the model were performed, and the results discussed below correspond to the second year of the simulations. The discussion of the results will focus on the impact of biomass burning on the global distribution of two important ozone precursors, CO and nitrogen oxides, as well as on ozone and on the main tropospheric oxidant, the OH radical.

3.1 Impact of biomass burning on the distribution of CO

Most previous studies have compared the results provided by two simulations to study the impact of surface emissions on the global distribution and budget of CO. In the first simulation, considered as the reference case, all surface emissions are included. In the second run, one of the surface emissions is excluded. The impact of this specific surface emission is then calculated from the difference between the two model results. In the present study, we have applied to carbon monoxide (see Granier *et al.*, 1998) the technique discussed by Lamarque *et al.* (1996) for the study of the global nitrogen budget. In the simulations, each CO molecule is "tagged" or "coloured", according to its source or origin, and is considered as an individual species in the model. This technique enables more accurate quantification of the contribution of each source for a given compound, while the other chemical species remain unaffected.

Surface CO mixing ratios have been measured at various sites over the past few years, enabling an accurate determination of its seasonal variation. The model results have been compared with observations collected within the CMDL cooperative network stations, as described in Novelli *et al.* (1998). An example of comparisons between model results and observations is shown in Figure 1, for 3 different sites, Alert (Canada), Ascension Island (Atlantic Ocean), and Crozet Island in the southern Indian Ocean. The monthly average of the observations is indicated by the solid line, and the shaded area represents the minimum and maximum concentrations observed for each month during the measurement period. The model results, shown in broken lines are in rather good agreement with observations, except for the high-latitude station, where the computed CO concentration is lower than observed during winter, which could reflect problems in the representation of transport processes at high latitudes in the northern hemisphere. At high- and mid-latitudes of the northern hemisphere, most stations show a similar seasonal variation, with the highest concentrations being observed in winter and spring. As shown in the figure, these high values result mostly from the contribution of industrial emissions. Emissions of CO from industrial activities are only slightly higher (20% or less) in winter than in summer. The main sink of CO molecules corresponds to their oxidation by the OH radical, whose concentration is strongly dependent on the amount of sunlight available. Most of the seasonal variation of the CO concentration at Alert can therefore be attributed to the higher summertime destruction of industrially produced CO. At this site, biomass burning contributes to about 12 and 19% of the surface concentration in winter and summer, respectively.

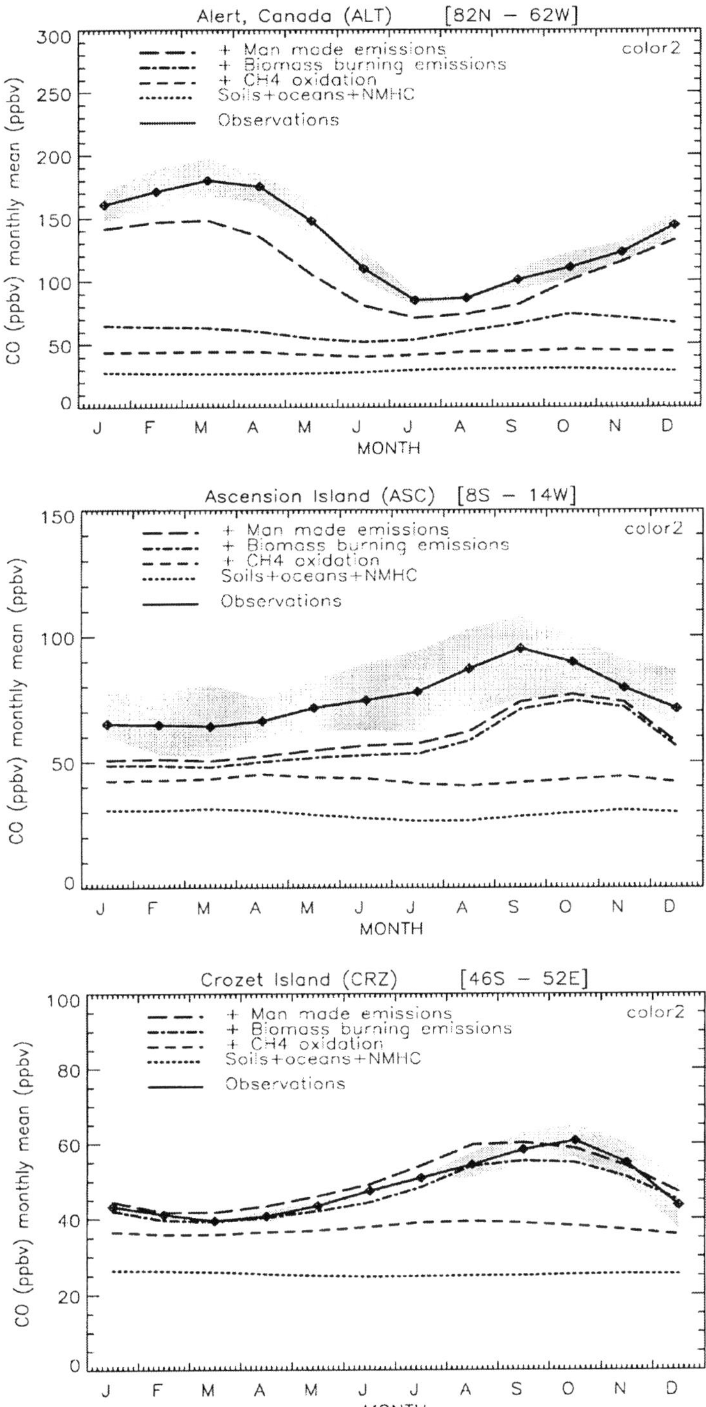

Figure 1. Comparisons between model results and observations for surface CO mixing ratios

The seasonal variation of CO is quite different at the other stations shown on Figure 1 and located at low and mid-latitudes in the southern hemisphere. As seen in Figure 1, the highest concentrations are observed between September and November, and most of the seasonal variation results from the contribution of biomass burning. According to the model, from January to July, most of the CO in this part of the world results from the oxidation of methane and non-methane hydrocarbons. The contribution of biomass burning emissions represents only 9% of the surface CO concentration at this period. During the austral spring, when biomass burning is at its maximum, the contribution of fires represents more than 30% of the CO concentration. At Ascension Island, the calculated seasonal maximum occurs from September to November, later than in the observations, which seems to show that the seasonal variation of biomass burning emissions is not well represented in the surface emissions data base.

Tables 5 and 6 provide the average CO surface mixing ratio (in ppbv) and tropospheric CO column abundance for different areas of the world, and the corresponding contributions from biomass burning, industrial and other natural (soils and oceanic) emissions, and the mixing ratio and column abundance of CO resulting from hydrocarbons (methane and non-methane hydrocarbon) oxidation.

Table 5. Modeled yearly average CO surface concentration (ppbv)
for different areas of the world

	Surface mixing ratio	CO from biomass burning		CO from industrial emissions		CO from hydrocarbon oxidation		CO from soils and oceans	
		ppbv	%	ppbv	%	ppbv	%	ppbv	%
Europe	155	25	16	82	53	39	25	9	6
Asia	127	28	22	50	39	41	32	8	7
Africa	113	39	34	20	18	46	41	8	7
North Amer.	121	24	20	50	41	39	32	8	7
South Amer.	100	34	34	9	9	48	48	9	9
Australia	61	13	21	6	10	37	61	5	8
North Atlan.	96	19	20	31	32	41	43	9	5
South Atlan.	48	9	19	3	6	33	69	3	6
North Pacific	81	15	18	23	28	39	48	4	6
South Pacific	45	6	13	3	7	32	71	4	9
Indian Ocean	47	8	17	3	6	32	68	4	9
Global aver.	87	18	21	25	29	37	42	7	8

The average surface concentration calculated by the model is 87 ppbv, of which 37 ppbv (42%) results from the oxidation of methane and non-methane hydrocarbons. Of the 50 ppbv of CO provided by surface emissions, 18 and 25 ppbv (36 % and 50%) result from industrial and biomass burning, respectively. These numbers should be compared to the values given in Table 4, which show that industrial emissions represent only 37% of the total surface emissions, while biomass burning represents 49% of these emissions. The difference in the contributions of the surface emissions and the surface concentrations reflects the fact that CO is emitted by biomass burning mostly at low latitudes, where the OH concentration is higher than at mid- and high latitudes, where most industrial activities take place. The highest destruction of CO emitted in the tropics by biomass burning is also illustrated by the model results shown in Table 6: the tropospheric CO burden resulting from biomass burning emissions and industrial emissions is similar, despite the large difference in surface emissions.

Table 6. Modeled yearly average tropospheric CO column abundance (Tg CO) for different areas of the world

	CO column	CO column from biomass burning		CO column from industrial emissions		CO column from hydrocarbon oxidation		CO column from soils and oceanic emissions	
		total	%	total	%	total	%	total	%
Europe	19	3	16	8	42	6	31	2	11
Asia	54	11	20	16	30	23	42	4	8
Africa	33	9	27	5	15	17	51	2	7
North Ameri.	31	6	19	10	32	12	39	3	10
South Amer.	20	5	25	1	5	12	60	2	10
Australia	7.5	1.5	20	0.5	7	5	67	.5	6
North Atlan.	22	5	23	6	27	10	45	1	5
South Atlan.	17	3	18	1	6	11	65	2	11
North Pacific	40	8	20	9	22	21	52	2	6
South Pacific	31	5	16	2	6	22	71	2	7
Indian Ocean	25	4	16	2	8	17	68	2	8
Total amount	299.5	60.5	20	60.5	21	156	52	22.5	7

3.2   Impact of biomass burning on the distribution of NO_x.

The "tagging" technique implemented to directly quantify the impact of the different surface sources on the atmospheric distribution has only been applied to CO within the present study. The quantification of the importance of biomass burning on the budget of other chemical species has been determined from the results of two simulations performed with the IMAGES model. In the standard run, all surface sources are taken into account and, in a second run, all emissions resulting from biomass burning are excluded. The comparison between the two runs provides an evaluation of the importance of biomass burning on the behaviour of the chemical species. Note that this technique is not equivalent to the "tagging" procedure described above, where the chemical feedbacks (e.g. influence on the hydroxyl concentrations) are not taken into account.

Table 7 gives the yearly average surface concentration and total tropospheric abundance (in Tg N) obtained in both runs for NO_x, and the percentage difference between the two simulation results. About 13% of the average surface concentration and 8% of the tropospheric total abundance of nitrogen oxides result from biomass burning. Because of the small lifetime of nitrogen oxides, the contribution of biomass burning to surface NO_x near burning areas is much larger. Biomass burning accounts for example for more than 40% of the surface concentration of nitrogen oxides in Africa and South America. Over other areas of the world, the main contributors to the distribution of nitrogen oxides are emissions related to industrial activities, natural soils and agricultural activities.

Table 7. Yearly average NOx surface concentration (pptv)
and total tropospheric NOx amount (in Tg N)

	NO_x surface concentration			tropospheric total NO_x		
	with biomass burning	without biomass burning	% difference	with biomass burning	without biomass burning	% difference
Europe	870	855	2	0.0156	0.0151	3
Asia	230	200	15	0.0270	0.0245	10
Africa	250	175	43	0.0205	0.0174	18
North America	330	309	7	0.0183	0.0172	6
South America	245	170	44	0.0109	0.0092	18
Australia	162	143	13	0.0061	0.0058	5
North Atlantic	45	40	12	0.0105	0.0097	8
South Atlantic	7.7	6.5	18	0.0077	0.0070	10
North Pacific	15	14	7	0.0140	0.0132	6
South Pacific	6.3	6	5	0.0156	0.0150	4
Indian Ocean	8	7	14	0.0101	0.0095	6
Global average/ total amount	170	150	13	0.156	0.144	8

3.3   Impact of biomass burning on the distribution of ozone and of the OH radical.

The photochemical formation of ozone strongly depends on the concentration of its precursors, particularly nitrogen oxides, CO and hydrocarbons. No "tagging" technique similar to the one applied to CO was available to study the ozone budget, and the impact of biomass burning on ozone has been quantified using the same simulations performed to study the impact of biomass burning on the distribution of nitrogen oxides. Figures 2a and 2b display the surface distribution of ozone in October, for the two simulations, one accounting for all surface emissions, and the second excluding the biomass burning source. Both Figures show a strong spatial variation in surface ozone concentrations; these are very different from one case to the other. When biomass burning is ignored, large ozone concentrations are found mostly over the mid-latitude regions of the northern hemisphere, resulting from the large industrial emissions over these areas. Low ozone values are found in all other regions, specifically in tropical regions. When biomass burning is taken into account, the large ozone concentrations over northern mid-latitudes are still present, and large ozone concentrations reaching 50 ppbv are also found over continental tropical and sub-tropical areas. In addition, significant export towards oceanic areas from biomass burning is also seen on the Figure.

Table 8 provides the yearly average ozone surface mixing ratio (in ppbv) and total tropospheric ozone amount (in Tg O_3), calculated with and without biomass burning emissions for different regions of the world. The global increase in surface ozone concentration is 7%, with large regional variations, as already seen in Figure 2. Over the biomass burning areas, ozone increases by more than 17% as a result of its precursors emissions, but the increase is smaller, between 4 and 8% in other regions of the world. The model results show that the net global photochemical production of ozone increases by about 25% in response to biomass burning. However, the change in tropospheric total ozone (given here as the vertical column from the surface to the tropopause) is relatively smaller as the ozone concentrations are high near the tropopause and as they are less affected by chemical processes than surface concentrations.

The hydroxyl radical, which acts as the main oxidant for compounds emitted at the surface, regulates the lifetime of a large number of gases, including several greenhouse gases. The distribution of OH depends on the distribution of most tropospheric species, including its main precursor, ozone, but also nitrogen oxides, CO, methane and non-methane hydrocarbons. The impact of biomass burning on OH has also been quantified, as indicated in Table 9, which gives for different areas of the world the yearly average surface concentrations (in molecules cm^{-3}), and tropospheric column amount (in molecules cm^{-2}). This Table shows that surface OH is significantly affected by biomass burning emissions in biomass burning areas. Over burning areas, as seen in Table 8, ozone is significantly increased, which results in the calculated increase in surface OH. In some areas, the hydroxyl concentration is slightly lower when biomass burning emissions are taken into account. The OH column is less affected by biomass burning than the surface concentrations, an increase being predicted only over burning areas, and a slight decrease being calculated elsewhere. The global abundance of OH is slightly smaller when biomass burning is taken into account. The small dependence of OH on

                                                          Claire Granier et al.

biomass burning and its significant regional dependence have already been indicated by Bonsang *et al.* (1994), who have calculated with the MOGUNTIA 3–D model a global decrease of 5% in tropospheric OH in response to biomass burning.

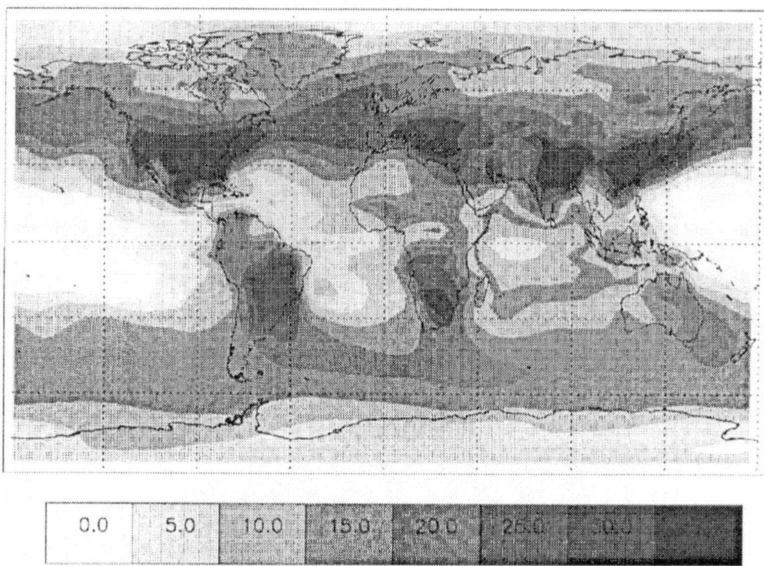

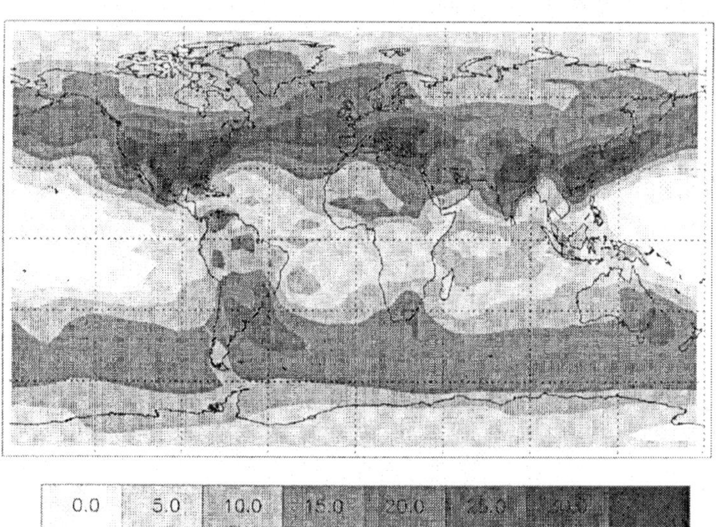

Figure 2. Simulated surface distribution of ozone in October

Table 8. Yearly averaged ozone mixing ratio (ppbv) and total tropospheric ozone amount (in Tg O3) including and excluding biomass burning, for different areas of the world

	ozone surface concentration			tropospheric total O_3		
	with biomass burning	without biomass burning	% difference	with biomass burning	without biomass burning	% difference
Europe	30.5	29.2	4	18.8	18	4
Asia	25.9	24.0	8	58	53.3	9
Africa	24.5	20.8	18	42.8	40	7
North America	27.7	26.1	6	33.9	32.6	4
South America	22.0	18.8	17	22.9	21.4	7
Australia	21.4	20.0	7	15.7	15.2	3
North Atlantic	26.0	24.7	5	27.6	26.3	5
South Atlantic	15.5	14.7	5	27.1	25.8	5
North Pacific	17.6	16.8	5	46.0	44.2	4
South Pacific	13.7	13.2	4	51.9	50.3	3
Indian Ocean	14.68	14.0	4	37.2	35.7	4
Global average/ total amount	20.8	19.4	7	381.9	362.8	5

Table 9. Yearly averaged surface concentration (cm-3) and tropospheric column abundance of OH (cm-2) for the different simulations, for different areas of the world

	OH surface concentration			tropospheric total OH		
	with biomass burning in 10^5 cm^{-3}	without biomass burning in 10^5 cm^{-3}	% difference	with biomass burning in 10^{12} cm^{-3}	without biomass burning in 10^{12} cm^{-3}	% difference
Europe	5.79	5.92	-2.2	1.81	1.91	-5.5
Asia	8.33	7.95	+4.8	3.09	3.13	-1.3
Africa	1.46	1.31	+11.4	4.13	4.03	+2.5
North America	6.44	6.25	+3	2.58	2.65	-2.7
South America	7.42	6.41	+16	3.14	2.99	+5
Australia	1.09	1.05	+3.8	3.37	3.41	-1.2
North Atlantic	1.177	1.181	-0.3	3.35	3.48	-3.9
South Atlantic	4.05	4.01	+1.0	2.00	2.03	-1.5
North Pacific	9.41	9.62	-2.2	2.88	3.00	-4.2
South Pacific	4.48	4.50	-0.4	2.02	2.07	-2.5
Indian Ocean	4.17	4.18	-0.2	1.88	1.93	-2.6
Global average	7.03	6.82	+3	2.68	2.71	-1.1

3.4 Sensitivity study

A sensitivity study has been performed with the model, in order to evaluate the dependence of the calculated ozone photochemical production on the emission ratios, which are not yet well quantified. A simulation has been performed, in which the emission factor of NO_x from biomass burning has been increased by a factor of 2, and the results have been compared with the standard run. Figure 3 shows the surface ozone distribution with the new emission ratio, as well as the difference with the reference run.

O3 surface mixing ratio (ppbv) — October — NOx biomass burning emissions *2

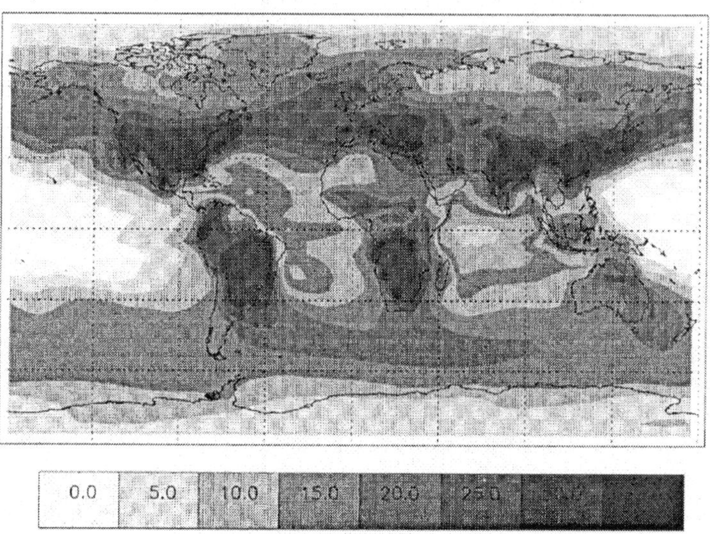

O3 % difference — [NOx biomass burning emissions *2 — reference]

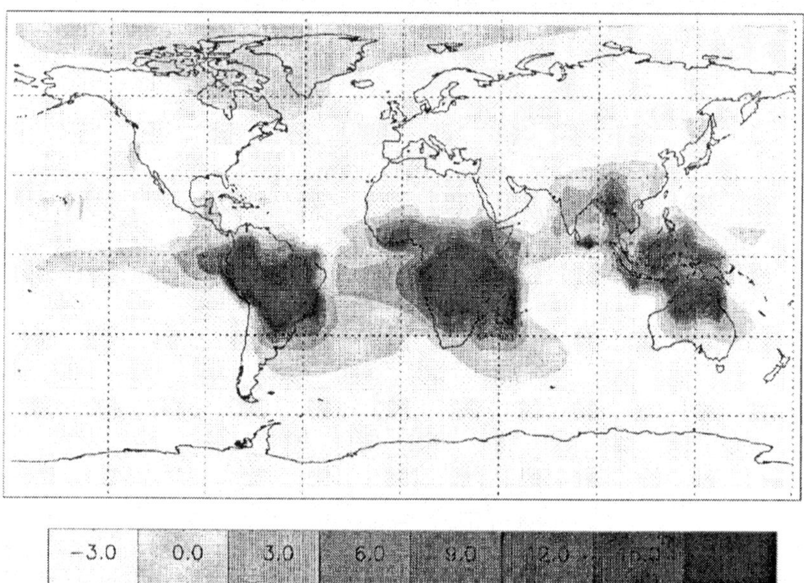

Figure 3. Simulated distribution of surface ozone using the new mixing ratios

Large differences in the ozone concentration are calculated, mostly over the burning areas. As the lifetime of ozone is relatively large (2–3 months), ozone can be formed or transported relatively far from the location of formation of its precursors, which explains the 4 to 8% increase in ozone concentration over oceanic areas. The average calculated ozone concentration and tropospheric ozone column abundance are also compared with the results from the reference run in Table 10. A global increase in surface ozone of about 2.9% is calculated when the NO_x emissions from biomass burning are doubled. The corresponding increase in the average total ozone is 1.7%. Increases in the surface ozone concentration of up to 9% are predicted over burning areas. Such results show that the large uncertainty which remains with the evaluation of the emission ratios for the different trace gases can affect the calculated distribution of gases directly emitted by biomass burning, but also other species such as ozone produced by photochemical processes.

Tableau 10. Yearly average surface ozone and total tropospheric ozone amount when the NOx emission ratio to CO2 from biomass burning is increased by a factor of 2. The percentage difference refers to the reference run (table 8)

	Ozone surface concentration for emission ratio of NO_x multiplied by 2		total tropospheric amount for emission ration of NO_x multiplied by 2	
	ppbv	% with reference	Tg O_3	% with reference
Europe	30.9	1.3	19.0	1.1
Asia	26.8	3.5	59.1	1.9
Africa	26.5	8.2	44.1	3.0
North America	28.4	2.5	34.3	1.2
South America	24.0	9.1	23.7	3.5
Australia	22.1	3.3	16.0	1.9
North Atlantic	26.6	2.3	28.0	1.4
South Atlantic	15.9	2.6	27.6	1.8
North Pacific	17.9	1.7	46.5	1.1
South Pacific	13.9	1.5	52.5	1.2
Indian Ocean	14.9	2.0	37.8	1.6
Global abundance	21.4	2.9	388.6	1.7

Large uncertainties exist over the temporal and spatial changes in the emission ratios. There is also uncertainty over the quantitative evaluation of biomass burning emissions, more specifically over the distribution of the amount of biomass burned each year and the interannual variability in these emissions. For example, the rate of deforestation has increased by about 40%

during the last decade, consumption of fuelwood has increased by about 3% yr^{-1}, and the time period between fires in savanna areas has been reduced (FAO, 1991; FAO, 1993). Year-to-year changes in emissions could affect substantially the distribution of all tropospheric chemical species. Future studies of the evolution of tropospheric chemical compounds should therefore take into account possible changes in biomass burning emissions.

## 4.		CONCLUSIONS

This study has confirmed the significance of biomass burning for the distribution of tropospheric chemical species. Biomass burning is estimated to represent approximately 49, 8, 21, and 10 of the surface emissions of CO, CH_4, NO_x and non-methane hydrocarbons, respectively. It contributes to 20% of the globally averaged surface concentration and total amount of CO, with values of about 35% over the biomass burning regions of South America and Africa. In the case of nitrogen oxides, an average of 13% of the surface concentration and 8% of the tropospheric total amount is produced by biomass burning. About 25% of the net chemical production of ozone results from biomass burning. An increase of 7% of the globally averaged surface concentration and 5% of the total tropospheric ozone is produced by biomass burning emissions. The amount of the hydroxyl radical OH is not significantly affected by biomass burning on the global scale, but significant regional differences are predicted by the model.

The large uncertainties which remain over the distribution of the amount of biomass burned each year and on the emissions ratios limit our ability to predict accurately the contribution of biomass burning to the distributions of tropospheric chemical species and to their global budgets.

## 5.		REFERENCES

Anderson, B.E., G.L. Gregory, J.D.W. Barrick, J.E. Collins, G.W. Sachse, C.H. Hudgins, J.D. Bradshaw, and S.T. Sandholm, 1993. Factors influencing dry season ozone distributions over the tropical South Atlantic. J. Geophys. Res., 98, 23491–23500.

Andreae, M.O., 1991. Biomass burning: its history, use, and distribution and its impact on environmental quality and global climate. in Global Biomass burning: Atmospheric , climatic, and biospheric implications. ed. J.S. Levine, MIT Press, Cambridge.

Baldy, S., G. Ancellet, M. Bessafi, A. Badr, and D. Lan Sun Luk, 1996. Field observations of the vertical distribution of tropospheric ozone at the Island of Reunion (southern tropics). J. Geophys. Res., 101, 23835–23850.

Benkovitz, C.M., J. Dignon, J. Pacyna, T. Scholtz, L. Tarrason, E. Voldner, and T.E. Graedel, 1996. Global inventories of anthropogenic emissions of SO_2 and NO_x. J. Geophys. Res., 101, 29, 239–250.

Bonsang, B., M. Kanakidou and C. Boissard, 1994. Contribution of tropical biomass burning to the global budget of hydrocarbons, carbon monoxide and tropospheric ozone. in Non-CO_2 greenhouse gases, J. van Ham *et al.* eds, Kluwer Academic Publishers, The Netherlands.

Browell, E.V., M.A. Fenn, C.F. Butler, W.B. Grant, M.B. Clayton, J. Fishman, A.S. Bachmeier, B.E. Anderson, G.L. Gregory, H.E. Fuelberg, J.D. Bradshaw, S.T. Sandholm, D.R. Blake, B.G. Heikes, G.W. Sachse, H.B. Singh, and R.W. Talbot, 1996. Ozone and aerosol distributions and air mass characteristics over the South Atlantic Basin during the burning season. J. Geophys. Res., 101, 24043–24068.

Food and Agriculture Organization (FAO), 1991. Yearbook of forests products 1978–1989. United Nations Organization, Rome.

Food and Agriculture Organization (FAO), 1993. Forests resources assessment 1990: Tropical countries. United Nations Organization, Rome.

Fishman, J., and V.G. Brackett, 1997. The climatological distribution of tropospheric ozone derived from satellite measurements using version 7 of Total Ozone Mapping Spectrometer and Stratospheric Aerosol and Gas Experiment data sets. J. Geophys. Res., 102, 19275–19278.

Friedl, R., ed., 1997. Atmospheric effects of subsonic aircraft: Interim assessment report of the Advanced Subsonic Technology Program. NASA Reference Publ. 1400, 143 pp, Washington D.C.

Granier, C., J.F. Müller, S. Madronich and G. Brasseur, 1996. Possible causes for the 1990–1993 decrease in the global tropospheric CO abundance: a three-dimensional study. Atmos. Env., 30, 1673–1682.

Granier, C., J.F. Müller, G. Pétron and G. Brasseur, 1998. A three-dimensional study of the global CO budget. Chemosphere, in press.

Guenther, A., C.N. Hewitt, D. Erickson, R. Fall, C. Geron, T. Graedel, P. Harley, L. Klinger, M. Lerdau, W.A. McKay, T. Pierce, B. Scholes, R. Steinbrecker, R. Tallamraju, J. Taylor, and P. Zimmerman, 1995. A global model of natural volatile organic compound emissions. J. Geophys. Res., 100, 8873–8892.

Hao, W.M., and D.E. Ward, 1993. Methane production from global biomass burning. J. Geophys. Res., 98, 20657–20661.

Hao, W.M., and M.H. Liu, 1994. Spatial and temporal distribution of tropical biomass burning. Global Biogeochem. Cycles, 8, 495–503.

Hudson, R.D., and A.M. Thompson, 1998. Tropical tropospheric ozone from total ozone mapping spectrometer by a modified residual method. J. Geophys. Res., 103, 22129–22145.

Lamarque, J.F., G.P. Brasseur, P.G. Hess, and J.F. Müller, 1996. Three-dimensional study of the relative contribution of the different nitrogen sources in the troposphere. J. Geophys. Res., 101, 22955–22968.

Levine, J.S., R.C. Wesley, D.R. Cahoon, and E.L. Winstead, 1995. Biomass burning: a driver for global change. Environ. Sci. and Tech., 29, 120–125.

Müller, J.F., 1992. Geographical distribution and seasonal variation of surface emissions and deposition velocities of atmospheric trace gases. J. Geophys. Res., 97, 3787–3804.

Müller, J.F., and G. Brasseur, 1995. IMAGES: a three-dimensional chemical transport model of the global troposphere. J. Geophys. Res., 100, 16445–16490.

Novelli, P., K.A. Masarie, and P.M. Lang, 1998. Distributions and recent changes of CO in the lower troposphere. J. Geophys. Res., in press.

Pham, M., J.F. Müller, G. Brasseur, C. Granier, and G. Mégie, 1995. A three-dimensional study of the tropospheric sulfur cycle. J. Geophys. Res., 100, 20061–20092.

Weller, R., R. Lilischkis, O. Schrems, R. Neuber, and S. Wessel, 1996. Vertical ozone distribution in the marine atmosphere over the Central Atlantic Ocean (56°S – 50°N). J. Geophys. Res., 101, 1387–1399.

Impact of the 1997 Indonesian Fires on Tropospheric Ozone and its Precursors

D.A. HAUGLUSTAINE[1], G.P. BRASSEUR[2] and J.S. LEVINE[3]

[1]*Service d'Aéronomie du CNRS, Université de Paris 6, Paris, France*
[2]*National Center for Atmospheric Research, Boulder, CO, USA*
[3]*Atmospheric Sciences Division, NASA Langley Research Center, Hampton, VA, USA*

Abstract: A global chemical transport model, called MOZART, is used to investigate the impact of the 1997 Indonesian biomass fires on the distribution of tropospheric ozone and its precursors. Due to the high release of methane, carbon monoxide, non-methane hydrocarbons, and NO_x by the forest fires and, more importantly, the peat fires, ozone increases significantly over the source regions (Sumatra and Kalimantan), and the tropospheric O_3 column increases by 25–30 DU over Indonesia in November. Our results indicate that the model reproduces with some accuracy the impact of the biomass fires on ozone when the magnitude of the emissions are properly accounted for in the simulations. Because of rapid venting of pollutants out of the boundary layer in the tropics, and their subsequent redistribution by large-scale transport processes, the simulations indicate a significant impact of this local event on the composition of the free troposphere on a regional scale.

1. INTRODUCTION

Deforestation in the tropics plays an important role in releasing large amounts of CO_2 into the atmosphere (Seiler and Crutzen, 1980; Crutzen *et al.*, 1989). Biomass burning is also a significant source of chemically active trace gases (e.g., CO, CH_4, non-methane hydrocarbons, NO_x) leading to the photochemical production of tropospheric ozone (Crutzen *et al.*, 1979). Satellite, aircraft and ground based measurements have identified plumes of elevated tropospheric ozone concentration emanating from South America and Africa during the biomass burning season (Fishman *et al.*, 1991; Andreae *et al.*, 1992; Thompson *et al.*, 1996; Blake *et al.*, 1999).

In 1997, unprecedented widespread fires occurred in Indonesia, starting in June–July and lasting for several months. Most of the fires were set by land owners, commercial loggers and small farmers in attempts to clear and cultivate the land. However, the severe drought induced by a strong El Niño Southern Oscillation event exacerbated the fires and smoke and haze. As a result of the release of significant amount of trace gases and particles, these fires profoundly affected the chemical composition of the troposphere on a regional scale. In particular, tropical tropospheric ozone maps from the Earth-Probe/TOMS instrument indicate an increase from about 40 Dobson Units (DU) in October 1996 to about 60–70 DU in October 1997 over the Indonesian area (Hudson and Thompson, 1998).

In this paper, we use a global three-dimensional chemical transport model, called MOZART, to investigate the sensitivity of the calculated distribution of tropospheric ozone and its precursors to estimated biomass burning emissions associated with the 1997 Indonesian fires. The calculations are aimed mainly at testing the ability of a state-of-the-art global tropospheric chemical transport model to reproduce the effects of biomass burning emissions on ozone in the troposphere. A brief description of the model is provided in section 2. The emission scenarios and the model simulations are described in section 3, and the results are discussed in section 4. Concluding remarks are given in section 5.

## 2.	GLOBAL CHEMICAL TRANSPORT MODEL

MOZART (Model for OZone And Related chemical Tracers) is a three-dimensional chemical transport model of the global troposphere, which has been described and evaluated by Brasseur *et al.* (1998) and Hauglustaine *et al.* (1998a). This model has been developed in the framework of the National Center for Atmospheric Research (NCAR) Community Climate Model (CCM). In MOZART (version 1), the time history of 56 chemical species is calculated on the global scale from the surface to the mid-stratosphere. The model accounts for surface emissions of chemical compounds (N_2O, CH_4, NMHCs, CO, NO_x, CH_2O, and acetone), advective transport (using the semi-Lagrangian transport scheme of Williamson and Rasch [1989]), convective transport (using the formulation of Hack [1994]), diffusive exchanges in the boundary layer (based on the parameterisation of Holtslag and Boville [1993]), chemical and photochemical reactions, wet deposition of 11 soluble species, and surface dry deposition. The chemical scheme includes 140 chemical and photochemical reactions and considers the photochemical oxidation schemes of methane (CH_4), ethane (C_2H_6), propane (C_3H_8), ethylene (C_2H_4), propylene (C_3H_6), isoprene (C_5H_8), terpenes (as α-pinene,

$C_{10}H_{16}$), and a lumped compound n-butane (C_4H_{10}) used as a surrogate for heavier hydrocarbons. The evolution of species is calculated with a numerical time step of 20 min for both chemistry and transport processes. The model is run with a horizontal resolution which is identical to that of CCM (triangular truncation at 42 waves, T42) corresponding to about 2.8 degrees in both latitude and longitude. In the vertical, the model uses hybrid sigma-pressure co-ordinates with 25 levels extending from the surface to the level of 3 mb. Dynamical and other physical variables needed to calculate the resolved advective transport as well as smaller-scale exchanges and wet scavenging are pre-calculated by the NCAR CCM (version 2, Ω0.5 library), and provided every 3 hours from pre-established history tapes. A preliminary version of the model was used by Brasseur *et al.* (1996) to investigate the budget of chemical compounds in the Pacific troposphere in conjunction with the MLOPEX measurements. More recent versions of the model were used by Hauglustaine *et al.* [1998b] in a study of ozone over the North Atlantic ocean, and by Emmons *et al.* [1997] for a comparison of nitrogen species distributions provided by various chemical transport models (CTMs) and observed climatologies.

Biomass burning emissions in MOZART take into account fires in tropical and non-tropical forests, savanna burning, fuel wood use, and agricultural waste burning. The spatial and temporal distributions of the amount of biomass burned is taken from Hao and Liu (1994) in the tropics and from Müller (1992) in other regions. The emission ratios of each species relative to CO_2 are taken from Granier *et al.* (1996) for each type of biomass fire except for the CO and NO_x (= NO + NO_2) from savanna where the values suggested by respectively Hao *et al.* (1996) and by Andreae *et al.* (1996) are used. Table 1 gives the global and annual emissions of CO_2 and ozone precursors used in the model standard simulation.

Table 1. Emissions considered in the model to account for the 1997 Indonesian fires (Experiment 1 assumes 100% of forest fires. Experiments 2 and 3 assume 70% of forest fires and 30% of peat fires) based on Levine *et al.* [1998]. Global and annual biomass burning emissions in the model are also given for comparison

Species	Experiment 1	Experiments 2–3	Base case global
CO_2 (TgC)	84.95	315.62	2963.2
CO (TgC)	7.22	51.60	283.6
CH_4 (TgC)	1.33	3.59	29.5
NMHCs (TgC)	1.74	6.18	37.6
NO_x (TgN)	0.21	8.91	7.5

3.   MODEL SIMULATIONS

The biomass burning emissions associated with the 1997 Indonesian fires are introduced in MOZART on the basis of the CO_2 emissions estimated by Levine *et al.* (1998). Liew *et al.* (1998) estimated a total burned area of 45,600 km^2 (4.5 million hectares) for the Sumatra and Kalimantan regions for the period of August 1997 to December 1997. Levine *et al.* (1998) estimated a release of 85 TgC of CO_2 if the area burned comprised tropical forest. If the biomass burned comprised 70% forest fires and 30% peat fires, then a higher CO_2 release of 316 TgC is estimated due to the higher carbon loading of the organic soil (Yokelson *et al.*, 1997). In our simulations, the two emission scenarios are considered and referred to as Experiment 1 (100% forest fires) and Experiment 2 (70% forest fires and 30% peat fires). Emissions of other species are calculated on the basis of the CO_2 released and the emission ratios relative to CO_2 (Yokelson *et al.*, 1997; Levine *et al.*, 1998). The estimated total emissions for the 1997 Indonesian fires are given in Table 1 and compared to the global and annual biomass burning emissions in the model.

For CO, the Indonesian fires contribute about 3% of the annual and global biomass burning emissions for Experiment 1 and and 20% for Experiments 2–3. For non-methane hydrocarbons (NMHCs), the Indonesian fires contribute about 5% of the annual emissions for Experiments 1 and 15% for Experiments 2–3. High NO_x release is estimated for the peat fires, leading to an emission during the 1997 fires 20% larger than the global and annual mean considered in the base case simulation in the model. In the model, these emissions are introduced in four model grid-cells, two grid-cells located in Sumatra and two in Kalimantan. In the tropics, the model resolution corresponds to a grid-cell area of about 300×300 km^2. The model is run for five months (August 1 to December 31), assuming constant emissions from the fires during the entire simulation. The results are compared to the base case scenario (no emission from the Indonesian fires).

Particulate levels over the Indonesian region were very high during the 1997 fires, leading to the formation of persistent smoke plumes and haze. The smoke plumes, mainly located in the boundary layer, cause direct radiative forcing by scattering solar radiation back to space. Consequently, the smoke particles affected the photolysis rates and potentially tropospheric ozone photochemical production and destruction rates. A discrete ordinates radiative transfer model (Madronich *et al.*, 1998) was used to illustrate the effects of smoke aerosols on NO_2 and O_3 (O^1D formation) photolysis rates.

We consider a uniform aerosol layer in the boundary layer (0 to 2 km). Single-scattering albedos (ω_0) for smoke particles have been measured by Hobbs *et al.* (1997) in regional hazes at three locations in Brazil. They report a range for ω_0 of 0.82 to 0.84 at 550 nm and 0.88 when a humidity factor is applied. In this study, we assume a wavelength independent value ω_0=0.88. Figure 1 shows the change in NO_2 and O_3 photolysis rates for aerosol optical depths (τ) of 1, 2, 3 and 4, averaged over the boundary layer height and in the free troposphere.

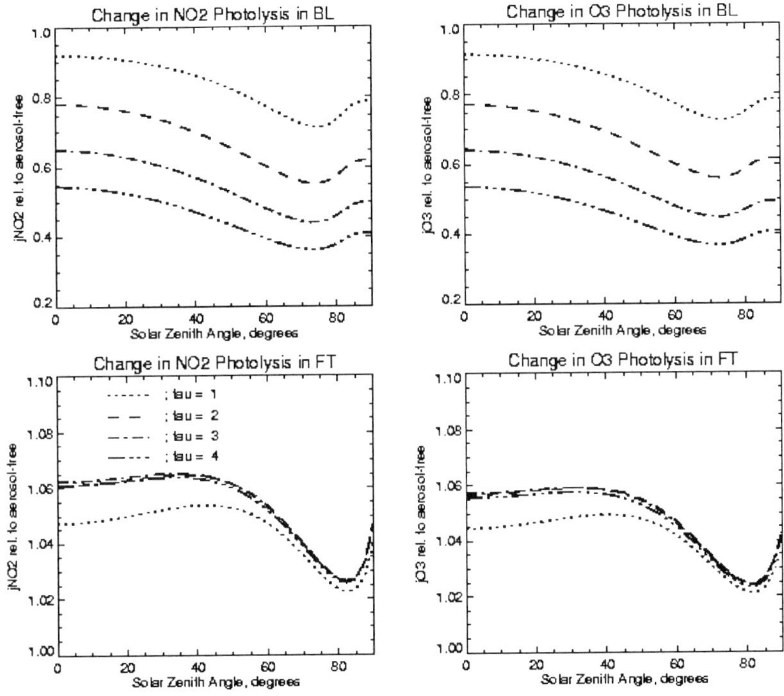

Figure 1. Calculated change in NO_2 and O_3 (O^1D formation) photolysis rates for an uniform smoke aerosol (ω_0=0.88) layer in the boundary layer (0–2km), relative to aerosol-free conditions. Different aerosol optical depths (τ=1, 2, 3, 4) are considered. The changes in j_{NO2} and j_{O3} are averaged over the boundary layer (BL) height (0–2km) or the free troposphere (FT).

Due to increased absorption of solar radiation in the boundary layer by aerosols, j_{NO2} and j_{O3} are decreased relative to aerosol-free conditions. For an optical depth of 3 (assumed conditions prevailing during the Indonesian fires), a decrease of the photolysis rates of about 35% is calculated for overhead sun conditions. In contrast, due to increased back scattering, j_{NO2} and j_{O3} are increased in the free troposphere by about 6 % for high-sun conditions (τ = 3). In MOZART, the photolysis rates are determined for clear sky and aerosol-free conditions from a pre-calculated lookup table. As in the case of clouds, a correction factor is applied to the clear sky and aerosol free value to account for the effect of smoke aerosols. In Experiment 3, we apply a correction factor to the photolysis rates of 0.7 in the boundary layer (surface up to 2 km) and of 1.06 in the free troposphere (2 km to 12 km) in the Indonesian region (7°S – 7°N in latitude and 100°E – 115°E in longitude) to account for the role played by aerosols on ozone photochemistry.

4.   RESULTS

The changes in the tropospheric ozone column (O_3 integrated from the surface up to the model tropopause) when Indonesian emissions are considered in the simulations are shown in Figure 2 for Experiments 1, 2, and 3 on November 25. In Experiment 1, the change in ozone reaches a maximum of only 12 DU (Dobson Unit) over Sumatra and Kalimantan, a value lower than the tropospheric column increase derived from Earth-Probe/TOMS. Typical TOMS tropospheric columns derived with the modified residual method (Hudson and Thompson, 1998) increase from about 25–35 DU in November 1996 to about 45–55 DU in November 1997 over the Indonesian region (increase of 20–30 DU). The model results from Experiment 1 clearly underestimate these values. In Experiment 2, because of greater emissions of ozone precursors to the atmosphere by the peat fires, the tropospheric O_3 column shows an increase of 20–30 DU over the Indonesian region with a local maximum of 32 DU. Transport of ozone and precursors in this version of the model driven by the NCAR CCM dynamics shows an eastward flow to the Pacific Ocean, and ozone increases by about 2–4 DU over the western Pacific Ocean. In Experiment 3, the results are characterized by a decreased photochemical activity in the boundary layer in comparison to Experiment 2 as a result of the reduction in photolysis rates. In this case, the ozone increase reaches about 20–25 DU over the Indonesian region with a local maximum of 30 DU. In Experiments 2 and 3, the magnitude of the ozone change is more in line with the TOMS measurements in the source region, suggesting that the emissions associated with the Indonesian fires and the ozone photochemistry in biomass burning regions in the model are reproduced with some accuracy in the simulations. However, the difference between the actual dynamical fields prevailing in 1997 and the CCM fields induces a discrepancy in the export of pollutants out of the source region. In the model, the transport is mainly dominated by export to the Pacific Ocean when strong export to India was actually observed in 1997.

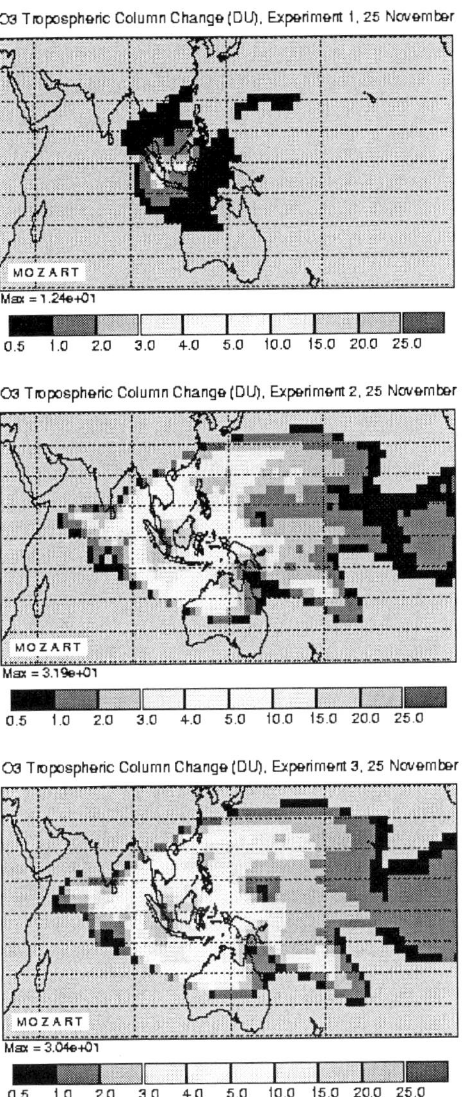

Figure 2. Calculated tropospheric ozone column increase (DU : Dobson Units) due to the Indonesian fire emissions of ozone precursors. The change in ozone are presented for November 25 conditions for Experiment 1 (top panel), Experiment 2 (middle) and Experiment 3 (bottom).

At the surface, a maximum increase in ozone mixing ratio reaching 40 ppbv is calculated in November over the Indonesian region for Experiment 3 (Fig. 3a). Due to the short lifetime of ozone at the surface in the tropics (about a week), the maximum increase is mainly localised over Sumatra and Kalimantan. Over the ocean, the ozone enhancement decreases rapidly due to ozone photochemical destruction. In the free troposphere (500 mb), the ozone increase reaches a maximum of about 50 ppbv over Indonesia (Fig. 3b). As a result of the longer lifetime of ozone at this altitude (about a month), export occurs towards the north-east and plumes with enhanced ozone (increase of 10–20 ppbv) reach the Indo-China peninsula. Export of ozone south of the source region is also visible. Carbon monoxide increases by up to 4500 ppbv at the surface over Sumatra and Kalimantan (Fig. 3c). A perturbation of 50–100 ppbv is calculated over the whole Indonesian region. This maximum is localised over the Indonesian region due to radical high concentrations and consequently short CO lifetime in the tropical boundary layer. In the free troposphere, the zonal winds are more intense and the transport as simulated by the CCM is characterized by a strong export to Indo-China and the Pacific Ocean. The CO increase reaches 1000 ppbv over Indonesia due to the rapid upward transport prevailing in this region and about 80–100 ppbv over Thailand (Fig. 3d). The calculated CO increase reaches about 20–40 ppbv over the western Pacific. However, the dynamics calculated by the NCAR CCM, and used to drive the transport in MOZART, represent mean conditions that are not necessarily representative of the conditions prevailing in 1997. Nevertheless, the results clearly emphasise the impact of the fires on the regional scale.

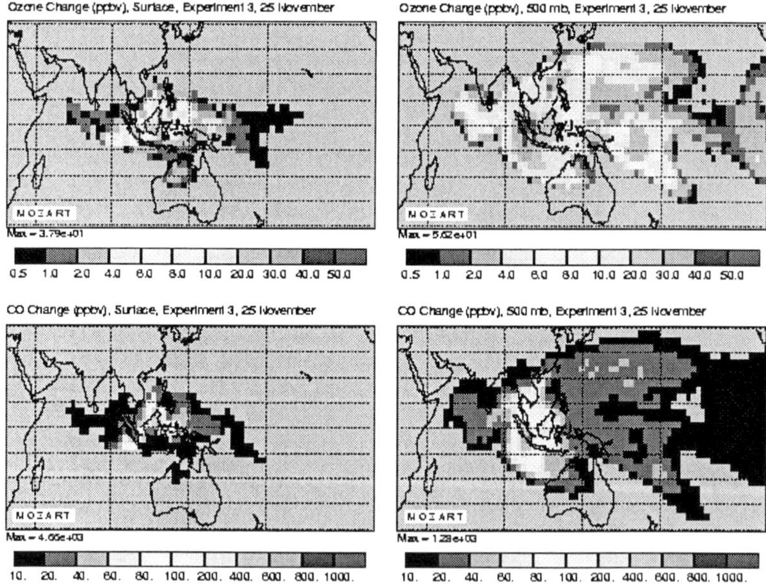

Figure 3. Change in ozone mixing ratio (ppbv) at the surface and at 500 mb calculated on November 25 for Experiment 3 (top), and corresponding change in CO mixing ratio at the surface and 500 mb (bottom). Note the different colour scales for O3 and CO

Figure 4 compares typical profiles of O_3, CO, NO_x and peroxyacetyl nitrate (PAN) calculated for the base case simulation together with the profiles simulated for Experiments 1, 2 and 3. The profiles are geographically averaged over the region 1.5°S – 1.5°N in latitude and 107°E – 110°E in longitude. In general, fairly small perturbations are calculated for Experiment 1, and clearly indicate an underestimate of the emissions by the fires. In Experiment 2, ozone reaches a maximum of 90 ppbv at the top of the boundary layer and 60 ppbv in the free troposphere in comparison to 30–40 ppbv in the base case simulation. Because of the lower photochemical activity in the boundary layer in Experiment 3, the calculated mixing ratios are lowered by about 5–10 ppbv in comparison to Experiment 2, and O_3 reaches a maximum of about 80 ppbv at 3 km. As a result of rapid vertical mixing, carbon monoxide is fairly constant with altitude in the base case simulation, with a value of about 100 ppbv throughout the troposphere. In Experiments 2 and 3, because of high CO release by the peat fires and photochemical production through NMHC oxidation, a mixing ratio of 400–500 ppbv is calculated at the top of the boundary layer. In the boundary layer, the CO oxidation by OH is more important and the mixing ratio reaches 300 ppbv at the surface. High mixing ratios (300 ppbv) are also calculated in the upper troposphere as a result of vigorous transport from the boundary layer to higher altitudes by convection.

The nitrogen budget was also considerably affected over the Indonesian fires. In the base case simulation (no emission from the 1997 fires), the NO_x profile shows a " C–shape " with values of about 200 pptv at the surface, decreasing to 50 pptv in the mid-troposphere and increasing with height above 9 km. In Experiments 2 and 3, higher values of 1000–1200 pptv are calculated in the boundary layer as a result of direct emissions by the fires. The mixing ratio rapidly decreases with height because of the conversion of NO_x to less reactive species (PAN, HNO_3). In the upper troposphere a secondary maximum of 1000 pptv is also calculated due to rapid vertical mixing through convective processes. Slightly higher NO_x mixing ratios are calculated in Experiment 3 in comparison to Experiment 2 as a result of decreased NO_2 photolysis in the boundary layer. PAN is formed by the reaction of peroxyacetyl radicals (issued from the NMHC oxidation chain) with NO_2. A significant increase of PAN mixing ratio is calculated and associated with both increased NO_2 and NMHC concentrations. The mixing ratio for Experiments 2 and 3 ranges from about 300 pptv at the surface to 1000 pptv at 3–4 km. A secondary maximum is also calculated in the upper troposphere (reaching 800 pptv) where PAN is more stable regarding its thermal decomposition.

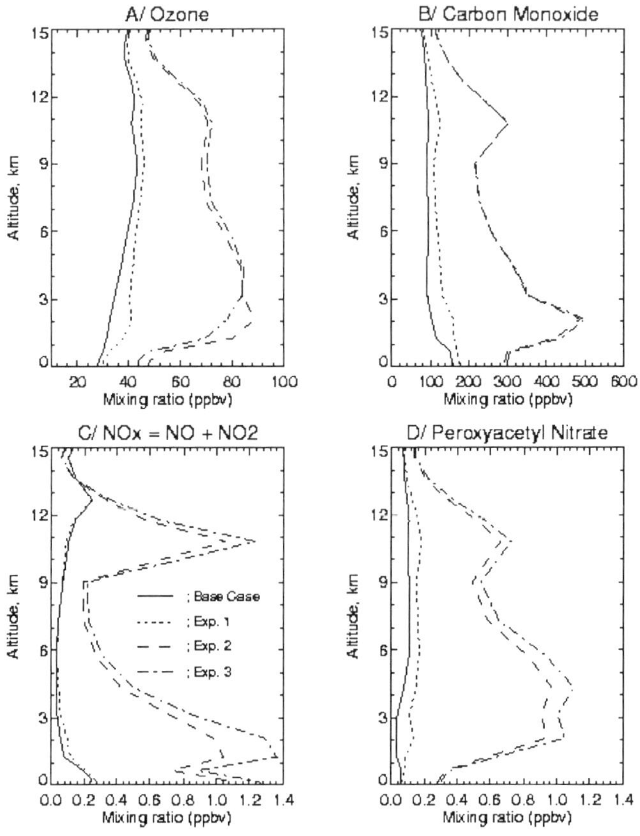

Figure 4. Vertical distributions of O_3, CO, NOx, and PAN mixing ratio (ppbv) simulated for the base case and compared to Experiments 1, 2, and 3 for November 25 conditions. The profiles are geographically averaged over the region 1.5°S – 1.5°N in latitude and 107°E – 110°E in longitude.

5.   CONCLUSIONS

In this study, a global three-dimensional chemical transport model (CTM) is used to assess the impact of widespread fires in the tropics on the composition of the troposphere. Sensitivity simulations are conducted to illustrate the changes in tropospheric ozone and its precursors associated with the 1997 Indonesian fires. This study is mainly aimed at testing the ability of a state-of-the-art CTM to reproduce the observed change in tropospheric ozone in a specific biomass burning case. A simple approach is adopted to account for the additional emissions from the Indonesian fires in the model based of the total CO_2 release estimated by Levine *et al.* (1998), and the effect of smoke particles on photolysis rates. Two emission scenarios are considered in our simulations and tested. In one case, we assume that the ecosystem is composed of 100 % of tropical forests, and in the second scenario 70% of forest fires and 30% of peat fires are assumed.

Our simulations show that the model results significantly underestimate the change in total ozone observed from space when the 100 % forest fire scenario is assumed. For the second emission scenario, because of the higher carbon content of peat, the ozone change is more pronounced and in line with actual observations over the source region (Sumatra, Kalimantan). The MOZART off-line chemical transport model used for the simulations is driven by the dynamical fields calculated by the NCAR Community Climate Model. Therefore, the simulations are characterized by a mean climatological state instead of actual meteorological conditions prevailing during the 1997 fire period. As a consequence, the transport of species in the model is mainly dominated by a strong export from the source region to the Pacific Ocean in the free troposphere when export to India was actually observed in 1997. A version of MOZART driven by assimilated dynamical and physical fields is currently under development and should help to address this issue.

Despite these limitations, the results clearly show that the budget of carbon monoxide, NMHCs, nitrogen species, and ozone are profoundly affected by the fires in the boundary layer, but also in the free troposphere due to rapid upward transport and subsequent redistribution of pollutants by large scale transport processes in the tropics. The simulations indicate a strong impact of this local event on the composition of the free troposphere on a regional scale, stressing the need for accurate estimates of biomass burning emissions (i.e., fire location and extent, burned area, type of ecosystem burned, timing of the fires) in chemical transport models. In the future, biomass fires determined by remote sensing techniques hopefully will provide more insight into this increasingly important aspect of the human impact on the global environment.

6.   ACKNOWLEDGEMENTS

Helpful discussions with S. Madronich and S. Walters are gratefully acknowledged. The National Center for Atmospheric Research is operated by the University Corporation for Atmospheric Research under the sponsorship of the National Science Foundation.

7.   REFERENCES

Andreae, M.O., A. Chapuis, B. Gros, J. Fontan, G. Helas, C. Justice, Y.J. Kaufman, A. Minga, and D. Nganga, 1992. Ozone and Aitken nuclei over equatorial Africa : airborne observations during DECAFE 88. *J. Geophys. Res.*, 97, 6137–6148.

Andreae, M.O., E. Atlas, H. Cachier, W.R. Cofer III, G.W. Harris, G. Helas, R. Koppmann, J.-P. Lacaux, and D.E. Ward, 1996. Trace gas and aerosol emissions from savanna fires. in *Biomass burning and global change,* J.S. Levine (ed.), pp. 278–295, MIT press, Cambridge, Mass.

Blake, D.R., N.J. Blake, I.J. Simpson, F.S. Rowland, E.V. Browell, M.A. Fenn, W.B. Grant, H.E. Fuelberg, G.W. Sachse, B. Anderson, G.L. Gregory, H.B. Singh, J.D. Bradshaw, S.T. Sandholm, R.W. Talbot, J.E. Dibb, J.A. Logan, M.G. Schultz, and D.J. Jacob, 1999. The pervasive influence of biomass burning on the mid-troposphere over the remote south central Pacific. Submitted for publication.

Brasseur, G.P., D.A. Hauglustaine, and S. Walters, 1996. Chemical compounds in the remote Pacific troposphere: Comparison between MLOPEX measurements and chemical transport model calculations. *J. Geophys. Res.*, 101, 14,795–14,813.

Brasseur G.P., D.A. Hauglustaine, S. Walters, P.J. Rasch, J.-F. Müller, C. Granier, and X.-X. Tie, 1998. MOZART: a global chemical transport model for ozone and related chemical tracers, Part 1. Model description. *J. Geophys. Res.*, 103, 28,265–28,289.

Crutzen, P.J., L.E. Heidt, J.P. Krasnec, W.H. Pollock, and W. Seiler, 1979. Biomass burning as a source of atmospheric gases CO, H_2, N_2O, NO, CH_3Cl, and COS. *Nature*, 282, 253–256.

Crutzen, P.J., W.M. Hao, M.H. Liu, J.M. Lobert, and D. Scharffe, 1989. Emissions of CO_2 and other trace gases to the atmosphere from fires in the tropics. in *Our changing atmosphere,* P. J. Crutzen, J.-C. Gérard and R. Zander (eds), pp. 449–471, Univ. of Liège, Liège, Belgium.

Emmons, L.K., M.A. Carroll, D.A. Hauglustaine, G.P. Brasseur, C. Atherton, J. Penner, S. Sillman, H. Levy, F. Rohrer, W.M.F. Wauben, P.F.J. Van Velthoven, Y. Wang, D. Jacob, P. Bakwin, R. Dickerson, B. Doddridge, C. Gerbig, R. Honrath, G. Hübler, D. Jaffe, Y. Kondo, J.W. Munger, A. Torres, and A. Volz-Thomas, 1997. Climatologies of NOx and NOy : a comparison of data and models. Atmos. Environ., 31, 1851–1904.

Fishman, J., K. Fakhruzzaman, B, Gross, D. Nganga, 1991. Identification of widespread pollution in the southern hemisphere deduced from satellite analysis. *Science,* 252, 1693–1696.

Granier, C., W.M. Hao, G. Brasseur, and J.-F. Müller, 1996. Land use practices and biomass burning: Impact on the chemical composition of the atmosphere. in *Biomass Burning and Global Change,* J. S. Levine (ed.), pp. 140–198, MIT Press, Cambridge, Mass.

Hack, J.J., 1994. Parameterization of moist convection in the NCAR community climate model (CCM2). *J. Geophys. Res.,* 99, 5551–5568.

Hao, W.M., and M.-H. Liu, 1994. Spatial distribution of tropical biomass burning in 1980 with 5° X 5° resolution. *Global Biogeochem. Cycles*, 8, 495–503.

Hao, W.M., D.E. Ward, G. Olbu, and S.P. Baker, 1996. Emissions of CO_2, CO, and hydrocarbons from fires in diverse African savanna ecosystems. *J. Geophys. Res.,* 101, 23,577–23,584.

Hauglustaine D.A., G.P. Brasseur, S. Walters, P.J. Rasch, J.-F. Müller, L.K. Emmons and M.A. Carroll, 1998a. MOZART: a global chemical transport model for ozone and related chemical tracers, Part 2. Model description. *J. Geophys. Res.*, 103, 28,291–28,335.

Hauglustaine D.A., G.P. Brasseur and S. Walters, 1998b. A three-dimensional simulation of ozone over the North Atlantic ocean. in *Atmospheric Ozone,* R. Bojkov and G. Visconti (eds), pp. 735–738, International Ozone Commission.

Hobbs, P.V., J.S. Reid, R.A. Kotchenruther, R.J. Ferrek, and R. Weiss, 1997. Direct radiative forcing by smoke from biomass burning. *Science*, 275, 1776–1778.

Holtslag, A., and B. Boville, 1993. Local versus nonlocal boundary-layer diffusion in a global climate model. *J. Clim.*, 6, 1825–1842.

Hudson, R.D., and A.M. Thompson, 1998. Tropical tropospheric ozone from total ozone mapping spectrometer by a modified residual method. *J. Geophys. Res.*, 103, 22,129–22,146.

Levine, J.S., T.D. Edwards, T.E. McReynolds, and C.W. Dull, 1998. Gaseous and particulate emissions from the fires in Kalimantan and Sumatra, Indonesia. *Report of WMO workshop on regional transboundary smoke and haze in South-East Asia,* World Meteorological Organization, Geneva, Switzerland.

Liew, S.C., O.K. Kim, L.K. Kwoh, and H. Lim, 1998. A study of the 1997 fires in South East Asia using SPOT quicklook mosaics. *Paper presented at the 1998 International Geoscience and remote sensing symposium,* Seattle, Washington, 3 pages, July 6–10.

Madronich, S., S.J. Flocke, J. Zeng, and I. Petropavloskyh, 1998. *Tropospheric Ultraviolet-Visible Model (TUV) Version 4.* NCAR Tech. Note, in preparation.

Müller, J.-F., 1992. Geographical distribution and seasonal variation of surface emissions and deposition velocities of atmospheric trace gases. *J. Geophys. Res.*, 97, 3787–3804.

Seiler, W., and P.J. Crutzen, 1980. Estimates of gross and net fluxes of carbon between the biosphere and the atmosphere from biomass burning. *Clim. Change*, 2, 207–247.

Thompson, A.M., K.E. Pickering, D.P. McNamara, M.R. Schoeberl, R.D. Hudson, J.H. Kim, E.V. Browell, V.W.J.H. Kirchhoff, and D. Nganga, 1996. Where did tropospheric ozone over sourthern Africa and the tropical Atlantic come from in October 1992 ? Insights from TOMS, GTE TRACE A, and SAFARI 1992. *J. Geophys. Res.*, 101, 24,251–24,278.

Williamson, D.L., and P.J. Rasch, 1989. Two-dimensional semi-Lagrangian transport with shape preserving interpolation. *Mon. Weather Rev.* 117, 102–129.

Yokelson, R.J., R. Susott, D.E. Ward, J. Reardon, and D.W.T. Griffith, 1997. Emissions from smoldering combustion of biomass measured by open-path Fourier transform infrared spectroscopy. *J. Geophys. Res.*, 102, 18,865–18,877.

The Relationship Between Area Burned by Wildland Fire in Canada and Circulation Anomalies in the Mid-Troposphere

W.R. SKINNER[1], B.J. STOCKS[2], D.L. MARTELL[3], B. BONSAL[1] and A. SHABBAR[1]

[1] *Climate Research Branch, Atmospheric Environment Service, Toronto, Ontario Canada*
[2] *Great Lakes Forestry Centre, Canadian Forestry Service, Sault Ste. Marie, Ontario Canada*
[3] *Faculty of Forestry, University of Toronto,Toronto, Ontario Canada*

Abstract: There is evidence that the area burned by wildland fire has increased in certain regions of Canada in recent decades. One cause for this increase is changes in the mid-tropospheric circulation at 500 hPa over northern North America. This study examines the physical links between anomalous mid-tropospheric circulation over various regions of Canada and wildland fire severity. Analysis of monthly and seasonal burned areas for the period 1953 to 1995 reveals a bimodal distribution with distinct low and extreme high burned area years. The high/low burned area years coincide with positive/negative 500 hPa height anomalies over north-western, western, west-central and east-central Canada. Total area burned and the 500 hPa height anomaly data are analysed for statistical relationships using the Spearman rank correlation non-parametric measure. Results for the May to August fire season indicate statistically significant correlations between regional total area burned and clusters of anomalous 500 hPa geopotential height values immediately over and immediately upstream of the affected region. For the north-western and west-central regions, significantly correlated clusters are also found in the central Pacific, providing evidence of the influence of a teleconnection structure on the summer climate of western and north-western North America. Two sample comparison tests show statistically significant differences in both the means and variances of the fire data populations during negative and positive phases of mid-tropospheric flow and the means of the height anomaly populations during extremely high and extremely low area-burned seasons. Increases in regional total area burned are related to increases in mean 500 hPa heights, taken from the significantly correlated clusters of height values, between two successive periods 1953–1974 and 1975–1995. For Canada as a whole, the five lowest area-burned seasons all occurred during the early period, while the five highest seasons occurred during the later period. The difference in the geopotential height fields between the two periods identifies an increase in 500 hPa heights over most of Canada with an amplification of the western Canada ridge and an eastward shifted Canadian Polar Trough (CPT).

# 1.	INTRODUCTION

Forests cover approximately 45% of the Canadian land area. Wildland fires, which burn millions of hectares of forests annually, are a major economic and environmental hazard. The reported area burned by wildland fires in some regions of Canada has tripled from 1980 to the present (Simard, 1997). In addition, the five worst Canadian fire years since 1918 have taken place in the past 15 years. It is important to note however, that area burned may not have been monitored consistently in all regions prior to 1970. Stocks *et al.* (1996) examined the spatial distribution of large fires in Canada during the 1980s when an average of almost 10,000 fires burned over 2.8 million hectares annually. They found that by far the greatest area burned occurred in the boreal region of west-central Canada. This was attributed to a combination of factors including fire-prone ecosystems, extreme fire weather, lightning activity, and varying levels of protection in this region. In the wake of the severe 1995 fire season in terms of area burned, there has been renewed effort to better understand the processes that influence wildland fire in Canada.

The fire season over much of Canada is relatively short with the majority of activity from May to September in the south, and from June to August in the north, when higher temperatures and thunderstorms with lightning strikes occur most frequently. Canadian fire regimes range from dry (<50 cm annual precipitation) in west-central, interior western and north-western regions of Canada to moist (>50 cm annual precipitation) in west coastal, east-central and eastern regions of Canada (Simard, 1997). Short droughts of up to one month can have significant effects on regions with drier climates, while longer duration droughts are normally required to similarly affect wetter regions. Precipitation frequency during the warm summer season is critically important to Canadian fire regimes. A moisture deficit for a sustained period of several weeks, coupled with one or more strong wind events, often results in short periods of extreme fire activity with large burned areas that can dominate fire season statistics. Extended sequences of days without rain strongly influence the provincial total area burned for all of Canada (Flannigan and Harrington, 1988). However, not all areas show significant correlations between annual temperature departures and annual total area burned in Canada (Kurz *et al.*, 1995). This study provides a method of integrating these surface climatological conditions on large spatial and temporal scales in terms of the large-scale mid- tropospheric circulation at 500 hPa.

The circulation of the middle and upper troposphere in the northern hemisphere meanders from west-to-east through a series of equatorward-extending troughs and poleward-extending ridges. These features, known as standing waves, often remain quasi-stationary in space for extended periods of time. The origin of these standing waves is linked to topography and to thermal influences related to the sources and sinks of heat such as the abrupt thermal differences between oceans and continents. The Rocky Mountains, which lie perpendicular to the zonal flow, cause the formation of an upper ridge over western North America and a deep trough over eastern North America. In addition, sea surface temperature (SST) anomalies in the North Pacific have been shown to be associated with atmospheric circulation patterns and thus ridging over western North America in spring and summer (Bonsal *et al.*, 1993). This affects the weather downstream across much of western and central North America.

The location, magnitude and persistence of the ridges and troughs greatly affects the weather and climate of North America. Normal climatological summer flow conditions produce a relatively dry climate in non-coastal western, north-western and west-central Canada. A stronger than normal ridge over western North America often produces abnormally warm and dry conditions for these regions. Occasionally, these standing waves become so pronounced that warm and dry pools of air can remain stagnant in north-western Canada for several days. If the deep trough over eastern North America shifts eastward, west-central and east-central Canada experience upper level convergence and subsidence on the western edge of the trough. Storm development is suppressed with an extended dry period. If the trough shifts westward, these regions are on the eastern edge where south-west to north-east flow produces uplift and cyclonic development. The result is an abnormally wet and generally cooler than normal period.

Variations in large-scale atmospheric flow cause variations in the frequency of surface air masses and frontal systems which dominate the climate of a region during any season. Previous studies have linked large-scale climatic mechanisms to smaller-scale conditions of surface temperature, precipitation, forest fuel moisture, and total area burned by wildland fire (Johnson and Wowchuk, 1993). The probability that large wildland fires will occur is increased significantly by weather conditions associated with stable anticyclones that promote drying of the forest fuels through low rainfall and higher than normal air temperature, but still generate enough convective activity to produce lightning to ignite fires, and wind of sufficient strength to spread them. Surface high pressure systems are transitory while long-wave ridging, at the 500 hPa geopotential height level in the mid-troposphere, can remain quite persistent, with characteristic oscillations in strength (Newark, 1975). Fire behaviour has been associated with persistent mid-tropospheric ridges for individual fire case studies (Finklin, 1973; Alexander *et al.*, 1983; Janz and Nimchuk, 1985; Street, 1985; Flannigan and Harrington, 1986) and for regional studies within Canada (Schroeder *et al.*, 1964; Street and Birch, 1986).

The purpose of this study is to establish baseline associations between variations in large spatial and temporal scale mid-tropospheric circulation patterns and wildland fire severity by examining the hemispheric 500 hPa record in conjunction with the more recent Canadian historical area-burned fire record. Total area burned and 500 hPa height anomaly data are analysed for statistical relationships for the May to August fire season for the common 43–year period of record 1953 to 1995. Hemispheric circulation at 500 hPa is analysed during extremely high and extremely low area-burned seasons. Key significantly correlated clusters of anomalous 500 hPa geopotential height values immediately over and immediately upstream of the affected fire region are identified. For the north-western (not shown) and west-central regions, significantly correlated clusters are found in the central Pacific as well. Two sample comparison tests during negative and positive phases of upper atmospheric flow and during extremely high and extremely low area-burned seasons are performed. A two-sample temporal analysis compares the regional total burned areas between two successive periods from 1953 to 1995. The difference in these geopotential height fields identifies an increase in 500 hPa heights over most of Canada with peaks over north-western and north-eastern Canada.

Until recently, there have been no attempts to identify and explain the links between mid-tropospheric circulation anomalies and wildland fire severity in Canada on large spatial and temporal scales. This information would aid in predicting wildland fire in Canada. Also, an improved knowledge of current climate–fire relationships would be useful for examination in conjunction with GCM and higher resolution Regional Climate Models (RCMs) (Caya *et al.*, 1995) outputs in order to assess potential changes in wildland fire regimes due to future climate change.

2.   DATA

2.1   Canadian Fire Statistics

Area burned by wildland fire is defined as "all land on which wildfires occur, including forest, cutover forest, grasslands, scrub, etc." (Flannigan and Harrington, 1988). For this study, the area analysed consists of all Canadian provinces and territories with the exception of the four eastern Atlantic provinces (Figure 1). Area-burned data are available for the four eastern provinces; however, the very low fire activity there, relative to the other provinces, excluded their use in this type of large spatial and temporal scale analysis. The following aggregation scheme is used to produce four large regions (see Figure 1); western Canada (British Columbia and Alberta) and west-central Canada (Saskatchewan, Manitoba, and north-western Ontario) are analysed with figures in this study; north-western (NW) Canada (Yukon and Northwest Territories) and east-central Canada (north-eastern Ontario, and Quebec)) are discussed but more fully treated with accompanying figures in Skinner *et al.* (1998).

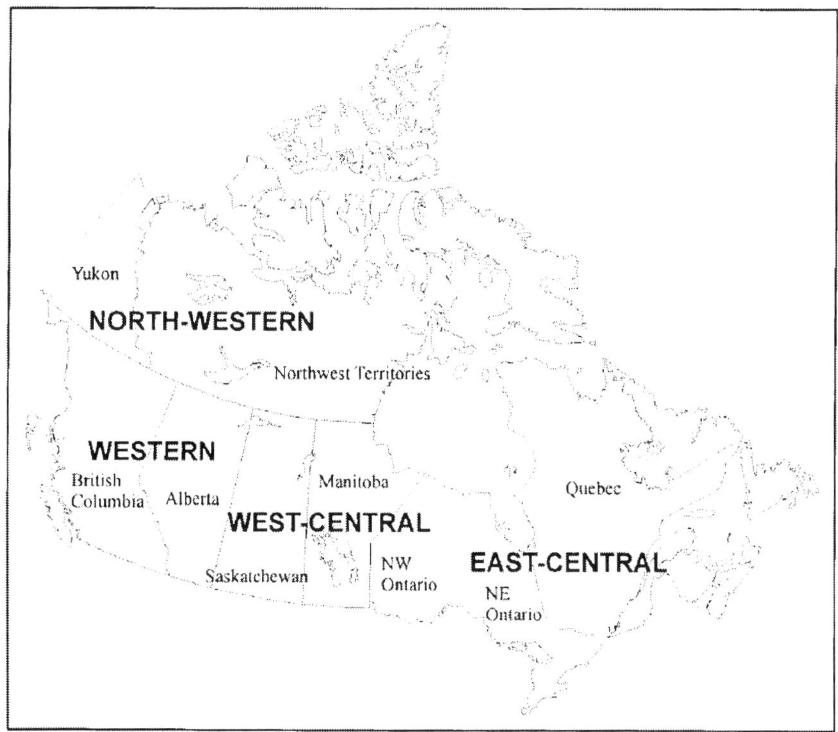

Figure 1. Provincial total area-burned data have been combined to yield information for four large Canadian regions: North-western (NW) Yukon/Northwest Territories (NWT); Western - British Columbia (BC) and Alberta; West-Central - Saskatchewan, Manitoba, and NW Ontario; and East-Central - NE Ontario and Quebec.

This regional framework is selected since area-burned data are only available on a provincial basis with Ontario being subdivided. Higher resolution studies are possible but not with the current data available for this time period and are beyond the scope of this study. The provincial data are aggregated into large climatic regions that encompass two or more provinces or territories as in Flannigan and Harrington (1988) and Harrington (1982).

Provincial area-burned data are compiled in calendar month summaries with no option to break them down any other way. It is not likely that mid-month to mid-month summaries, or any other temporal configuration, would yield vastly different totals. Figure 2 shows the monthly percentage of regional and national total area burned by wildland fire in Canada during the fire seasons from 1953 to 1995.

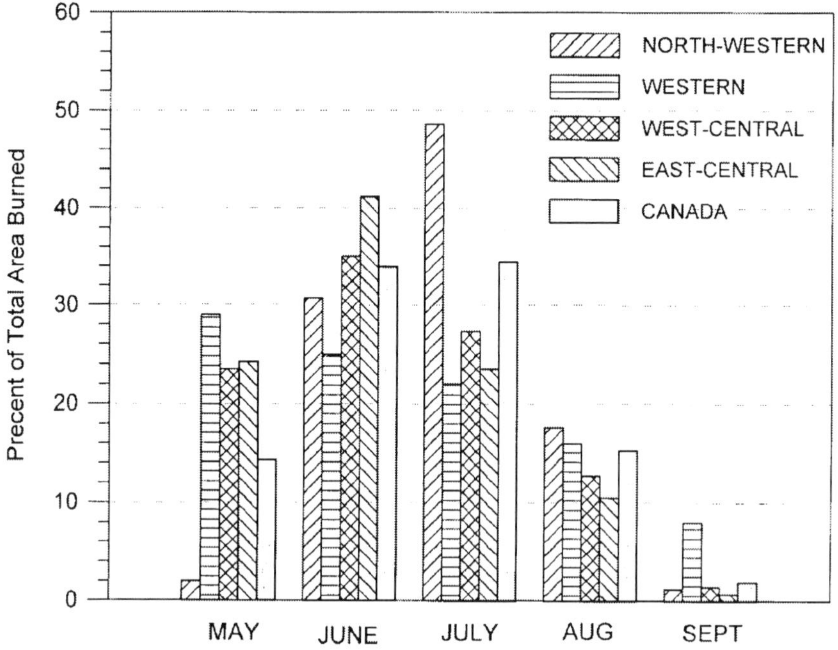

Figure 2. Regional total area burned by wildland fire in Canada by month (percent of total) during the fire season from 1953 to 1995

The Canadian national statistic represents the sum of all regions used in this analysis. September data represent less than 2% of the total annual area burned in all of Canada and thus are not considered in this analysis. Similarly, May data were not included in the north-west region fire season. Approximately 84% of the total area burned in Canada occurs during June, July and August. Nearly 78% of the total area burned in Canada over the May to August period occurs in the dry climatic regimes of west-central and north-west Canada. Approximately 12% occurs in the mainly dry western region while only about 10% takes place in the more humid climatic regime of east-central Canada. Total area burned is generally distributed evenly throughout the May to August fire season in the three regions south of 60°N with increases in June and slight decreases in August.

Figure 3 shows the regional total area-burned time series for the May to August period. Each series is smoothed by a low order polynomial trend line to provide an indication of long-term temporal variability. Detectable increases in total area burned by wildland fire since the mid-1970s are clearly evident in the north-west and west-central regions. However, this may in part be influenced by incomplete fire records prior to 1970. These two regions (see Figure 2) contain the largest percentages of the area burned during the 1953–95 period. The western region, with a much lower area-burned regime, shows no evidence of change in recent decades. East-central Canada, with even less of the total area burned, shows indication of slight increases since the mid-1970s similar to the north-west and west-central regions, but this is mainly because of the very high 1989 fire season.

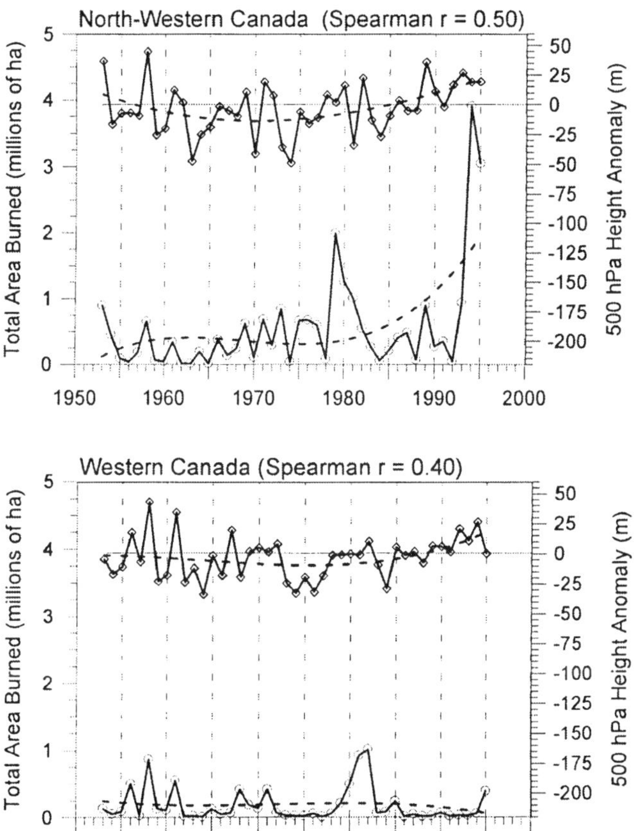

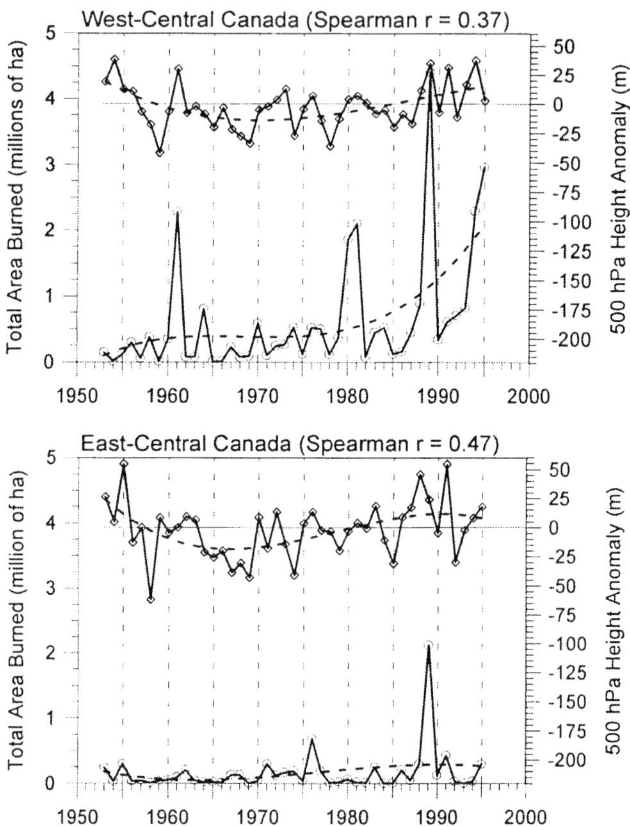

Figure 3 Regional time series of total area burned (bottom curve, left axis; dashed lines represent a low order polynomial that provides an indication of long-term temporal variability), and 500 hPa height anomalies (top curve, right axis) for the May–August fire season (June to August in NW Canada). The 500 hPa height anomalies are the average of the significantly correlated cluster of grid points ($\alpha = 0.01$) as derived from the hemispheric Spearman rank correlation maps (Figure 6).

2.2   Northern Hemisphere 500 hPa Heights

The upper air data base consists of daily (1200 UTC) 500 hPa geopotential height values for the Northern Hemisphere from 15° N to the pole on a 455–point grid. The grid has a spatial resolution of 5° latitude by 10° longitude from 15° to 65°N, and 5° latitude by 20° longitude from 70°N to the pole. The data from 1953 to 1981 were obtained from the National Center for Atmospheric Research (Jenne, 1975) and the balance, 1982 to 1995, from the Canadian Meteorological Centre in Montreal. A detailed description of the 500 hPa height data set is given in Knox *et al.* (1988). The daily 500 hPa heights were averaged by month for each grid point. 500 hPa height anomalies, expressed in decametres (dams = meters * 10), were calculated for each calendar month to correspond with the calendar month total area-burned data, at each grid point based on the full period (1953–1995) average.

The average 500 hPa heights from 1953 to 1995 for the May–August Canadian fire season over the Northern Hemisphere are shown in Figure 4. The main features of this climatology which affect northern North America are indicated on the map. They include from west to east: a weak southward-extending trough located adjacent to the west coast of North America; a parallel strong northward-extending ridge over western and north-western Canada ranging from the west coast of North America to approximately 110°W and extending from mid-latitudes to Alaska; and a strong southward-extending trough over north-eastern North America. This eastern trough has been termed the Canadian Polar Trough (CPT) during winter (Shabbar *et al.*, 1997). In summer, the spatial domain of the CPT is much weaker. However, it still dominates north-eastern North America extending from and including the closed circulation over northern Baffin Island, to the mid-latitudes where it broadens to occupy a sector from approximately 100°W to 50°W.

During normal summer conditions, moisture-bearing systems from the Pacific Ocean are generally deflected away from the forested north-west, western and west-central regions of Canada thus producing drier conditions. The east-central Canada region is normally subjected to an abundance of moisture-bearing systems, generally originating from the south and south-west. Surface system development is greatest in the baroclinic zone where the jet stream is strongest, in the southern areas of CPT. This region therefore normally experiences higher precipitation than its north-west, western and west-central counterparts.

The standard deviation map of May–August 500 hPa heights, as calculated over the period 1953 to 1995 at each grid point (not shown) reveals three major centres of variability affecting northern North America, including north-western Canada, north-eastern Canada, and a third large centre over the Arctic Ocean. These centres are somewhat transitory for individual months during summer and are generally stronger earlier and later in the season. The first centre occupies a large area over northern British Columbia/southern Yukon/western North-West Territories and has an amplitude of approximately 5.0 decametres (dams) or 50 meters. It is associated with the variability in the locations and magnitudes of the northern portions of the continental ridge and coastal trough described in Figure 4. The second centre, located over southern Baffin Island/Ungava, is associated with the CPT. It is slightly weaker with an amplitude of approximately 4.0 dams. The third centre, located over the Arctic Ocean off the coast of northern Greenland and Ellesmere Island, has an amplitude exceeding 5.0 dams. While the trough-ridge-trough (Figure 4) represents the normal summer flow configuration across northern North America, there is considerable variability in both the monthly and seasonally averaged patterns.

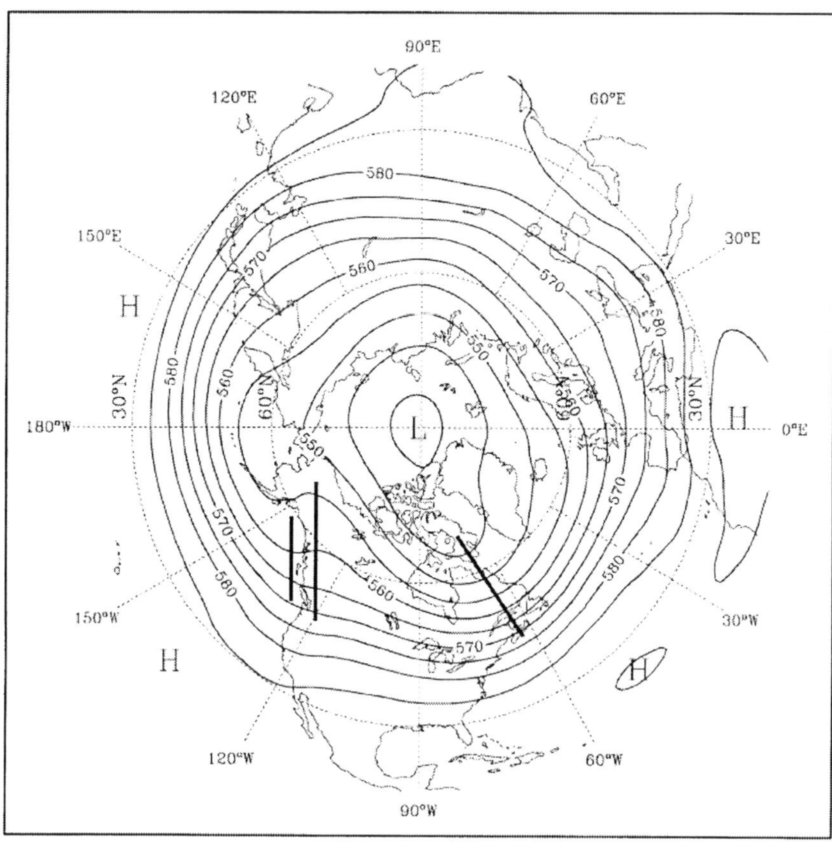

Figure 4. Average May–August 500 hPa circulation in decametres (dams = meters * 10)(1953 to 1995) over the Northern Hemisphere. Vertical lines represent the main features affecting northern North American summer climate. From west to east these are: west coast trough; continental ridge; Canadian Polar Trough (CPT). Contour interval is 5.0 dams.

3.   DATA ANALYSIS

3.1   Northern Hemisphere 500 hPa Height Composites

Figure 5a–5b shows the May–August 500 hPa height anomaly composites for the five highest and five lowest total area-burned seasons from 1953 to 1995 for the western and west-central regions, respectively. These extremes represent approximately the upper and lower 10% of the area-burned data population. Analysis of extremes is important because it is the extremely high area-burned seasons that have become more numerous and frequent as indicated by the observed increases in recent decades (Figure 3). A more detailed analysis of the other two regions (north-western and east-central) is provided in Skinner *et al.*, 1998.

Composites for the extremely high and low regional area-burned fire seasons are shown for the western region (Figure 5a(i) and (ii)) and for the west-central region (Figure 5b(i) – (ii)). For the extremely high fire composite for western Canada, there is a peak of anomalous ridging immediately over the forested areas of British Columbia and Alberta (2.0 – 2.5 dams). This is accompanied by deep troughs over the north Pacific (-2.0 dams) and eastern North America (CPT) (-2.0 dams). The entire hemispheric flow resembles the normal configuration (Figure 4), but is highly accentuated and anomalously meridional. In this case, there would be little moisture transport into the affected region. The extremely high fire composite for the west-central region shows anomalous ridging immediately over the forested areas of Saskatchewan, Manitoba and north-western Ontario (2.0 – 3.0 dams). The CPT is deeper than normal (-1.5 dams) but shifted to the northeast while the trough off the west coast is in its normal position and slightly deeper than normal (-0.5 dams). Figures for north-west and east-central Canada are not shown although similar peaks of anomalous ridging at 500 hPa are clearly evident. For north-west Canada, highly anomalous ridging occurs immediately over the forested areas of the North-West Territories and Yukon (2.0 – 2.5 dams) with a peak upstream over the Gulf of Alaska (3.0 dams). The CPT is located near its normal position although it is deeper in the north (-1.0 dams) and weaker near the southern limits (2.0 dams). For east-central Canada the extremely high fire composite is associated with a peak of anomalous ridging immediately over and upstream of the forested areas of north-eastern Ontario and Quebec (3.5 – 4.0 dams). The extensive ridging extends much further east into the region of normal upper troughing (see Figure 4). The CPT is also much deeper than normal (-2.5 – -3.0 dams) but is displaced well to the north-east in the Baffin Bay/Davis Strait region. The west coast trough is in its normal position and is deeper than normal (-1.0 dams).

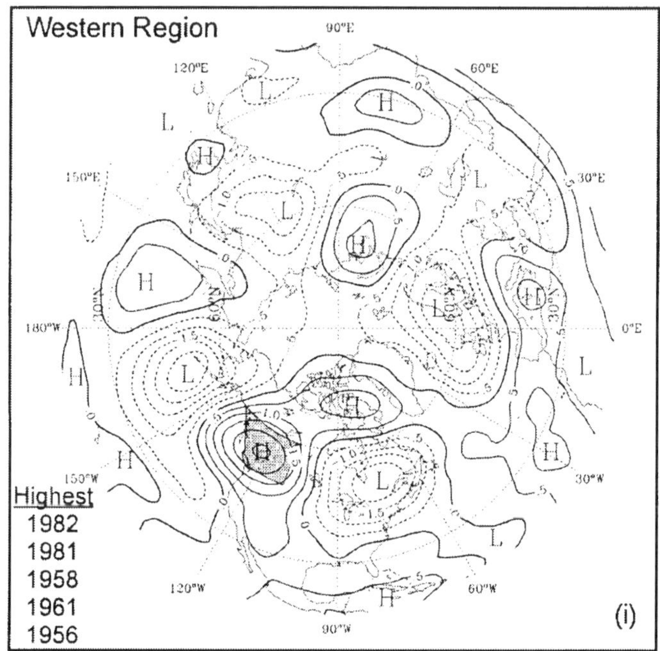

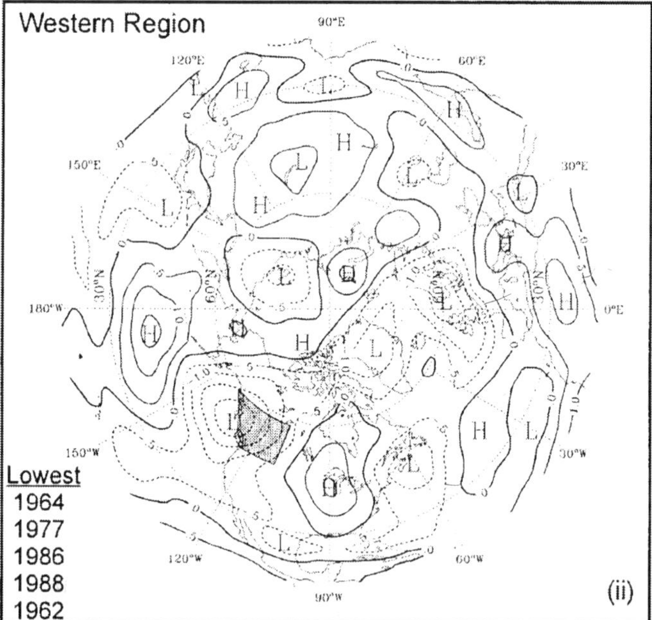

Figure 5a. 500 hPa height anomaly composites (in dams) for (i) the five highest and (ii) the five lowest total area-burned years for the May–August fire season for Western Canada. The five years used in the composites are listed. Contour interval is 0.5 dam. Negative contours are dashed. The region is shaded.

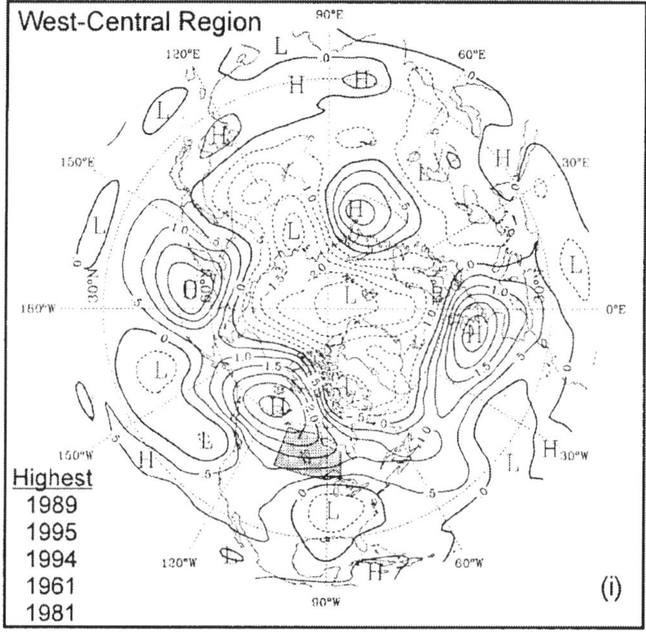

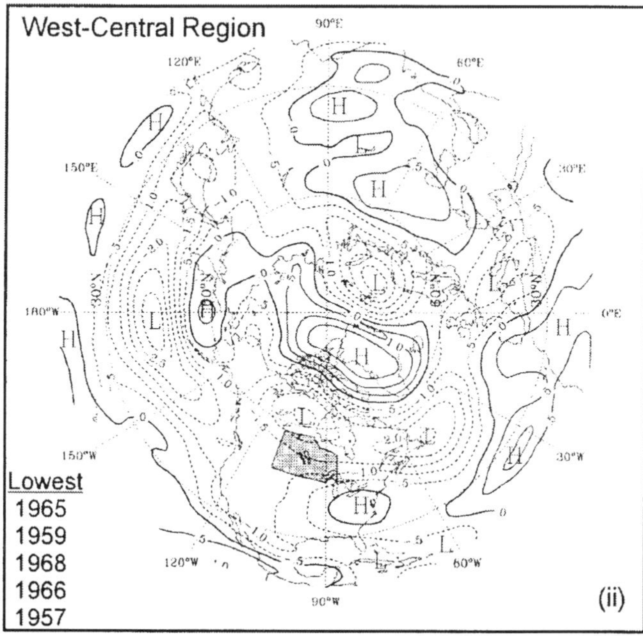

Figure 5b. 500 hPa height anomaly composites (in dams) for (i) the five highest and (ii) the five lowest total area-burned years for the May–August fire season for West-Central Canada. The five years used in the composites are listed. Contour interval is 0.5 dam. Negative contours are dashed. The region is shaded.

The extremely high regional area-burned fire seasons are clearly associated with highly anomalous ridging immediately over and upstream of the affected region. During these periods, 500 hPa circulation over northern North America is more meridional than normal with an amplification and northward displacement of the continental ridge and a deeper west coast trough and CPT. Additionally, the entire hemispheric flow at mid-latitudes is more meridional than normal. Under these conditions, the jet stream is displaced well to the north and often splits with a southern arm across the continental U.S. As a result, moisture-bearing systems are deflected well north and south of the affected regions. This is accompanied by high pressure subsidence with surface evaporation and thus dry conditions on a large regional basis. This configuration can dominate the affected region for extended periods of several weeks to months. Individual monthly 500 hPa height anomaly composites during the May–August fire season for all four Canadian regions (not shown) also indicate similar atmospheric flow patterns to the seasonal configurations.

The extremely low regional area-burned fire season composites, as shown in Figure 5a(ii) and 5b(ii) for the most part, show quite different 500 hPa circulation anomalies over northern North America. In all cases, there are either normal, or lower than normal, heights over western and north-western North America. The continental ridge is much weaker than normal suggesting a more zonal flow over North America during these periods. As a result, the jet stream follows a relatively uninterrupted west-to-east pattern across mid-continent. Moisture bearing systems from the Pacific Ocean are free to move directly across the continent thus affecting the low burned-area totals. This configuration can also dominate the affected region for extended periods of several weeks to months. Individual monthly 500 hPa height anomaly composites during the May–August fire season for all four Canadian regions (not shown) also indicate similar atmospheric flow patterns to the seasonal configurations.

3.2   Correlation Analysis

Frequency histograms (not shown) of the regional total area-burned seasonal time series (Figure 3) indicate non-normal bimodal data distributions. This is because of the monthly (not shown) and fire season series being typically represented by a large number of low burned areas, and a smaller number of large burned areas (Figure 3). To address the problem of non-normality in the distributions, the total area-burned data are analysed using the non-parametric Spearman rank correlation, which measures the correlation between the ranks of data. Spearman rank correlation coefficients between the 500 hPa anomaly series at each of the

455 Northern Hemisphere grid points and regional total area burned are calculated and mapped for the entire study period 1953–1995. The maps are used to illustrate the spatial associations between regional total area burned and 500 hPa height anomalies.

The Spearman rank correlation maps for the May–August fire season are shown in Figures 6a and 6b for the western and west-central regions. A more detailed analysis of the other two regions (north-western and east-central) is provided in Skinner *et al.* (1998). Each regional map clearly indicates statistically significant positive correlations (α = 0.01 level; r >= +0.40 is shaded) between regional total area burned and a cluster of 500 hPa anomalies immediately over or upstream of the affected region. The significance level α = 0.01 indicates that it is expected to have five of the 455 hemispheric grid points significantly correlated when in fact they are not and these should be randomly distributed across the entire hemispheric field. The significantly correlated clusters are spatially consistent with the regional area-burned configurations. Also, significantly correlated clusters of grid points in the central Pacific are identified with the north-west region (not shown) and the west-central region (Figure 6b) burned areas.

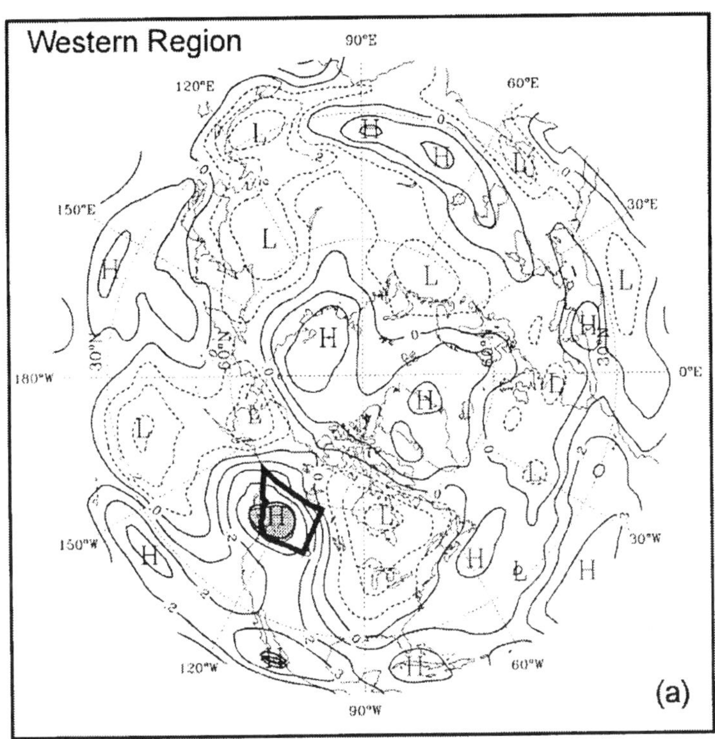

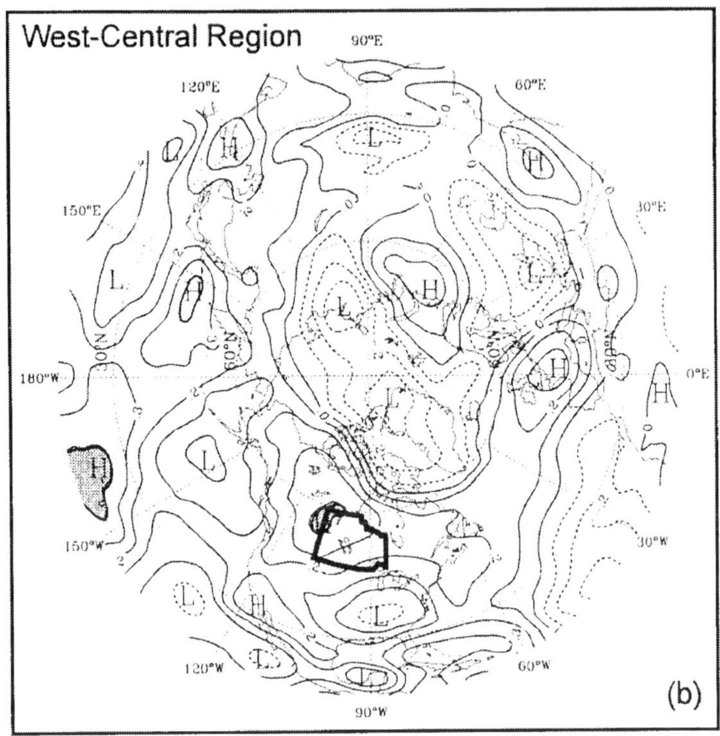

Figure 6. Spearman rank correlation maps for the Northern Hemisphere between regional total area burned and 500 hPa height anomalies at each of the 455 grid points for the May–August fire season for; (a) western Canada, and (b) west-central Canada. Contour interval is 0.5 dam. Clusters of statistically significant correlations at the $\alpha = 0.01$ significance level ($r = 0.40$) are shaded. The letter "H" indicate areas of high positive correlation while the letter "L" indicates areas of high negative correlation. The regions are outlined.

3.3   Two-Sample Comparisons

Table 1 presents the locations of the significantly correlated continental grid points($\alpha = 0.01$ level) as identified in Figure 6b for the west-central Canada region and their anomalous 500 hPa geopotential height values (in dams) that are in common to both the five highest total area-burned fire seasons (Figure 5b(i)) and the five lowest total area-burned seasons (Figure 5b(ii)). The west-central region is examined here because it is normally a high area-burned region (Figure 2) and also shows evidence of increases in total area burned over the past few decades (Figure 3). Also, it is located directly downstream to the normal position of the strong northward-extending ridge (Figure 4) which can be highly variable. During the highest fire seasons (Table 1), average cluster anomalous height is 2.0 dams, ranging from an average of -0.1 dams in 1995 to 3.4 dams in 1961. For the lowest fire seasons (Table 1), values range from -0.2 dams in 1966 to -3.3 dams in 1959 with an average of -1.9 dams. This example is typical of the other regions and the shorter monthly time periods (not shown).

Table 1. Locations of the significantly correlated grid points at the α = 0.01 level and their anomalous 500 hPa geopotential height values (in dams) that are in common to the five highest and five lowest total area-burned fire seasons for the west-central Canada region as shown in Figure 5b(i)

| Lat. °N | Long. °W | Highest Fire Seasons | | | | | | Lowest Fire Seasons | | | | | |
		1989	1995	1994	1961	1981	Highest Composite	1965	1959	1968	1966	1957	Lowest Composite
60.0	110.0	3.7	-0.3	4.9	4.5	1.8	2.9	-1.2	-5.2	-2.8	-0.7	-1.5	-2.3
60.0	100.0	4.2	-0.2	3.8	3.6	0.8	2.4	-2.5	-4.9	-3.9	1.1	-2.4	-2.5
60.0	90.0	3.7	0.1	1.7	1.5	0.0	1.4	-4.2	-3.1	-3.7	1.7	-2.1	-2.3
55.0	110.0	1.4	-0.6	2.7	5.1	1.9	2.1	-0.6	-3.7	-2.1	-1.3	-0.1	-1.5
55.0	100.0	3.5	-0.1	2.3	3.4	1.1	2.0	-2.0	-2.8	-3.2	-0.5	-0.6	-1.8
55.0	90.0	4.0	1.0	1.6	1.3	0.6	1.7	-3.4	-0.8	-3.2	-0.5	-1.1	-1.8
50.0	110.0	0.2	-0.8	2.4	4.3	0.4	1.3	-1.6	-2.3	-2.2	-1.0	0.3	-1.4
	Ave.	2.9	-0.1	2.8	3.4	1.0	2.0	-2.2	-3.3	-3.0	-0.2	-1.1	-1.9

Figure 7a is a box plot comparing the May–August total area-burned data from 1953 to 1995 associated with periods of only positive (> zero) and only negative (< zero) 500 hPa height anomalies from the grid points identified in Table 1 for west-central Canada. The analysis shows statistically significant differences between both means using the student's t test, and between both variances using the F test (α = 0.01). Area burned is extremely low during negative height anomalies when surface climatic conditions are more humid and favour low fire danger. However, when there are positive height anomalies only, there can be extremely high variability in area-burned totals. Surface climatic conditions are drier than normal and fire danger is high. Varying levels of protection would contribute greatly to the highly variable area-burned totals in a given year. This example is typical of the other three geographical regions as well as for the shorter monthly time periods (not shown).

Figure 7b is a box plot comparing the May–August anomalous 500 hPa heights for west-central Canada for each of the five high and five low extreme area-burned cases outlined in Table 1. The means are significantly different (α = 0.01); however, the variances are not. 500 hPa height anomalies are exclusively positive during the extreme high, and exclusively negative during the low, area-burned cases. This example is typical of the other regions and the shorter monthly time periods (not shown).

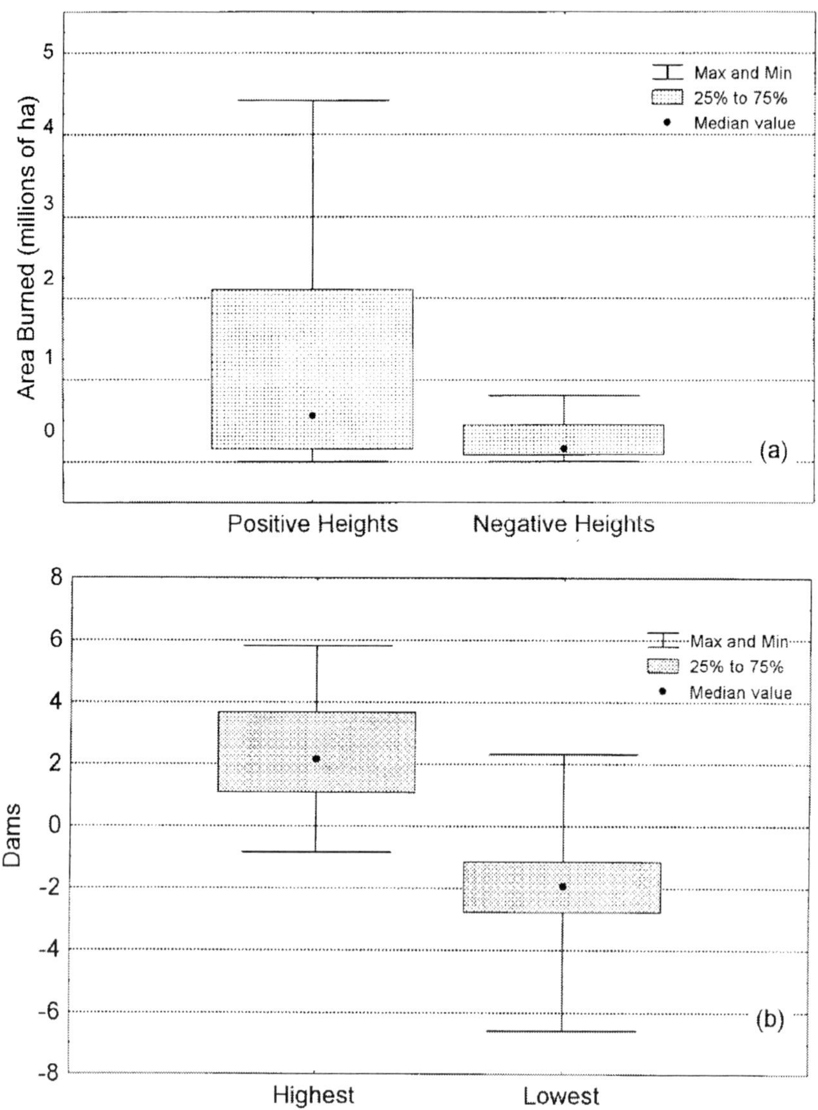

Figure 7. (a) Box plot showing the differences between the two samples of total area burned by wildland fire at times of positive (greater than normal), and negative (less than normal), 500 hPa height anomalies during the May–August fire season in the west-central Canada region. To test for difference between the two sample means the *t* statistic was used (*t* = 2.88, α = 0.01); to test for difference between two sample variances the F statistic was used (F = 29.71, α = 0.01), and (b) box plot showing the differences between the two samples of anomalous 500 hPa geopotential height values at times of the five highest and five lowest total area burned May–August fire seasons in the west-central Canada region. To test for difference between two sample means the t statistic was used (t = 12.09, α = 0.01); to test for difference between two sample variances the F statistic was used (F = 0.945, not significant).

3.4   Temporal changes in wildland fire activity and associated 500 hPa heights

All of the preceding analyses show strong support for 500 hPa heights being related to total area burned in Canada. The next step is to determine the nature of the temporal trends in these variables. As previously stated, area burned in Canada appears to be increasing since the mid-1970s especially in the north-west and central regions. Figure 3 shows the regional time series of total area burned and the associated average 500 hPa height anomalies in the statistically significant clusters identified in Figure 6 for 1953 to 1995. Each series is smoothed by a low-order polynomial to provide an indication of temporal trend. All regional 500 hPa series show generally higher than normal heights in the 1950s, generally lower than normal heights during the 1960s and 1970s, and a return to higher than normal heights since the 1980s. Detectable increases in wildland fire activity since the mid-1970s in the north-west and west-central regions, and to a lesser extent in the east-central region, are accompanied by increases in 500 hPa height anomalies over the past two decades.

Based on the detectable increase in west-central and NW Canada burned areas since the mid-1970s, the 43 year record of the May–August 500 hPa height field is divided into two periods of similar length, the 22 year period from 1953 to 1974 and the 21 year period from 1975 to 1995. The selection of these two periods is also supported by recent findings by Mantua *et al.* (1997) of an interdecadal shift in the winter climatic regime in 1977 in the Pacific basin that also includes interdecadal climate variability in the tropical Pacific. Other researchers have identified this abrupt shift in large-scale winter circulation in the mid-1970s (Trenberth, 1990; Graham, 1994). The mid-decade point (1975) was arbitrarily selected. The student's t test indicates that the means of the areas burned of each period are significantly different ($\alpha = 0.01$) in the north-west and west-central regions. Figure 8a shows the difference height field obtained by subtracting the earlier period from the later period. The difference pattern identifies an increase in 500 hPa heights over most of Canada. There is greater ridging over noth-west Canada (1.0 dam) and the northern limit of the CPT is strengthened (-1.0 dam). There is a weakened southern CPT over eastern Canada (1.0 dam). There has been more meridional flow over northern Canada north of 60°N. South of 60°N, the western Canada ridge is amplified possibly leading to increases in upper blocking situations.

To better identify the temporal change signal, the May–August total area-burned data are summed for all regions of Canada and the five highest and five lowest area-burned years are identified. The analysis shows that all five highest total area-burned years in Canada have occurred since 1980, while all five lowest years occurred prior to 1966. Figure 8b shows the difference height field obtained by subtracting the mean field of the five lowest area-burned years from that of the five highest area-burned years. The western Canada ridge is amplified possibly leading to increases in upper blocking situations. The pattern shows an amplification of the western Canada ridge with an eastward shifted CPT. North-west, western and west-central Canada experience more frequent upper blocking situations and thus drier surface conditions.

4.   DISCUSSION

Our analysis has demonstrated a strong and consistent statistical association between high total area-burned wildland fire seasons and the presence of anomalous ridging at 500 hPa immediately over and upstream of the affected region of Canada. During these periods, 500 hPa circulation over northern North America is more meridional than normal, with an amplification and northward displacement of the continental ridge and a deeper west coast trough and CPT. Additionally, the entire hemispheric flow at mid-latitudes is more meridional than normal. Under these conditions, the jet stream is displaced well to the north and often splits with a southern arm across the continental U.S. As a result, moisture-bearing systems are deflected well north and south of the affected regions. This is accompanied by high pressure subsidence with surface evaporation and thus dry conditions on a large regional basis. This configuration can dominate the affected region for extended periods of several weeks to months.

The 500 hPa height composites for the extreme cases from the five highest total area-burned seasons for the western and west-central Canada regions indicate a structure similar to the PNA (Pacific–North American) pattern (Wallace and Gutzler, 1981) during individual months (not shown), and for the entire May–August fire season, as shown in Figures 5a(i) and 5b(i). In addition, the Spearman rank correlation maps identify the same structure during extremely high area-burned seasons for the western and west-central regions (Figures 6a and 6b).

This enhancement of the positive phase of the west-to-east structure is termed "high index", and has a negative anomaly in the north-central Pacific, a positive anomaly over western North America, and another negative anomaly over eastern/south-eastern North America. Significantly correlated clusters of 500 hPa height anomalies are identified in the north-central Pacific, as well as over the continent (Figure 6b). This evidence supports the influence of this teleconnection structure on the summer climate of western and north-western North America.

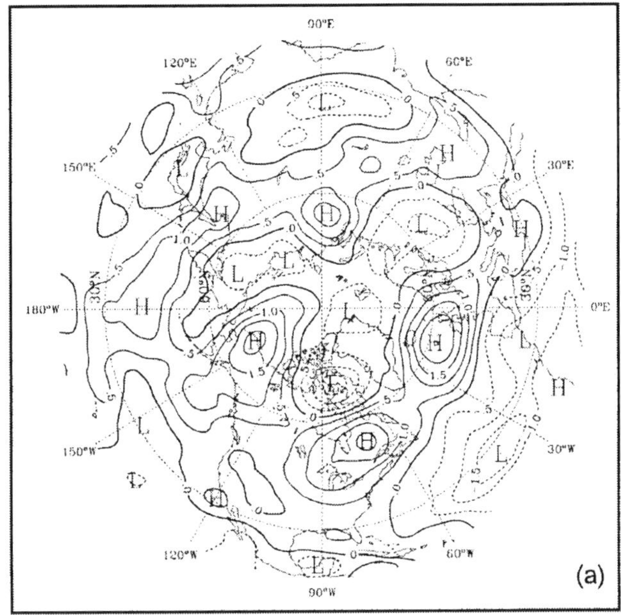

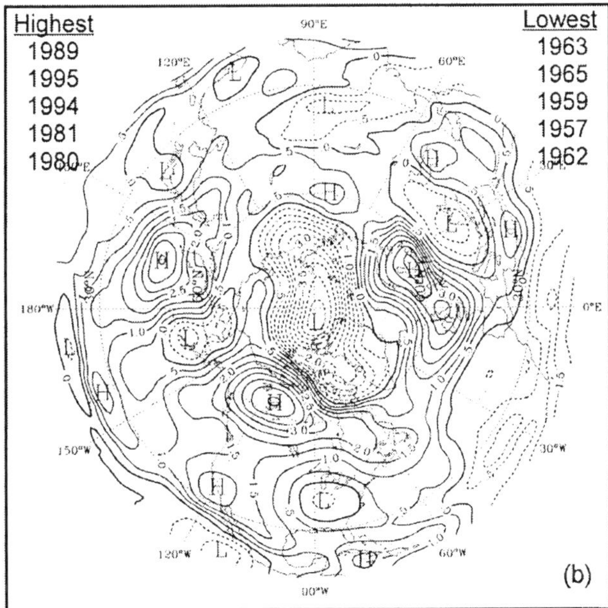

Figure 8. (a) Northern Hemisphere summer May–August 500 hPa geopotential height change derived by subtracting the mean height field of Period 1 (1953–1974) from that of Period 2 (1975–1995). Contour interval is 0.5 dam. Negative contours are dashed, and (b) Northern Hemisphere May–August 500 hPa geopotential height change derived by subtracting the mean height field of the five lowest total area-burned seasons for Canada from the five highest total area-burned seasons for Canada. The 5 years used in each extreme case are listed. Contour interval is 0.5 dam. Negative contours are dashed.

This structure produces warm and dry weather in western and west-central Canada during summer due to a northward displacement of the polar jet stream. This analysis also identifies a slightly northward shifted positive phase of a PNA-like pattern with lower heights over the north-central Pacific Ocean, higher heights over north-western North America, and lower heights over the south-eastern USA. This is consistent with the findings of others for the winter season (Shabbar *et al.*, 1997). Further examination of this structure in time and space and its relationship to Pacific Ocean forcing mechanisms is necessary for a better understanding of surface moisture transport and continental surface climatic conditions during summer.

Normally, the PNA pattern is thought to be a mode that has full amplification during the colder months and which attenuates with decreasing wavelength as the summer season approaches (Barnston and Livezey, 1987; Horel, 1981). However, Knox and Lawford (1990), found that the anomaly field over the east Pacific Ocean and North America during Canadian Prairie dry months in summer had the geometrical characteristics of the 3-cell west-to-east structure of the PNA. Recently, Johnson and Wowchuk (1993) have identified a link between PNA and severe fire years in the southern Canadian Rockies. These results are consistent with the results of others which have associated North American summer drought in the context of circulation anomalies (Namias, 1955; Namias, 1982; Trenberth *et al.*, 1988; Knox and Lawford, 1990).

While the summer structure of PNA is similar, the three main centres have somewhat different locations than those previously outlined for non-summer analysis (Yarnal and Diaz, 1986; Leathers *et al.*, 1991; Henderson and Robinson, 1994). The three main centres, as designed by Knox (1991) are; 45°N, 160°W; 55°N, 110°W; 35°N, 80°W. Figure 5a(i), the composite for the most severe fire seasons in the western region, identifies the first two centres in similar locations. However, the third centre is situated much further north than during its non-summer position at 50°N, 80°W. Figure 5b(i), the composite for the most severe seasons in the west-central region, identify a PNA-like structure more similar to the non-summer anomaly field with the third centre located closer to 35°N.

Temporal analysis shows a ridge over western and north-western Canada and Alaska that is oriented north-west to south-east with a maximum height rise of 2.0 dams since the mid-1970s, and 4.0 dams for the extreme cases since 1980. This height rise is consistent with that found in the same geographic locations for winter (Shabbar *et al.*, 1997). Also, it is consistent with the surface warming observed in the western half of Canada in recent decades, in general, and in north-west Canada in summer, in particular (Skinner and Gullett, 1993).

The identified clusters in the mid-tropospheric flow are critically located with respect to the movement of moist air to an affected region and thus to the development of regional surface moisture stress during summer. They can ultimately affect large regions and impact the potential for wildland fire severity. It is important to concentrate on these clusters and their varying magnitudes from non-severe to severe fire seasons for a more accurate assessment of future fire regimes under changing climatic conditions

5.   SUMMARY

This represents the first study of this kind to identify and understand the links between atmospheric circulation anomalies in the mid-troposphere and wildland fire severity in Canada on large spatial and temporal scales. Statistically significant relationships have been demonstrated. There appears to be increases in both area burned, especially in north-west and west-central Canada, and 500 hPa heights in key geographical areas. Further research is required at finer spatial and temporal scales.

Current climate change projections (IPCC, 1995), along with recent assessments of projected climate change impacts on forest fire activity (e.g. Stocks *et al.*, 1998) clearly indicate that increases in regional seasonal fire weather severity will result in large increases in the areal extent of extreme fire danger at northern latitudes. The accurate prediction of future Canadian fire regimes therefore requires the identification of relationships between synoptic-scale climate variability and forest fire impacts (for example, 500 hPa ridging frequency and strength and area burned) that will be reconcilable with future Regional Climate Model (Caya *et al.*, 1995) results. Also, there is a need to improve the spatial and temporal resolution of the Canadian fire burned area database and to identify in aggregate terms, the spatial and temporal variability of the levels of fire protection that have been brought to bear on our study regions over time.

6.   REFERENCES

Alexander, M.E., Janz, B., Quintillo, D., 1983: Analysis of extreme wildfire behaviour in east-central Alberta: a case study. In *Proceedings of the 7th Conference on Fire and Applied Meteorology*, 25–28 April 1983, Fort Collins, Colorado. American Meteorological Society, Boston. Mass., 38–46.

Barnston, A.G., Livezey, R.E.,1987: Classifications, seasonality and persistence of low frequency atmospheric circulation patterns. *Monthly Weather Review*, **115**, 1083–1126.

Bonsal B.R., Chakravarti, A.K., Lawford, R.G., 1993: Teleconnections between north Pacific SST anomalies and growing season extended dry spells on the Canadian Prairies. *International Journal of Climatology*, **13**, 865–878.

Caya D., Laprise, R., Giguere, M., Bergeron, G., Blanchet, J.P., Stocks, B.J., Boer, G.J., McFarlane, N.A., 1995: Description of the Canadian Regional Climate Model. In Price, D.T., Wisniewski, J., (eds.), *Boreal Forests and Global Change*, Kluwer Academic Publishers, Dordrecht, 477–482.

Finklin, M.D., 1973: Meteorological factors in the Sundance fire run. *USDA Forestry Service General Technical Report* INT–6.

Flannigan, M.D., Harrington, J.B., 1986: Synoptic weather conditions during the Porter Lake experimental fire project. *Climatological Bulletin*, **20**, 19–40.

Flannigan, M.D., Harrington, J.B., 1988: A study of the relation of meteorological variables to monthly provincial area burned by wildfire in Canada (1953–80). *Journal of Applied Meteorology* **27**,441–452.

Graham, N.E., 1994: Decadal–scale climate variability in the tropical and North Pacific during the 1970s and 1980s: observations and model results. *Climate Dynamics*, **10**, 135–162.

Harrington, J.B., 1982: A statistical study of area burned by wildfire in Canada 1953–1980. Environment Canada, Canadian Forestry Service, Petawawa National Forest Institute *Information Report PI–X–16*, 32p.

Henderson, K.G., Robinson, P.J., 1994: Relationships between the Pacific/North America teleconnection patterns and precipitation events in the south-eastern USA. *International Journal of Climatology*, **14**, 307–232.

Horel, J.D., 1981: A rotated principal component analysis of the interannual variability of the Northern Hemisphere 500 mb height field. *Monthly Weather Review*, **109**, 2080–2092.

IPCC 1995: *Climate change 1995: impacts, adaptations and mitigation of climate change: scientific-technical analysis.* Watson, R.T., Zinyowera, M.C., R.H. Moss, R.H., (eds), Cambridge University Press, Cambridge, UK., 878p.

Janz, B., Nimchuk, N., 1985: The 500 mb chart – a useful fire management tool. In *Proc. of the 8th Conf. on Fire and Appl. Met.*, 29 April – 2 May 1985, Detroit, Michigan. Society of American Foresters, Washington, D.C., 233–238.

Jenne, R., 1975: *Data Sets for Meteorological Research*, NCAR–TN/JA111, 194 p.[Available from Order Department, US Department of Commerce, National Technical Information Service, Springfield Virginia, US 22161].

Johnson, E.A., Wowchuk, D.R., 1993: Wildfires in the southern Canadian Rocky Mountains and their relationship to mid-tropospheric anomalies. *Canadian Journal of Forest Research*, **23**, 1213–1222.

Knox, J.L., Higuchi, K., Shabbar, A., Sargent, N.E., 1988: Secular variations of the Northern Hemisphere 50 kPa geopotential height. *Journal of Climatology*, **5**, 500–511.

Knox, J.L., Lawford, R.G., 1990: The relationship between Canadian prairie dry and wet months and circulation anomalies in the mid-troposphere. *Atmosphere-Ocean*, **28**, 189–215.

Knox, J.L., 1991: *Northern Hemisphere Circulation*. Internal Report, Canadian Climate Centre, Downsview, Ontario, 29pp.

Kurz, W.A., Apps, M.J., Stocks, B.J., Volney, W. JA., 1995: Global climate change: disturbance regimes and biospheric feedbacks of temperate and boreal forests. In Woodwell, G.M., Mackenzie, F.T., eds. *Biotic feedbacks in the global climatic system.* New York: Oxford University Press, 119–13.

Leathers, D.J., Yarnal, B., Palecki, M.A., 1991: The pacific/North American teleconnection pattern and United States climate. Part I: Regional temperature and precipitation associations. *Journal of Climatology*, **4**, 517–528.

Mantua, N.J., Hare, S.R., Zhang, Y., Wallace, J.M., Francis, R.C., 1997: A Pacific interdecadal climate oscillation with impacts on salmon production. *Bulletin of the American Meteorological Society*, **78**, 6, 1069–1079.

Namias, J., 1955: Some meteorological aspects of drought with special reference to the summers of 1952–54 over the United States. *Monthly Weather Review*, **83**, 199–205.

Namias, J., 1982: Anatomy of Great Plains protracted heat waves (especially the 1980 U.S. summer drought). *Monthly Weather Review*, **110**, 824–838.

Newark, M.J., 1975: The relationship between forest fire occurrence and 500 mb longwave ridging. *Atmosphere*, **13**,1,26–33.

Schroeder, M.J., Glovinsky, M., Hendricks, V.H., *et al.*, 1964: Synoptic weather types associated with critical fire weather. USDA Forest Service. Pacific Southwest Forest and Range Experimental Station, Berkeley, California.

Shabbar, A., Higuchi, K., Skinner, W., Knox, J.L., 1997: The association between the BWA index and winter surface temperature variability over eastern Canada and west Greenland. *International Journal of Climatology*, **17**, 1195–1210.

Simard, A.J., 1997: *National workshop on wildland fire activity in Canada*. Science Branch, Canadian Forestry Service, Natural Resources Canada, 38p.

Skinner, W.R., Gullett, D.W., 1993: Trends of daily maximum and minimum temperature in Canada during the past century. *Climate Bulletin*, **27**, 2, 543–557.

Skinner, W.R., Stocks, B.J., Martell, D.L., Bonsal, B., and Shabbar, A.. 1998: The association between circulation anomalies in the mid-troposphere and area burned by wildland fire in Canada. *Theoretical and Applied Climatology*, in press.

Stocks, B.J., Lee, B.S., Martell, D.S., 1996: Some potential carbon budget implications of fire management in the boreal forest. p. 89–96 in *Forest Ecosystems, Forest Management and the Global Carbon Cycle*. Apps, M.J., Price, D.T., (eds.), NATO ASI Series, Subseries 1, Global Environmental Change . Vol.40, Springer-Verlag, Berlin, Germany.

Stocks, B.J., Fosberg, M.A., Lynham, T.J., Mearns, L., Wotton, B.M., Yang, Q., Jin, J-Z., Lawrence, K., Hartley, G.R., Mason, J.A., McKenney, D.W., 1998: Climate change and forest fire potential in Russian and Canadian boreal forests. *Climate Change* **38**, 1–13.

Street, R.B., 1985: Drought and synoptic fire climatology of the boreal forest region of the Canadian prairie provinces. In *Proc. of the 8th Conf. on Fire and Appl. Met.*, 29 April – 2 May 1985, Detroit, Michigan. Society of American Foresters, Washington, D.C., 108–112.

Street, R.B., Birch, E.C., 1986: Synoptic fire climatology of the Lake Athabasca – Great Slave Lake area, 1977–1982. *Climate Bulletin*, **20**, 3–18.

Trenberth, K.E., Branstator, G.W., Arkin, P.A., 1988: Origins of the 1988 North American drought. *Science*, **242**, 1640–1645.

Trenberth, K.E., 1990: Recent observed interdecadal climate changes in the Northern Hemisphere. *Bulletin of the American Meteorological Society*, **71**, 7, 988–993.

Wallace, J.M., Gutzler, D.S., 1981: Teleconnections in the geopotential field during the Northern Hemisphere winter. *Monthly Weather Review*, **109**, 784–812.

Yarnal, B., Diaz, H.F., 1986: Relationships between extremes of the Southern Oscillation and the winter climate of the Anglo-American Pacific coast. *Journal of Climatology*, **6**, 197–219.

Underestimation of GCM-Calculated Short-Wave Atmospheric Absorption in Areas Affected by Biomass burning

MARTIN WILD

Swiss Federal Institute of Technology, Department of Geography, Zurich, Switzerland

Abstract: Many current General Circulation Models (GCMs) exhibit a common problem, namely that their atmosphere is too transparent to solar radiation. The underestimation of atmospheric short-wave absorption by these models is particularly large in areas and seasons where extensive biomass burning takes place. This is shown using surface radiation measurements combined with co-located satellite observations at sites affected by biomass burning in Equatorial Africa. The observed absorption of solar radiation in the atmosphere is significantly larger during the dry season where biomass burning takes place than during the wet season. In contrast, GCMs, which do not account explicitly for aerosols arising from biomass burning, calculate a maximum absorption in the wet season and a minimum absorption in the dry season, as a result of the dominance of water vapour absorption in the calculations. This indicates that the inadequate representation of absorbing aerosols in GCMs significantly contributes to the underestimated short-wave atmospheric absorption in regions affected by biomass burning. This can reach up to 30–40 Wm^{-2} regionally and seasonally, and more so locally. For a realistic simulation of the radiation budget in these regions, an adequate spatial and temporal representation of aerosols is therefore crucial. Thus, in addition to possible underestimates in water vapour and cloud absorption, the lack of absorbing aerosols should be accounted for in the ongoing discussion of the problem of underestimated short-wave absorption in the GCM atmospheres. Evidence is presented that an appropriate representation of aerosol and water vapour absorption may be sufficient to close the gaps between model-calculated and observed estimates of short-wave atmospheric absorption. Hence, a substantial enhancement of cloud absorption in GCMs, as claimed in other studies, seems not immediately necessary from the present investigation.

1.   INTRODUCTION

The sun is the only significant source of energy needed to account for the climate of the Earth. The distribution of solar radiation in the global climate system is therefore of key importance for the thermal, dynamical and hydrological conditions on our planet. Due to recent satellite programs such as ERBE, the total amount of solar energy absorbed by the global climate system is well established (Barkstrom *et al.*, 1990). Much larger uncertainties, however, still exist with respect to the relative fraction of solar energy absorbed in the atmosphere and at the surface. Large differences are therefore also found in the simulated short-wave radiation budgets in different General Circulation Models (GCMs) (e.g., Wild *et al.*, 1995a). More insight into this problem was recently gained through the availability of a comprehensive dataset of the world-wide measured surface energy fluxes compiled at the author's institute, namely, the Global Energy Balance Archive (GEBA, Ohmura *et al.*, 1989; Gilgen *et al.*, 1998).

Combining the surface measurements from GEBA with co-located top-of-atmosphere (TOA) fluxes from ERBE at more than 700 sites, Wild *et al.* (1995a, 1998) concluded that current GCMs typically overestimate the solar irradiance at the surface, and attributed this to a lack of absorption in the atmosphere. The principal absorbers in the atmosphere are the radiatively active gases (water vapour, carbon dioxide, ozone), clouds, and absorbing aerosols (such as black carbon aerosols originating from biomass burning). Wild *et al.* (1995a, 1998), Arking (1996) and Li *et al.* (1997) presented evidence that an underestimated water vapour absorption contributes to the lack of atmospheric absorption in GCMs. It has further been argued that the GCM-calculated absorption of solar radiation by clouds is underestimated (Cess *et al.*, 1995; Ramanathan *et al.*, 1995). Less emphasis has so far been put on a possible inadequate representation of absorbing aerosols in GCMs as a further source for the lack of atmospheric absorption. Aerosols are to date only very crudely represented in GCMs. The lack of absorption in the GCM atmospheres has been found to be particularly large at low latitudes (Wild *et al.*, 1995a, 1998; Wild and Ohmura, 1999). A similar deficiency in a satellite-based radiation climatology has been noted by Li (1998), which he related to the absence of absorbing aerosol in the retrieval algorithm. In many low latitude areas, the atmosphere is periodically loaded with large amounts of aerosols from human-induced vegetation fires, which are used as a tool to clear savannas and forests. The bulk aerosol originating from such biomass burning is principally composed of black carbon and organic carbon. While organic carbon is mainly a scatterer of solar radiation, black carbon is a strong absorber.

The present study assesses the extent to which the neglect of these absorbing aerosols in GCMs can explain the underestimated absorption in the model atmospheres. It will focus on Equatorial Africa because this region has the highest density of surface observations amongst those affected by biomass burning. A map with the global distribution of observations sites with long-term monitoring records available from GEBA is given in Wild *et al.* (1995a).

2.   MODELS AND EXPERIMENTS

Three GCMs are considered in this study, namely the ECHAM3 GCM of the Max Planck Institute for Meteorology, Hamburg (Roeckner *et al.*, 1992), the ARPEGE GCM (Déqué *et al.*, 1994) from Météo-France, Toulouse, and the HadAM2b GCM (Stratton, 1999) from the Hadley Centre for Climate Prediction and Research, Bracknell. All models include broad-band radiation schemes with two-stream approximation, as typically used in GCMs. The effects of aerosol are very crudely represented in these GCMs, as is the case in many others. The ECHAM3 and ARPEGE GCMs only include two aerosol profiles, one for continental and the other for maritime conditions, which are constant in time and space over all land- and sea-points. Scattering and absorption coefficients are taken from a dataset provided by Shettle and Fenn (1976). The HadAM2b model does not include any aerosol effects. Two additional simulations were available from the Hadley Centre with the next generation model version HadAM3, one with a simple aerosol climatology as in ECHAM and ARPEGE, and the other with a switched off aerosol effect (Cusack *et al.*, 1998).

In the framework of the EU project HIRETYCS (High Resolution Ten Years Climate Simulation), high resolution simulations of present day conditions have been performed with these models (T106 for the spectral models ECHAM and ARPEGE, or 1.25° x 0.83° for the grid point model HadAM2b). The same simulations were also carried out at standard resolution (T42, or 3.75° x 2.5°, respectively). All experiments used prescribed AMIP SST and sea ice boundary conditions representative for the period 1979–1988 (Gates, 1992). The results shown below are from the high resolution experiments only, which allows a more appropriate comparison between the gridded model fluxes and point observations (Wild, 1997). The results, however, are also consistently found at lower resolution and are thus not resolution dependent (cf. Wild *et al.*, 1995a).

3.   OBSERVATIONAL DATA

The observational data used in this study are retrieved from a database containing the world-wide instrumentally measured surface energy fluxes, the Global Energy Balance Archive (GEBA, Ohmura *et al.*, 1989; Gilgen *et al.*, 1998). This database currently possesses 220,000 monthly mean fluxes for approximately 1600 sites and has been used in a number of studies to assess model-calculated and satellite derived estimates of surface energy fluxes (e.g., Garratt, 1994; Li *et al.*, 1995; Wild *et al.*, 1995a, 1995b, 1997, 1998; Konzelmann *et al.*, 1996; Rossow and Zhang, 1995). Gilgen *et al.* (1998) estimated the relative random error (root mean square error / mean) of the incoming short-wave radiation values in the GEBA at 5% for the monthly means and 2% for annual means.

The satellite climatologies of the radiative fluxes at the top-of-atmosphere (TOA) are ensemble averages from the Earth Radiation Budget Experiment (ERBE, Barkstrom, 1990) over the period 1985–1989, with a resolution of 2.5° x 2.5°. The uncertainties in the monthly averaged scanner data are estimated within +/- 5 Wm^{-2}.

4.   GCM SHORT-WAVE RADIATION BUDGETS IN EQUATORIAL AFRICA

4.1   Seasonal and latitudinal discussion

The distribution of the sites available for this study from GEBA in Equatorial Africa is shown in Fig. 1. Only sites with long-term monitoring records (longer than 5 years) have been selected in order to ensure the representativity of the observations for climatological mean conditions. Interpolation of the gridded GCM and ERBE fluxes to these GEBA sites has been done by taking into account the four surrounding grid points weighted by their inverse spherical distance. The observed short-wave absorption within the atmospheric column above the sites has been determined as difference between the net short-wave flux at the TOA from ERBE, and the net short-wave flux at the surface. The net short-wave flux at the surface has been obtained by combining the observed values of the incoming short-wave radiation from GEBA together with the collocated values of a surface albedo climatology as given in the Surface Radiation Budget Project (SRB, Darnell *et al.*, 1992).

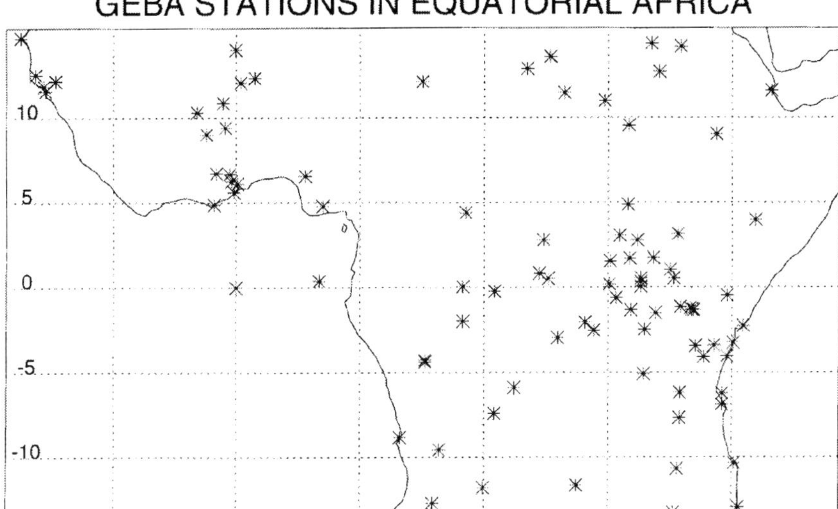

Figure 1. Geographical distribution of the observation sites in Equatorial Africa with long-term observations of short-wave downward radiation used in this study from the Global Energy Balance Archive (GEBA).

In the following, the model-calculated short-wave absorption is compared with the GEBA derived observations at the surface, to the co-located ERBE estimates at the TOA, and to the difference between ERBE and GEBA within the atmosphere. The resulting annual mean model biases in Equatorial Africa at the TOA, within the atmosphere and at the surface are shown in Fig. 2. Biases are displayed as a function of latitude, which were obtained by averaging over the sites located within latitudinal bands of 5°. At the TOA, the three models agree well with the ERBE estimates in Equatorial Africa, with mean model biases generally smaller than 10 Wm^{-2} (top panel of Fig. 2). This indicates that the net solar energy absorbed in the climate system in Equatorial Africa is well represented in the models. Note that the global mean planetary albedo in the GCMs has been tuned to match the ERBE global mean value, so this good agreement is not unexpected. At the surface, however, the models largely overestimate the absorption of solar radiation compared to the GEBA derived estimates, on the order of 20–40 Wm^{-2} (bottom panel of Fig. 2). This is caused by a significant lack of absorption in the atmosphere, of equivalent magnitude (20–40 Wm^{-2}, centre panel of Fig. 2). The problem common to all these models is thus not so much one of the reproduction of the net amount of solar energy absorbed in the climate system as a whole, but rather one of the partitioning of the absorbed energy between atmosphere and surface: too little solar energy is absorbed in the model atmosphere while too much is absorbed at the surface.

SW ABSORPTION ANNUAL

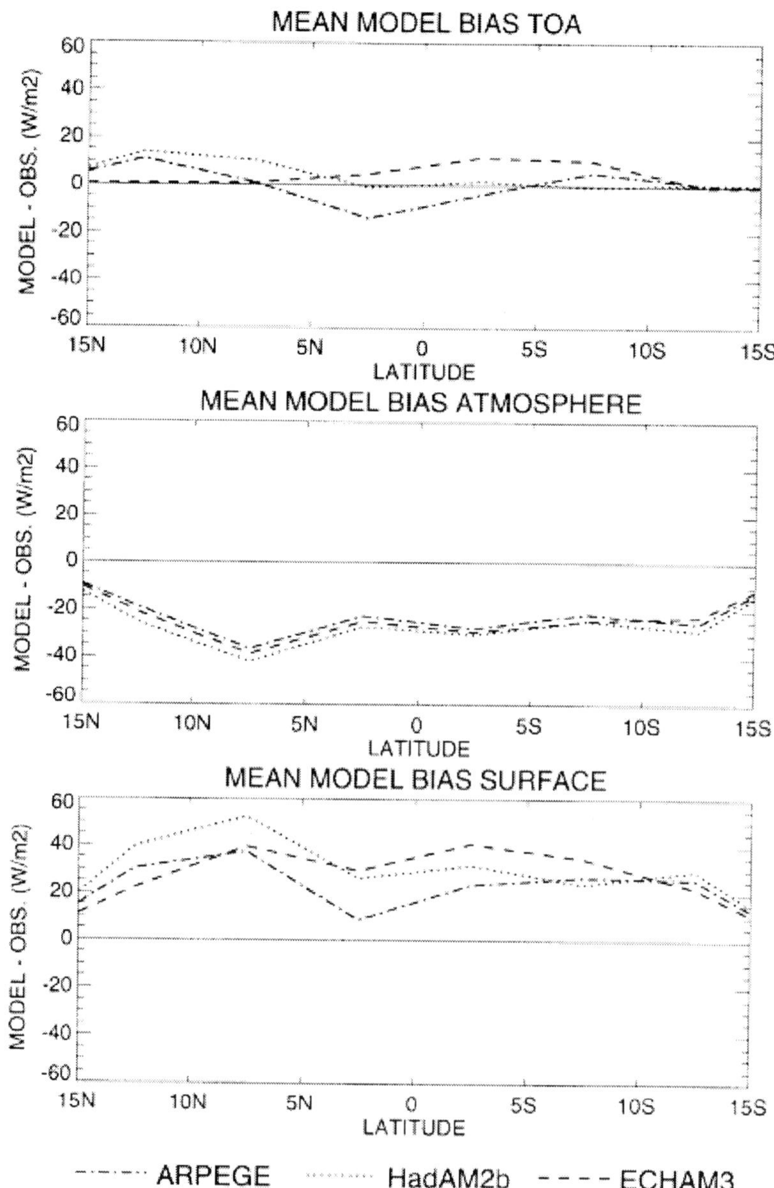

Figure 2. Annual mean model bias of short-wave absorption in the total surface/atmosphere system (TOA, upper panel), within the atmosphere (centre panel) and at the surface (lower panel), at observational sites in Equatorial Africa as function of latitude, for the 3 GCMs ARPEGE, HadAM2b and ECHAM3. Biases averaged over sites within 5° latitudinal bands.

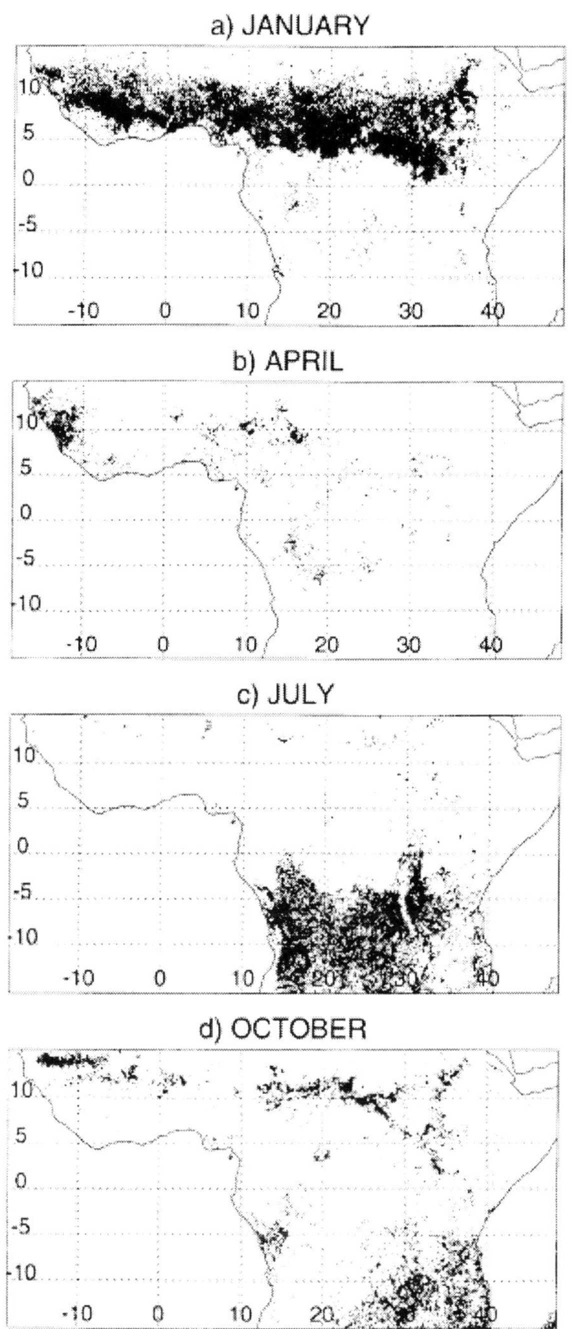

Figure 3. Occurrence of vegetation fires in Equatorial Africa as given
in the Global Fire Atlas (Arino and Melinotte 1998) for the months January 1993 (a),
April 1993 (b), July 1993 (c), October 1993 (d).

To estimate the contribution of aerosols from biomass burning, a discussion on the seasonal scale is necessary, as human-induced biomass burning is a very distinct seasonal phenomenon: It is predominately practised during the dry seasons as a tool to clear savannas and forests. This is reflected in the seasonal occurrence of fires in Equatorial Africa as shown in the four mid-season months in Fig. 3, based on data provided by the ESA Fire Atlas (Arino and Melinotte, 1998): in the northern hemispheric winter (January), with the dry season to the north of the Equator, fires are largely concentrated in the area between 5°N and 10°N (Fig. 3a). In contrast, in the northern hemisphere summer (July), with the dry season shifted to the southern hemisphere, the fire activity is dominantly to the south of the Equator, with a maximum concentration between 5°S and 15°S (Fig. 3c). The fire activity is significantly less pronounced during northern hemispheric spring (April, Fig. 3b) and autumn (October, Fig 3d), i.e. during the rainy seasons over Equatorial Africa (see also Stroppiana *et al.*, this book). The fire distributions shown in Fig. 3 are based on data from the year 1993, but are qualitatively similar in other years (cf. e.g., Cahoon *et al.*, 1992, for the seasonal fire distributions in the year 1987).

The model-calculated biases in the Equatorial African radiation budgets in the four mid-season months are given in Figs. 4–7. At the TOA, the 3 GCMs are generally in good agreement with the ERBE data for all seasons, hence not only in the annual mean as noted above (cf. Fig. 2 with Figs. 4–7). At the surface and in the atmosphere, however, substantial deviations from the annual mean biases can be noted as follows: In January (Fig. 4), all models show a significant peak overestimation of solar radiation at the surface to the north of the Equator, due to a maximum underestimation of short-wave absorption in the GCM atmospheres at the same latitudes. The underestimation of atmospheric absorption is largest between 5°N and 10°N, interestingly enough precisely at the latitudes of maximum fire activity (cf. Fig. 3a). In July (Fig. 5), the peak underestimation of short-wave absorption in the atmosphere is shifted to the south of the Equator, again coinciding with the maximum fire activity, which is also shifted to the southern hemisphere (cf. Fig 3c).

Thus, there is strong evidence for a correlation between biomass burning activity and underestimated atmospheric absorption in GCMs: Aerosols from biomass burning have a significant influence on atmospheric short-wave absorption not taken into account in GCMs, thus causing significant biases in simulated radiation budgets. Accordingly, during the periods with little fire activity (April, October), no such dominant peak can be found in the radiation budget biases (Figs. 6 and 7).

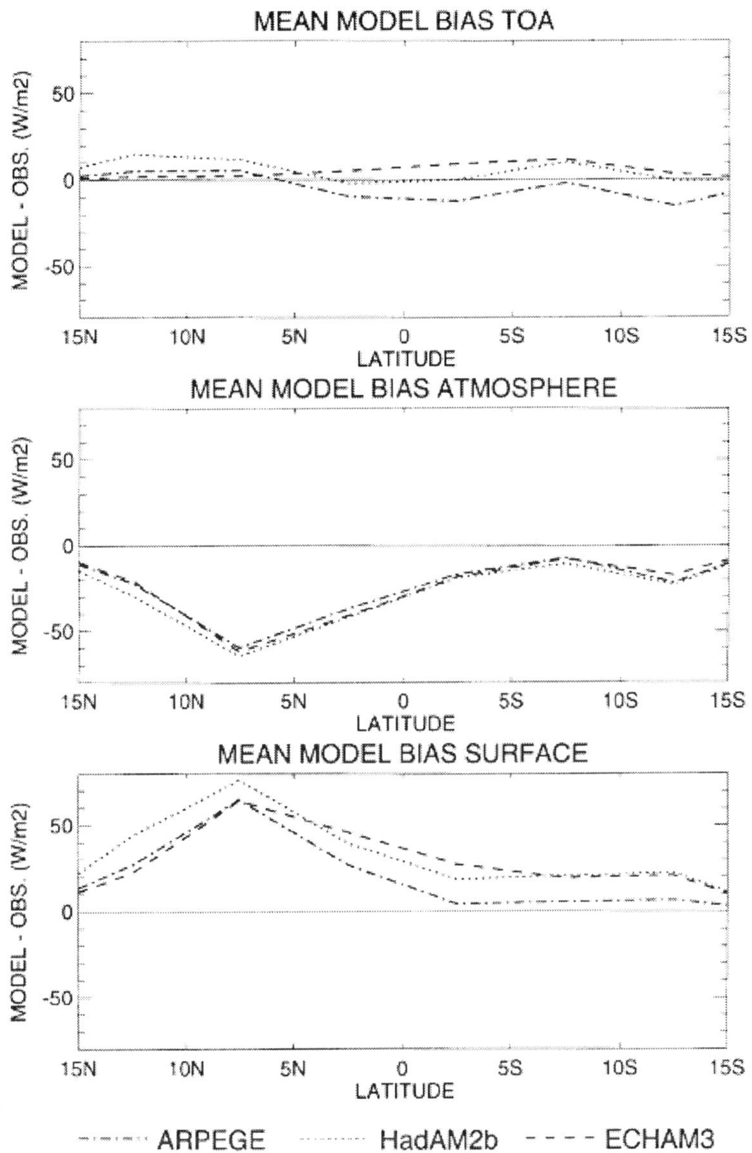

Figure 4. Mean January model bias of short-wave absorption a) in the total surface/atmosphere system (TOA, upper panel), within the atmosphere (centre panel) and at the surface (lower panel), at observational sites in Equatorial Africa as function of latitude, for the 3 GCMs ARPEGE, HadAM2b and ECHAM3. Biases averaged over sites within 5° latitudinal bands.

SW ABSORPTION JULY

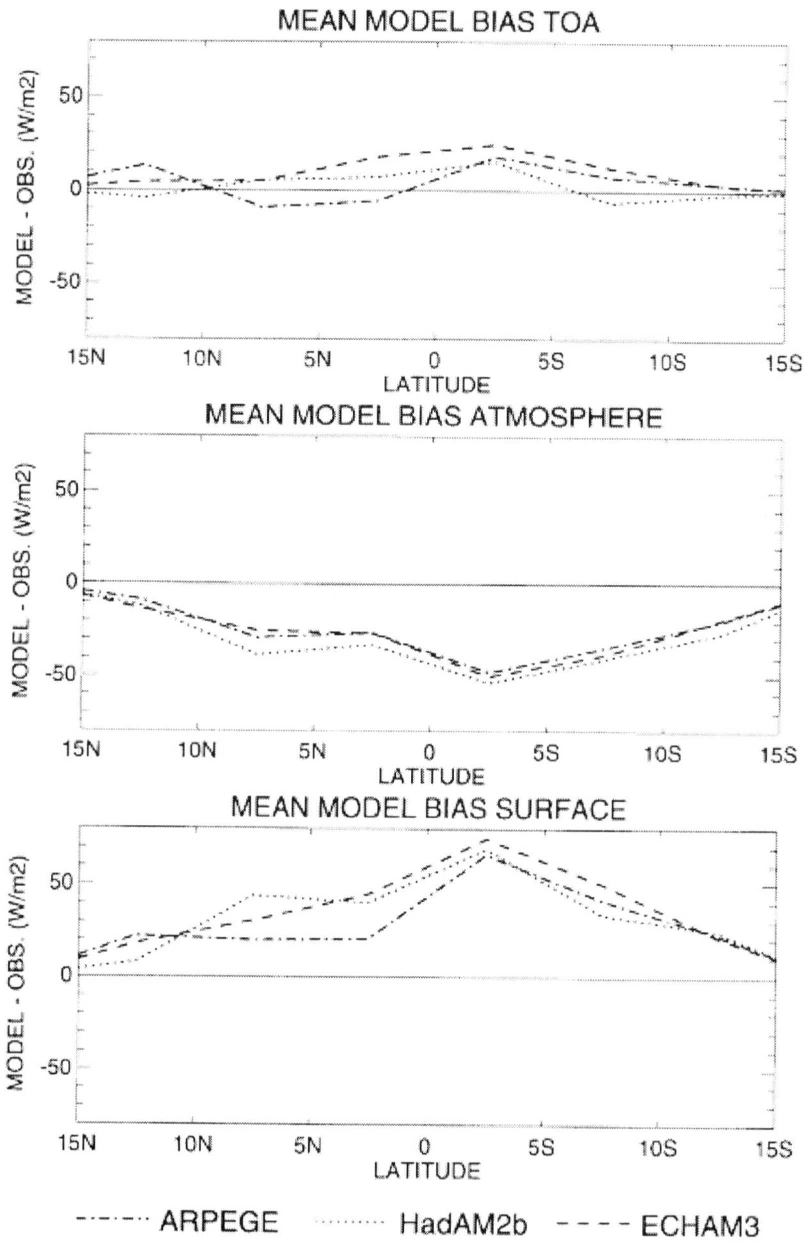

Figure 5. As Figure 4, but for July.

SW ABSORPTION APRIL

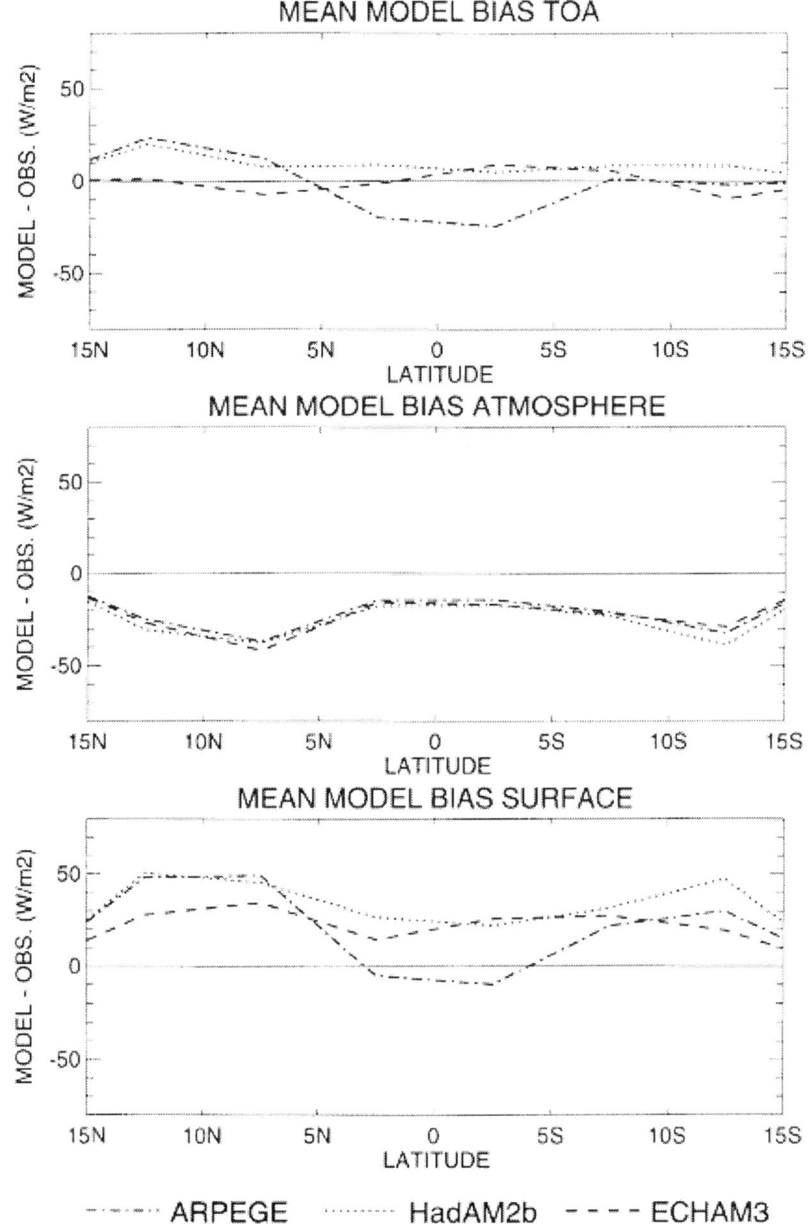

Figure 6. As Figure 4, but for April.

 Martin Wild

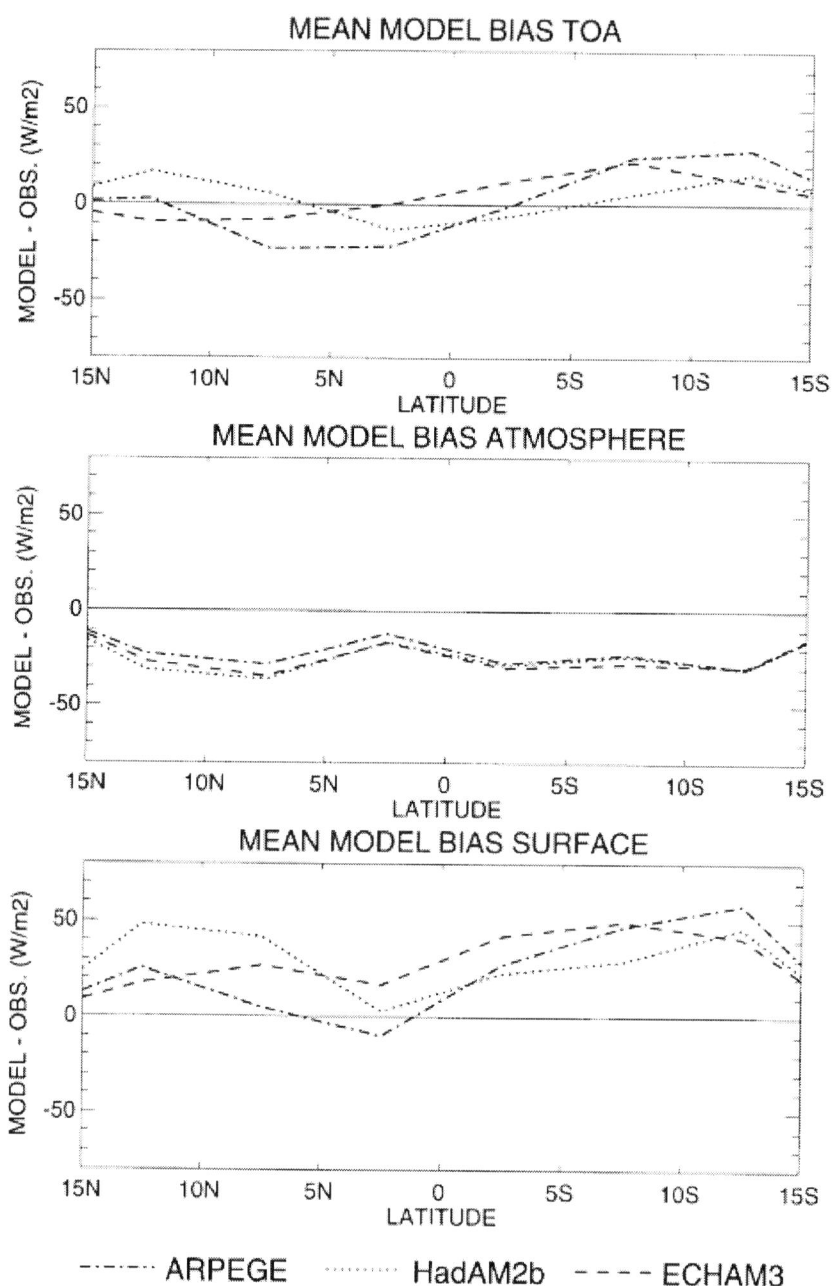

Figure 7. As Figure 4, but for October.

There is, however, still a remaining underestimation of atmospheric absorption in the models, on the order of 20 Wm^{-2}, which in these months can hardly be attributed to the neglect of aerosols from biomass burning. A lack of absorption of similar magnitude is also found in January to the south and in July to the north of the Equator, respectively, i.e. in the areas and seasons not affected by biomass burning (Figs. 4 and 5). This lack of absorption in the GCM atmospheres in areas with little aerosol loading from biomass burning can be related to the following causes:

a) lack of water vapour absorption

Water vapour is the principal absorber of solar radiation in the atmosphere. Vertically integrated water vapour calculated at the Equatorial African GEBA sites in the three GCMs is compared with an observational-based dataset in Fig. 8 as a function of latitude. Again the data have been averaged over the sites within latitudinal bands of 5°.

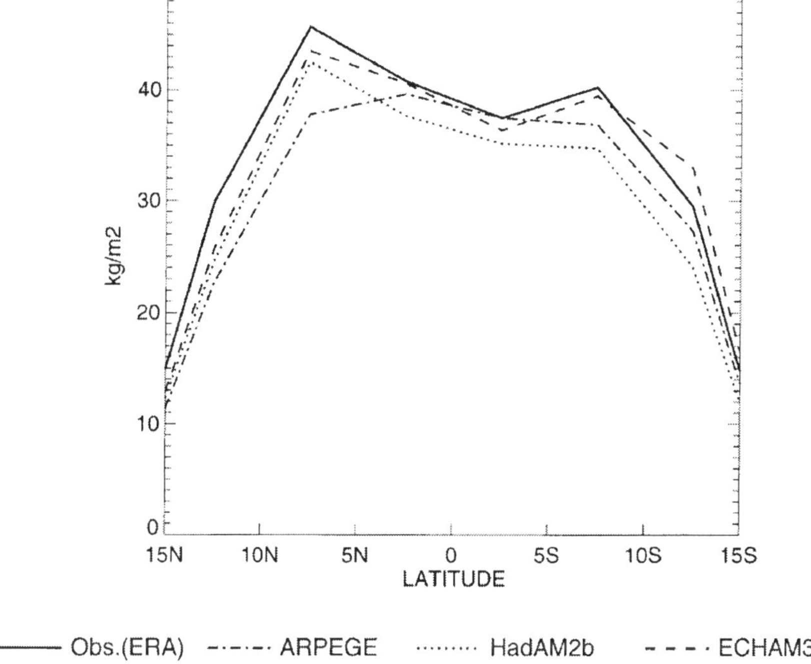

Figure 8. Annual mean vertically integrated water vapour at the GEBA sites in Equatorial Africa as function of latitude, for the 3 GCMs ARPEGE, HadAM2b and ECHAM3, and observational reference from the ECMWF Re-Analysis (Gibson *et al.* 1997). Values averaged over sites within 5° latitudinal bands.

The reference dataset is taken from the ECMWF Re-Analysis (ERA, Gibson *et al.*, 1997), which currently provides the best possible estimate of the atmospheric water vapour content. The models show a tendency to underestimate the water vapour content over Equatorial Africa, by the order of 5 to 10 %. This can, however, explain no more than 2–3 Wm^{-2} of the absorption biases, as shown in sensitivity studies (Wild, 1997; Wild *et al.*, 1997). More seriously, the GCM radiation codes tend to underestimate the short-wave absorption even given the correct amount of water vapour. This has been shown by Wild *et al.* (1995a) in stand-alone calculations with a GCM radiation scheme in isolation, where observed atmospheric water vapour and temperature profiles were prescribed from radiosondes. This enables the appropriate validation of the calculated short-wave fluxes at the surface with co-located surface measurements: In Fig. 9, surface insolation calculated with the ECHAM3 radiation scheme is compared with measured insolation at Payerne, Switzerland, for a number of clear-sky cases at noon. A substantial overestimation of 50 Wm^{-2} on average in the calculated insolation can be found. This corresponds to a daily mean overestimation of approx. 10–15 Wm^{-2}, and is related to an underestimation of the atmospheric water vapour absorption (Wild *et al.*, 1998). This suggests that the underestimated absorption and amount of water vapour are responsible for at least 10–15 Wm^{-2} of the above biases.

b) lack of absorption in clouds and aerosol of other origin than biomass
   burning
Several studies have stated that the absorption of solar radiation by clouds is substantially underestimated in GCMs (e.g., Cess *et al.*, 1995). This subject has been very controversial to date. The present study does not claim that the entire 20 Wm^{-2} underestimation in areas unaffected by biomass burning should necessarily be the result of underestimated water vapour absorption, and hence leaves the possibility that some remaining part is caused by underestimated absorption in clouds. This remaining fraction, however, is estimated at a maximum of 10 Wm^{-2}, if present at all (cf. paragraphs above and below). This is significantly smaller than given in Cess *et al.* (1995), who claimed an underestimation of the order of 25 Wm^{-2} in the GCM cloud absorption. Moreover, if a substantial underestimation of cloud absorption is present in the models, the largest underestimation may be expected during the wet seasons. This is not found in the models.

Furthermore, some of this remaining underestimate might be explained by effects other than clouds, e.g. by the neglect of absorbing aerosol from sources other than biomass burning in the models. The mineral aerosols from desert dust with their absorbing capacity should be particularly considered here, as they are likely to be underestimated in the spatially invariant aerosol climatologies used here. The lack of explicit treatment of desert dust aerosols may also contribute, in addition to the biomass burning aerosols, to the peak underestimation to the north of the Equator in winter (Fig. 4), since during this time of the year the "Harmattan" trade wind advects desert dust from the Sahara into these latitudes. As the atmospheric underestimation is significantly reduced in areas closer to the Sahara (north of 10°N) where the influence from biomass burning fades (cf. Fig 3a) and the influence of desert dust increases, it seems likely that the dominant contribution to these biases is from aerosols derived from biomass burning.

To summarize, assuming a "background" underestimation of the order of 20 Wm^{-2} in the models due to the reasons outlined above, the peak underestimations of 50–60 Wm^{-2} in the active fire areas (Figs. 4 and 5) suggest an error due to the neglect of absorbing aerosol of up to 30–40 Wm^{-2} on regional and seasonal scales. Locally these effects might even be larger as shown in the next section.

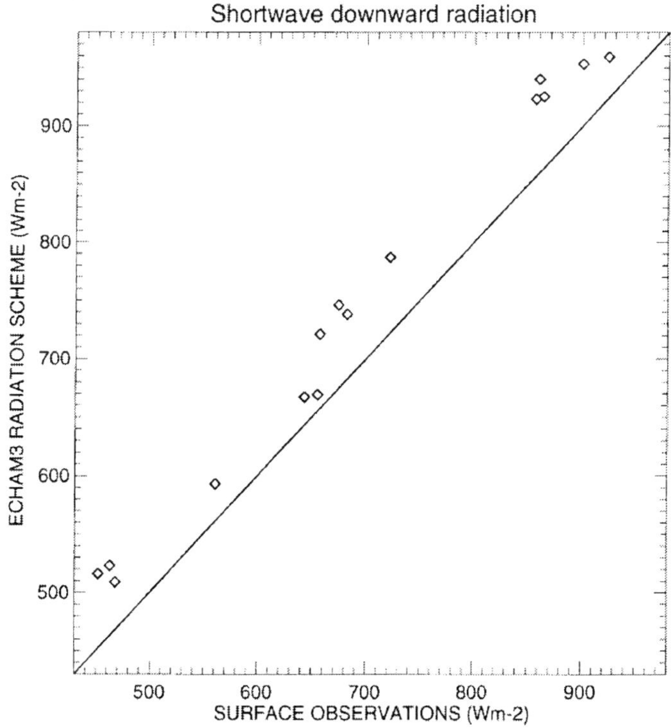

Figure 9. Offline calculations with the isolated radiation codes of the ECHAM3 GCM. Calculated surface insolation using prescribed profiles of atmospheric temperature and humidity from radiosonde data versus synchronous surface observations for clear-sky conditions. Radiation measurements and upper air soundings from the Swiss aerological station Payerne.

4.2   Case studies at four sites

To further illustrate the discrepancies between the calculated and observed fluxes on seasonal time-scales, we concentrate now on four equatorial sites, all of which are strongly influenced by biomass burning. These sites are located near to the Equator and have been used in the study of Konzelmann *et al.* (1996) for the assessment of the surface insolation in satellite-derived products.

They include the sites Mbanaka (0.1°N, 18.3°E), Inongo (1.1° S, 18.3°E), Kinshasa (4.4°S, 15.3°E), and Kananga (5.9°S, 22.4°E). While the study by Konzelmann *et al.* (1996) has focused on the surface insolation at these sites, the focus here is on the absorption of solar radiation in the atmospheric column above the sites, estimated as the residual between the measured TOA and surface absorption from ERBE and GEBA, as outlined before. Annual cycles of the observed estimates of atmospheric absorption are compared with the GCM estimates at the four sites in Fig. 10. It might be argued that discussions of the model performance at single sites are somewhat critical, due to the scale differences involved between the GCM grid and the local field observations. However, the biases found at the different sites show a very strong common signal, which is therefore likely to be of larger scale in nature. At all four sites, the observed and model-calculated seasonal cycles of atmospheric absorption are out of phase and exactly reversed. The observed atmospheric absorption reaches its maximum during the dry seasons (with peaks in January and July), where the GCMs calculate a minimum absorption. Vice versa, the maximum absorption in the models occurs during the wet seasons (with peaks in April and November), where the observational estimate indicates a minimum.

The models calculate a higher atmospheric absorption during the wet seasons as a result of higher water vapour and cloud amounts than during the dry seasons. This can also be seen in the annual cycle of vertically integrated water vapour in the GCMs, which are shown together with the observational estimate from ERA in Fig. 11. The annual cycle of integrated water vapour in the GCMs is in line with the annual cycle of atmospheric absorption in Fig. 10. Although there is again a tendency in the models to underestimate the atmospheric water vapour content, as noted in the above section, the differences show no distinct seasonal dependencies, and thus fail to provide an explanation for the strong seasonal signal in the atmospheric absorption bias.

The observed atmospheric absorption, in contrast, shows its maximum in the dry rather than in the rainy seasons suggesting that other absorbers over-compensate for the reduced amounts of water vapour and clouds. It thus seems highly plausible that such additional absorption in the atmosphere is related to the large aerosol loadings from the biomass burning in the dry seasons.

In January, the four sites are affected by aerosols advected by the trade winds from fires located to the north (cf. Fig 3a) (see Konzelmann *et al.*, 1996). These trade winds also contain significant loadings of desert dust, which may further contribute to the increased atmospheric absorption (and hence to the increased biases in the models).

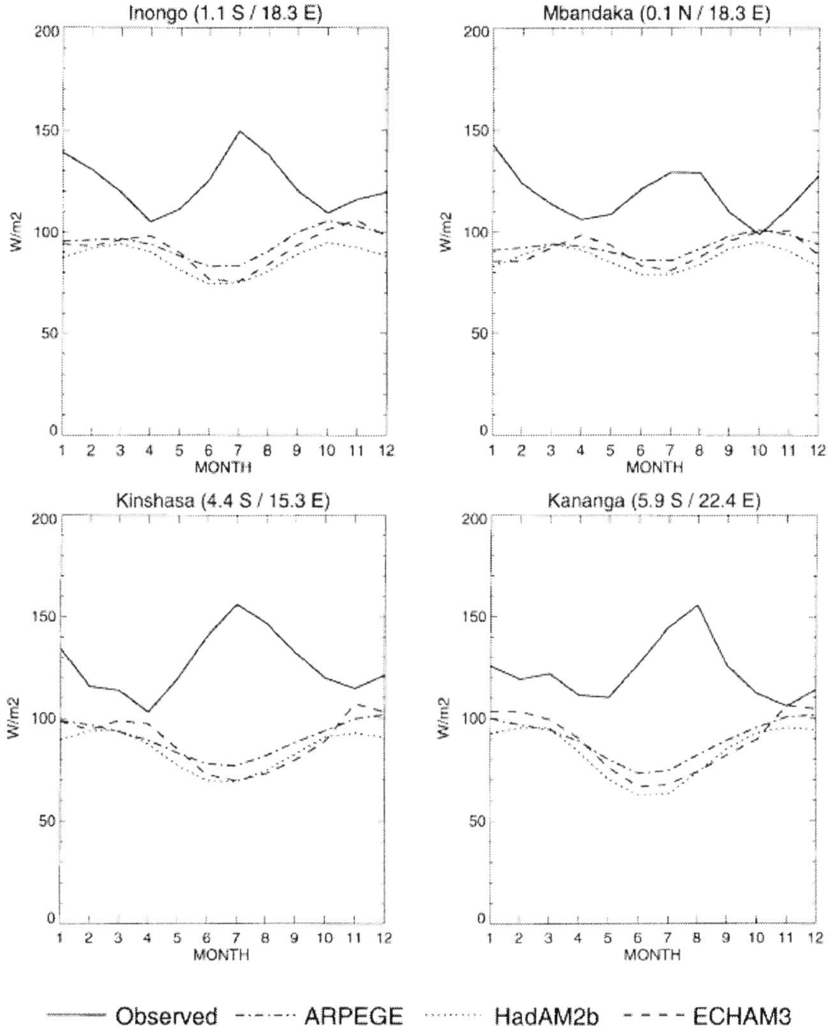

Figure 10. Annual cycles of model-calculated and observed atmospheric absorption of solar radiation at four observation sites in Equatorial Africa. Models are the GCMs ARPEGE, HadAM2b and ECHAM3. Observations derived from collocated surface measurements from GEBA and TOA measurements from ERBE at the sites Mbanaka (0.1°N, 18.3°E), Inongo (1.1° S / 18.3°E), Kinshasa (4.4°S, 15.3°E), and Kananga (5.9°S, 22.4°E).

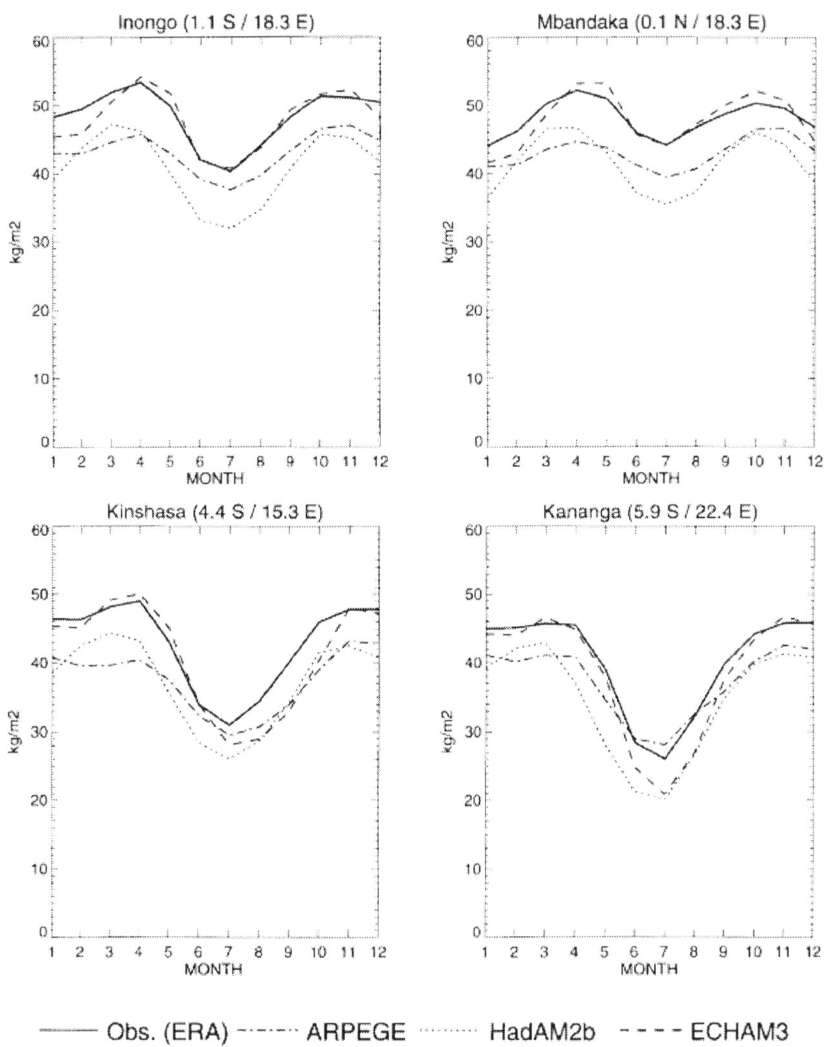

Figure 11. As Fig. 10 but for vertically integrated water vapour. Observational reference from the ECMWF Re-Analysis (ERA, Gibson *et al.* 1997).

In July, air masses are advected with high aerosol loadings from the fires to the south of the Equator (cf. Fig 3c). This also explains why the two stations located more to the south (Kinshasa and Kananga) show a larger peak in July than the two sites closer to the Equator (Inongo and Mbandaka). These in turn show a larger peak in January, due to their closer location to the fires, which are then predominantly located to the north (c.f. Fig 3a).

In the seasons with little aerosol loadings (around April and October) an overestimation of the order of 10–20 Wm^{-2} remains. As discussed above, this can firstly be attributed to the underestimated water vapour absorption in the radiation codes, and secondly to a lack of water vapour amount (cf. Fig. 11). A certain lack of absorption in clouds cannot be excluded, although it is quantitatively unnecessary to close the gaps between model and observational estimates in these seasons.

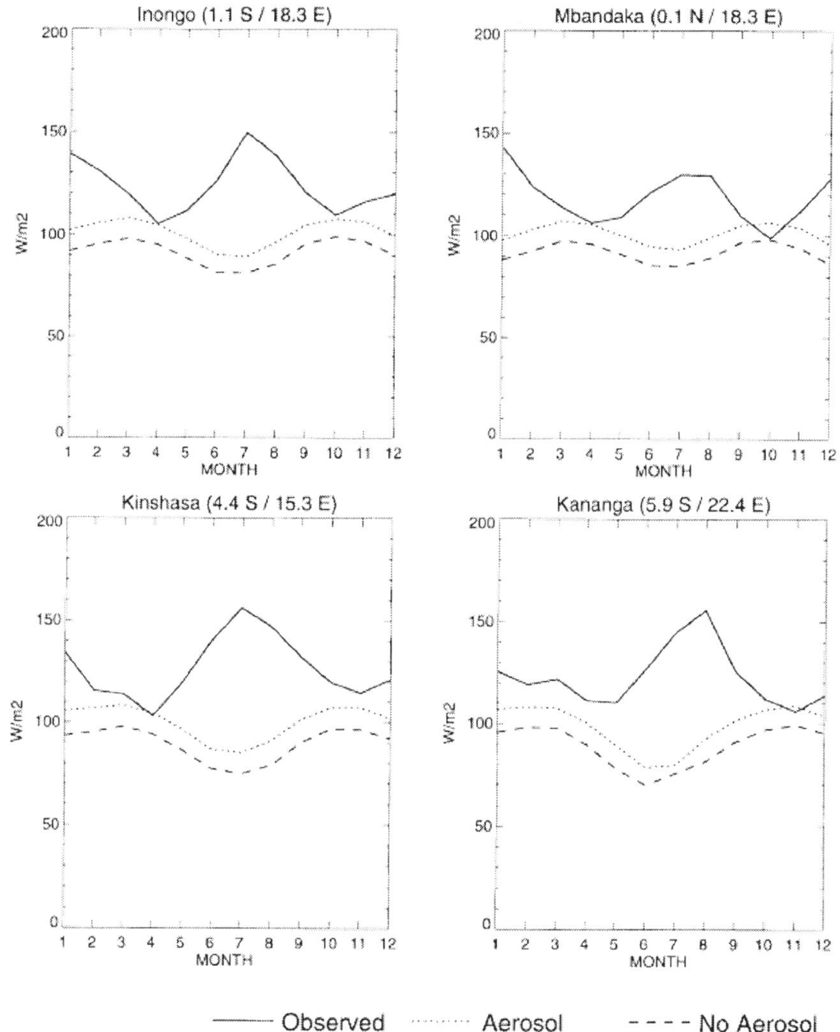

Figure 12. As Fig. 10 but for two experiments with the HadAM3 GCM, one using a simple aerosol climatology (with no seasonal variation) and one without any aerosols (Cusack *et al.* 1998).

Overall it can be concluded that simple aerosol climatologies with no seasonal variation cannot account properly for the forcing from biomass burning aerosols in the models. This is further illustrated using two experiments from the next generation Hadley Centre model HadAM3 (Cusack *et al.*, 1998), one without any aerosol effect and the other including a simple aerosol climatology (WMO, 1983) similar to the one in ECHAM3 and ARPEGE, with no seasonal variation. In Fig. 12, the simulated annual cycle of atmospheric absorption in the two experiments is compared to the observations at the four equatorial sites. The inclusion of the aerosol climatology reduces the biases in the atmospheric absorption in the annual mean. Seasonally, however, the aerosol climatology without any seasonal variation is not capable of removing the large biases. Note the very good agreement between the HadAM3 simulation and observations during the times with little aerosol loading (April/October). This indicates that the atmospheric absorption apart from the aerosol absorption is accurately captured in the HadAM3 model. The HadAM3 model benefits from the inclusion of the new radiation scheme of Edwards and Slingo (1996).

5. SUMMARY AND CONCLUSION

Current GCMs typically absorb too little solar radiation in the atmosphere, as shown in a comparison with a comprehensive set of co-located surface and satellite observations. The lack of atmospheric absorption is particularly strong at low latitudes. The present study has demonstrated that the largest gaps between model-calculated and observed atmospheric absorption coincide in time and space with the areas of most active biomass burning. Thus, there is an obvious connection between the lack of absorption and the misrepresentation of aerosol forcing in the GCMs, which do not properly take into account the spatial and seasonal effects of the absorbing aerosols such as from biomass burning. This neglect can amount up to 30–40 Wm^{-2} regionally and seasonally, and possibly more at local scales. However, this does not imply that all of the underestimated absorption in GCM atmospheres is due to an inadequate representation of absorbing aerosols, since an underestimation in atmospheric absorption of the order of 20 Wm^{-2} remains also in seasons and regions with little aerosol loadings. This is mainly related to the lack of water vapour absorption in the GCMs, as shown in previous studies (Wild *et al.*, 1995a, 1998; Wild and Liepert, 1998). A significant enhancement of the GCM cloud absorption, as claimed in other studies, may therefore not be necessary to close the gap between model-calculated and observational estimates of short-wave atmospheric absorption.

6. ACKNOWLEDGEMENTS

I am grateful to Prof. A. Ohmura for the support of this study and his valuable comments on the manuscript. This study would not have been possible without the many efforts of Dr. H. Gilgen regarding all aspects of the GEBA database. Thanks to Drs. M. Déqué, Météo-France, and R. Stratton, Hadley Centre, for making available the output of the ARPEGE and HadAM2b GCMs within the framework of the EU project HIRETYCS. The data of the fire occurrences stem from the web site http://shark1.esrin.esa.it established by Drs. O. Arino and J.-M. Melinotte from ESA/ESRIN. Special thanks to Drs. S. Cusack and A. Slingo, Hadley Centre, for providing the data from their HadAM3 experiments. The Swiss Scientific Computing Centre CSCS generously provided the necessary computer resources for the high resolution ECHAM T106 simulations. This study is part of the EU program environment, project HIRETYCS, and is supported by the Swiss Bundesamt für Bildung und Wissenschaft (BBW) Grant No. 95.0640. Thanks to Dr. M. Verstraete and an anonymous reviewer for their detailed and constructive reviews. Finally I would like to thank Prof. M. Beniston, University of Fribourg, for the organisation of the workshop on biomass burning in Wengen, Switzerland, which gave motivation to carry out the above study.

7. REFERENCES

Arino O., and J.-M. Melinotte, 1998: The 1993 Africa Fire Map, Cover Page. *Intern. J. Remote Sensing*, **19**, 2019–1023.

Arking, A., 1996: Absorption of solar energy in the atmosphere: Discrepancy between model and observations. *Science*, **273**, 779–782.

Barkstrom, B.R., E.F. Harrison, and R.B. Lee III, 1990: Earth Radiation Budget Experiment. *EOS*, **71**, 297–305.

Cahoon, D.R., Jr., B.J. Stocks, J.S. Levine, W.R. Cover III, and K.P. O'Neill, 1992: Seasonal distribution of African savanna fires. *Nature*, **359**, 812–815.

Cess, R.D., M.H. Zhang, P. Minnis, L. Corsetti, E.G. Dutton, B.W. Forgan, D.P. Garber, W.L. Gates, J.J. Hack, E.F. Harrison, X. Jing, J. T. Kiehl, C.N. Long, J.-J. Mocrette, G.L. Potter, V. Ramanathan, B. Subasilar, C.H. Whitlock, D.F. Young, and Y. Zhou, 1995: Absorption of solar radiation by clouds: observations versus models. *Science*, **267**, 496–499.

Cusack, S., A. Slingo, J.M. Edwards, and M. Wild, 1998: The radiative impact of a simple aerosol climatology on the Hadley Centre atmospheric GCM. *Quart. J. Roy. Meteor. Soc.*, **124**, 2517–2526.

Darnell, W.L., W.F. Staylor, S.K. Gupta, N.A. Ritchey, and A.C. Wilber, 1992: Seasonal variation of surface radiation budget derived from International Satellite Cloud Climatology Project C1 data. *J. Geophys. Res.*, **97**, 15741–15760.

Déqué, M., C. Dreveton, A. Braun, and D. Cariolle, 1994: The ARPEGE/IFS atmosphere model: a contribution to the French community climate modelling. *Climate Dynamics*, **10**, 249–266.

Edwards, J.M., and A. Slingo, 1996: Studies with a flexible new radiation code. I: Choosing a configuration for a large-scale model. *Q. J. R. Meteorol. Soc.*, **122**, 689–719.

Garratt, J. R., 1994: Incoming shortwave fluxes at the surface – a comparison of GCM results with observations. *J. Climate*, **7**, 72–80.

Gates, W. L., 1992: AMIP: The atmospheric model intercomparison project. *Bull. Amer. Meteor. Soc.*, **73**, 1962–1970.

Gibson, R., P. Kallberg, S. Uppala, A. Hernandez, A. Nomura, and E. Serrano, 1997: ECMWF Re-Analysis project report series, 1. ERA description, ECMWF, Reading, 72 pp.

Gilgen, H., M. Wild, and A. Ohmura, 1998, Means and trends of shortwave irradiance at the surface estimated from Global Energy Balance Archive data. *J. Climate,* **11**, 2042–2061.

Konzelmann, T., D.R. Cahoon, and C.H. Whitlock, 1996: Impact of biomass burning in Equatorial Africa on the downward surface shortwave irradiance: observations and calculations. *J. Geophys. Res.*, **101**(D1), 22833–22844.

Li, Z., C. H. Whitlock, and T. P. Charlock, 1995: Assessment of the global monthly mean surface insolation estimated from satellite measurements using Global Energy Balance Archive data. *J. Climate*, **8**, 315–328.

Li, Z., L. Moreau, and A. Arking, 1997: On solar energy disposition. *Bull. Amer. Meteor Soc.*, **78**, 53–70.

Li, Z., 1998: Influence of absorbing aerosols on the inference of solar surface radiation budget and cloud absorption. *J. Climate*, **11**, 5–17.

Ohmura, A., H. Gilgen, and M. Wild, 1989: Global Energy Balance Archive GEBA, World Climate Program – Water Project A7, Report 1: Introduction. Zürcher Geografische Schriften Nr. 34, Verlag der Fachvereine, Zürich, 62pp.

Ramanathan, V., B. Subasilar, G. Zhang, W. Conant, R. Cess, J. Kiehl, H. Grassl, and L. Shi, 1995: Warm pool heat budget and shortwave cloud forcing: a missing physics? *Science*, **267**, 499–503.

Roeckner, E., K. Arpe, L. Bengtsson, S. Brinkop, L. Dümenil, M. Esch, E. Kirk, F. Lunkeit, M. Ponater, B. Rockel, R. Sausen, U. Schlese, S. Schubert, and M. Windelband, 1992: Simulation of the present day climate with the ECHAM model: impact of model physics and resolution. Max Planck Institute for Meteorology Report No. 93, 171 pp.

Rossow, W.B., and Y. C. Zhang, 1995: Calculation of surface and top of atmosphere radiative fluxes from physical quantities based on ISCCP data sets. Part II: Validation and first results. *J Geophys. Res.*, **100** (D1), 1167 – 1197.

Shettle, E.P., and R. Fenn, 1976: Models of the atmospheric aerosols and their optical properties. AGARD Conference Proceedings No. 183, AGARD–CP–183.

Stratton, R.A., 1999: A high resolution AMIP integration using the Hadley centre model HadAM2b, *Climate Dynamics,* **15**, 9–28.

Wild, M., A. Ohmura, H. Gilgen, and E. Roeckner, 1995a: Validation of GCM simulated radiative fluxes using surface observations. *J. Climate*, **8**, 1309–1324.

Wild, M., A. Ohmura, H. Gilgen, and E. Roeckner, 1995b: Regional climate simulation with a high resolution GCM: surface radiative fluxes. *Climate Dynamics,* **11**, 469–486.

Wild M., 1997: The heat balance of the Earth in GCM simulations of present and future climate. Zürcher Geografische Schriften Nr. 68, Verlag der Fachvereine, Zürich, 188 pp.

Wild M., A. Ohmura, and U. Cubasch, 1997: GCM simulated surface energy fluxes in climate change experiments. *J. Climate*, **10**, 3093–3110.

Wild, M., A. Ohmura, H. Gilgen, E. Roeckner, M. Giorgetta, and J.J. Morcrette, 1998: The disposition of radiative energy in the global climate system: GCM versus observational estimates. *Climate Dynamics,* **14,** 853–869.

Wild, M., and B. Liepert, 1998: Excessive transmission of solar radiation through the cloud-free atmosphere, *Geophysical Research Letters*, **25**, 2165–2168.

Wild, M., and A. Ohmura, 1999: The role of clouds and the cloud-free atmosphere in the problem of underestimated absorption of solar radiation in GCM atmospheres. *Phys. Chem. Earth (B)*, **24**, 261–268.
World Meteorological Organisation, 1983: Report of the experts meeting on aerosols and their climatic effects, WCP–55, 107 pp.

Wildland Fire Detection from Space: Theory and Application

DONALD R. CAHOON, Jr.[1], BRIAN J. STOCKS[2], MARTIN E. ALEXANDER[3], BRYAN A. BAUM[4], JOHANN G. GOLDAMMER[5]
[1]*NASA-Langley Research Center, 21 Langley Boulevard, Hampton, USA*
[2]*Canadian Forest Service, Sault Ste. Marie, Ontario, Canada*
[3]*Canadian Forest Service, Edmonton, Alberta, Canada*
[4]*Langley Research Center, Hampton, VA USA*
[5]*Fire Ecology Research Group, University of Freiburg, Germany*

Abstract:
New satellite instruments are currently being designed specifically for fire detection, even though to date the detection of active fires from space has never been an integral part of the design of any in-orbit space mission. Rather, the space-based detection of fires during the last two decades has been exploiting measurements obtained for other objectives. The current fire products have proved to be of great benefit and interest, but their usefulness is not fully understood. Part of the confusion about the utility of these measurements stems from the lack of detailed knowledge about the data and its acquisition. The remote sensing research community has spent considerable time and effort trying to rationalize the usefulness of existing satellite imagery for active fire detection. Unfortunately, uncertainties about instrument capabilities pervades much of this research and the true limits of fire detection from space have not been fully evaluated and understood.

To analyze the active fire detection capability of any instrument, the flow of energy from the source to the instrument and the instrument's response to that energy must be considered. For this reason, an approach has been developed that models the energy emitted from surface fires, allowing for the fact that fire is itself a variable phenomenon. The energy transmission is then modelled along its path through the atmosphere and through the instrument's optical system. A fundamental concern is in the estimation of the total surface area that emits the energy which defines a single pixel in the image. Unfortunately, most of the fire detection modelling done to date is based on a misconception about the pixel and its actual size. Rather than using the radiometric footprint size, the instantaneous-field-of-view (IFOV) is used to describe the 'resolution' of the instrument. In fact, the radiometric footprint is considerably larger than the IFOV and greatly affects the energy modelling used to estimate the fire detection thresholds of a particular instrument. Based on knowledge of the radiometric footprint, the fire detection capability of AVHRR, DMSP–OLS, and MODIS are reviewed.

151

1.   INTRODUCTION

> "the pixel, the elementary unit of analysis in remote sensing..., is a
> delusion which may become a snare for the unwary", P. Fisher (1997)

Even though this original quotation by Fisher is aimed more at the concern of the use of image pixels in geographical information system (GIS) analyses, it captures the essence of a fundamental problem in most remote sensing studies. That is, the great majority of remote sensing investigators lack a basic understanding of what defines an image pixel on the ground. Typically, image pixels are treated as non-overlapping tiles in a two-dimensional array of values which, when displayed, renders itself into a digital scene. In the image's simplest sense, this is the case. However, this simple interpretation of an image avoids the realities of what the pixel really represents and how the image is constructed. And, unbeknownst to many, the results of our analyses are tainted with these artifacts of reality which lead to misinterpretations of our data and, even worse, show up as errors in our analyses. Often the limitations of our image data pass unnoticed, but even when they do not, there is no explanation available to explain or justify the seemingly mysterious results.

There is far more to understand about the pixel. In the strictest sense, the word pixel is short for 'picture element' and describes one data point in an image. Alternatively, the word footprint is used to describe the surface area on the ground whose integrated energy defines the value of the pixel. The people that use remote sensing data in the Earth Sciences are from a wide variety of academic backgrounds; largely the sciences, but not always. The people that most likely understand what an image pixel represents are the instrument engineers. These two groups, remote sensing scientists and instrument engineers, do not interact frequently enough. Hence, some of the essential information about a pixel is lost or neglected. The users of the image data do not always possess the academic tools to sift through the engineering data, analyses and publications that cloak the meaning of an image pixel. To make things worse, the engineers analyze data relative to its frequency components (wave domain), and scientists relative to its spatial variations (physical domain). In essence, the remote sensing community suffers from a mathematical communication problem between themselves and the instrument engineers. This paper will try to bridge the gap and convey a simple concept of an instrument, how the instrument defines the pixel, and what is the value in understanding the pixel through the application to fire detection modelling.

The satellite remote sensing of fire began in the late 1970s (Croft, 1978; Matson *et al.*, 1987). Since this time the global importance of fire has become internationally recognized. Fire has also escalated to the forefront in the global carbon budget discussions (e.g. Kurz *et al.*, 1995; Kasischke *et al.*, 1995; Stocks *et al.*, 1998). Even with this increased awareness and the global importance of fire, there has not been a satellite instrument specifically designed for fire detection that has been launched at the time of this writing. Two decades have passed since the launch of the first Advanced Very High Resolution Radiometer (AVHRR). Although there are numerous studies in the literature on fire detection, there is not a rigorous comparison in the literature that documents what the thresholds of fire detection are for the various existing instruments. For the AVHRR (Kidwell, 1991), the Defense Meteorological Satellite Program Operational Linescan System (DMSP–OLS) (Elvidge *et al*, 1996), and the Moderate Resolution Imaging Spectrometer (MODIS) (King *et al.*, 1992) instruments, this paper will apply the knowledge of the actual image pixel size on the ground (radiometric footprint), the knowledge of fire behaviour, the knowledge of the energy transmission through the atmosphere and instrument optics, and model the fire detection limitations for these three polar-orbiting instruments. At each step along the way, actual imagery will be used to assess the performance of the model and its components.

2.   INSTRUMENTS AND THE PIXEL

The simplest notion of a satellite imager can be seen in Figure 1. An imaging instrument can be sub-divided into three systems, the optics, the detector, and the electronics. The energy from the actual scene passes through the optics and impinges on the detector. The detector must respond to the energy that it receives and convert that energy into a signal. This signal is typically filtered and converted from analog to digital. However, some of the newer detector technology can acquire a digital signal directly. The final product of the instrument is to pass a digital value out which corresponds to a single observation: the pixel. It is easy to conceive how the definition of a pixel is completely reliant upon the system which acquires it. At each step through the instrument, the original input signal is modified slightly. This modification, or degradation, of the input signal occurs for a variety of reasons. Optical systems are not perfect and energy passing through the optics will not be perfectly focused on the detector because of effects like diffraction. Detectors have response delays that are a function of the detector material, the sampling time, and the rate of change in the input signal. As the signal passes through the electronic system, it is converted from analog-to-digital, filtered, and sampled. Each step further induces change to the input signal. In a mathematical sense, we are convolving the input signal with a numerical filter corresponding to each phase of the measurement process. The cumulative effect of these filters at each step through the instrument can be represented as a single mathematical filter for the entire system which defines how much the input signal has been modified. Thus, the resulting output digital signal, our image pixel value, has been numerically modified by each component of the instrument and can never fully resolve the original scene.

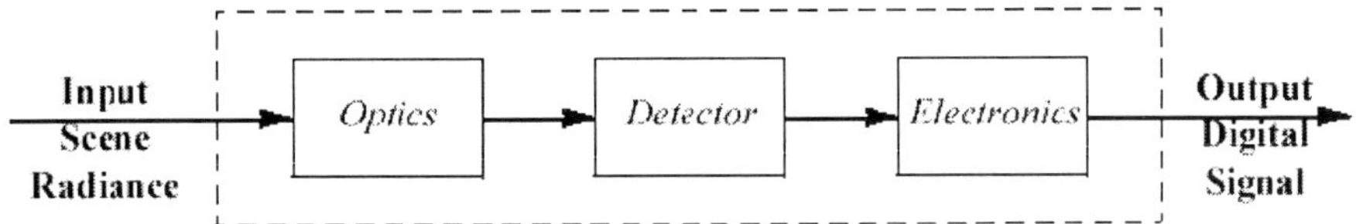

Figure 1. A simple schematic of an imaging instrument showing its three fundamental systems, the optics, the detector, and the electronics. The arrows show the direction of flow, from incoming energy on the left to the output signal on the right. After Breaker, 1990.

The performance of an instrument, its ability to reproduce the original scene, is measured by engineers and reported as the modulation transfer function (MTF) (Smith, 1966). The units of the MTF are those of spatial frequency (e.g., cycles per kilometer) and the MTF is usually reported in the engineering documentation for each individual imaging instrument. To more thoroughly understand the MTF, examine Figure 2. Figure 2a shows a pattern of black and white bars, which represents an input scene, that the instrument scans across. If the instrument were to reproduce the scene correctly, then each scan would result in a square wave whose amplitude would resolve all of the contrast between the black and white bars, or in other terms, the maximum scene contrast between black and white (Figure 2b). If the instrument were to scan slowly enough, allowing the detector time to respond to the input signal before it changes significantly, then the maximum scene contrast could almost be detected. However, as the scan rate increases in speed, the ability of the detector to capture the maximum scene contrast is lost and the amplitude of the output signal will be reduced (Figure 2c). The output signal then represents a blur of energy that emerges from both the black and white bars. This loss in the output signal's contrast occurs because the detector is integrating the contrast of the black and white bars together as it scans faster. The reason for the integrated signal from both the black and white bars is because the sampling interval of the instrument is fixed by design and the observation time over any location in the input scene is decreasing.

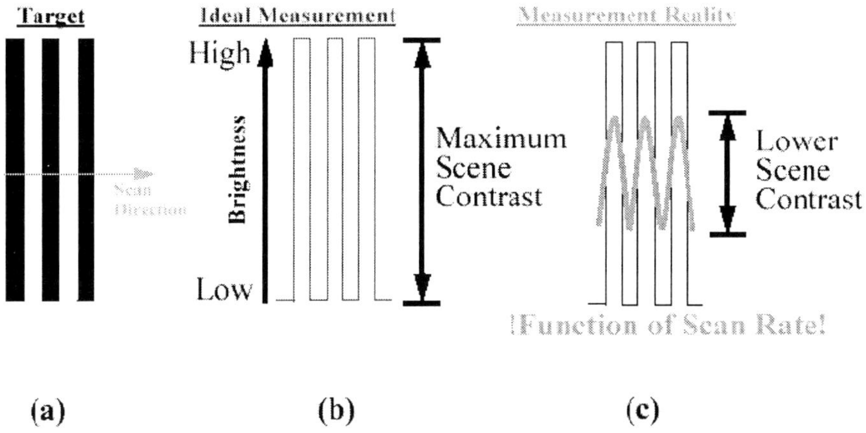

Figure 2. A three schematic drawing to demonstrate the concept of the modulation transfer function (MTF). The MTF provides a measure of how well an instrument can reproduce changes that occur in the input scene. (a) Shows an ideal scene of evenly spaced high contrast bars. (b) The ideal instrument, as it scans across the bars in 'a' will reproduce a perfect square wave showing the maximum contrast. (c) The reality is that something less than the ideal can be resolved by the instrument and the result is a loss of information from the input scene.

As mentioned in the introduction, there is a communication problem between the instrument engineers and the scientists regarding the nature of the pixel. This problem is not necessarily only attributed to language, but is also mathematical. Engineers are trained to and prefer to perform measurements and calculations in the mathematical frequency domain. Remote sensing scientists are trained to and prefer to work more in the spatial domain. Simply put, those of us working in the realm of remote sensing typically want to resolve spatial features of a landscape, a cloud field, and so on. The MTF, in units like cycles per kilometer, has the appearance of being a rather foreign quantity and can be easy to ignore. With the use of the inverse Fourier transform, the MTF can be mathematically transformed into its spatial domain representation. This representation has units of distance (e.g., kilometers) and is more intuitive to use for remote sensing scientists. The transformed MTF can be used to describe the size of each image pixel on the ground (or in other terminology the footprint) and is commonly referred to as the Point Spread Function (PSF) or sometimes as the Point Response Function (PRF). In this paper, the term PRF will be used.

How does everything that has been discussed in this section manifest itself in the imagery? This question can be answered by examining an AVHRR coastline crossing which is a clear example of a high contrast boundary (like the black and white bars) in a scene. If the signal is truly being degraded, then there should not be a sharp edge in the image across the water/land boundary. Figure 3 shows a coastline crossing which has a distinct water/land interface on the ground and which also has a water/land interface that is perpendicular (up–down) to the scan direction (left–right). It can be seen in the graph, for both the reflected energy channel (channel 2) and the thermal channel (channel 3), that there are as many as four pixels which contain intermediate values between the plateau of land values (on the upper left) and the plateau of water values (on the lower right). The sharp coastline is not reproduced in the image data where there is one in reality and there are several pixels, instead of just one pixel, that have mixed water/land observations. The figure clearly demonstrates that as pixels gain more distance from the water/land interface, the effect of the boundary crossing diminishes slowly until the plateau is reached. To summarize the crossing in terms of distance, the actual coastline boundary which occurs over a few meters on the ground is taking over 3 km to resolve in the imagery. To further restate this from a visual interpretation standpoint, a close examination of the image shows that the coastline is blurred and thus not well defined. This blurring phenomena can be explained and modelled by gaining a more detailed understanding of the PRF; the actual footprint size.

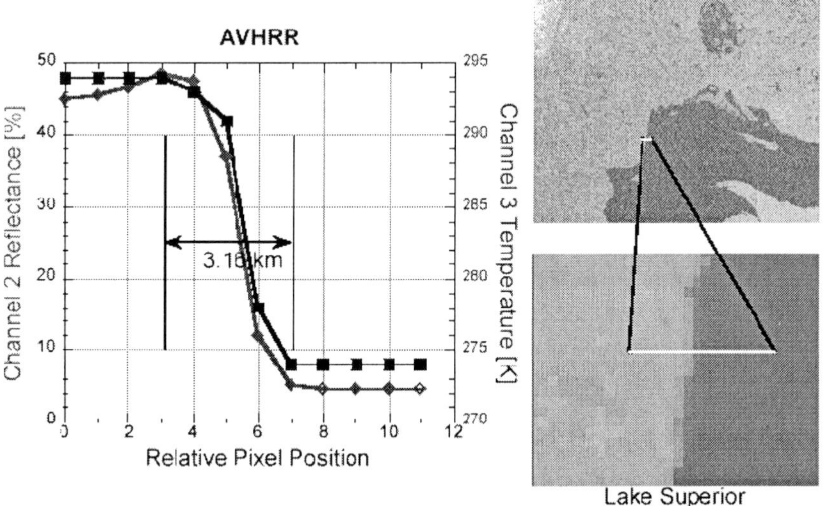

Figure 3. A coastline crossing along a transect (shown on the right) as seen in the AVHRR imagery. The sharp boundary of the coast is not being reproduced and it takes several pixels to complete the transition in both the reflected energy (diamonds) and the thermal energy (squares) channels.

3.   THE PRF AND THE PIXEL

Figure 4 shows the PRF for the AVHRR instrument whose imagery has been used in countless studies in many disciplines, including fire research. The first thing to note is that the PRF is a weighting function, for a single pixel, which describes the response of the instrument to the energy emerging from the footprint in the input scene. From Figure 4 it can be seen that the actual AVHRR footprint at nadir is over 5 km in diameter across. For this study, the full footprint is being defined as the input scene area from which 99% of the energy that constitutes a pixel is emerging. The nadir resolution of AVHRR is commonly calculated to be about 1.26 km (Cahoon *et al*, 1992b; Setzer and Malingreau, 1996). At a radius of 0.63 km (half of 1.26 km) it can also be seen that a surface area of the size typically quoted as the instrument's resolution only accounts for 28% of the total energy of the pixel. This means that 72% of the energy that makes up one AVHRR pixel (at nadir) is from outside the scene area that is commonly assumed to be the pixel. It should also be noted that 80% of the energy of each pixel is from a surface area that is almost 3 km in diameter at nadir. Thus, the AVHRR image pixels are broadly overlapping each other.

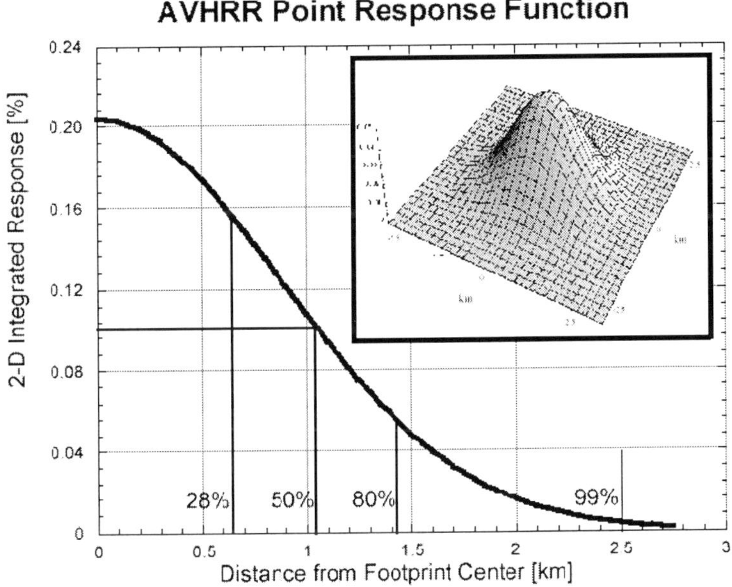

Figure 4. The cross-section of the AVHRR point response function (PRF) at nadir. The 3–D view of the PRF can be seen in the upper right-hand corner. The response of the instrument at the pixel level is essentially Gaussian and declines outward from the pixel centre. Integrating under the surface provides the estimates of the total energy at various radii outward from the centre.

Setzer and Malingreau (1996) present the common and largely accepted view of the overlapping pixel structure in the AVHRR imagery. This common representation can be seen to be inappropriate as a basis for interpreting the imagery in lieu of our knowledge of the PRF. To represent the information in Figure 4, Figure 5 shows the AVHRR PRF, derived from engineering data, superimposed on a horizontal plane which has the classical AVHRR footprint layout drawn for a sample of image pixels near the centre of the image. The tremendous footprint overlap is clearly demonstrated by the PRF. One may note that the total energy for any one pixel comes from the surface area represented by over a dozen image pixels (which of course all have the same PRF). This means that the images are extremely complex representations of the original input scenes. It is only with an understanding of the PRF that we can begin to fully comprehend our analyses and the limitations of the imagery. The most common perceptions of the image pixel are not sufficient to explain the inherent underlying subtleties of the imagery. Support for our definition of the PRF can be found in the imagery.

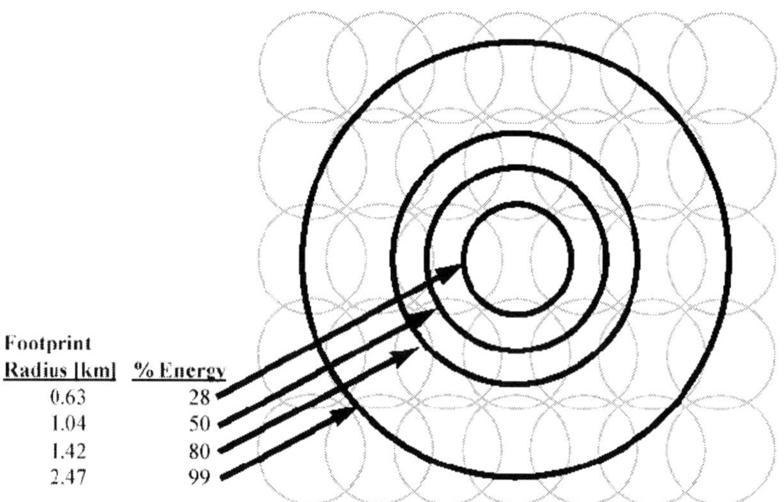

Figure 5. The set of thick concentric rings is a top-down view of the AVHRR PRF that has been superimposed on a set of smaller overlapping rings which coincide with the classical view of AVHRR image pixels. Each ring coincides with the integrated energy levels as shown in Figure 4. The number of image pixels that fall within the PRF is quite large, even at the 80% total energy level.

The AVHRR PRF is used to model a sharp land/water boundary. This modelled boundary can be seen in Figure 6 as the thick curve which has each individual pixel plotted as a diamond. It is similar to the coastline crossing in Figure 3 in that it has four pixels that are between the plateaus of land and water pixel values. Analysis is performed on a random AVHRR image from our inventory that contains clear-sky (no clouds) coastline crossings around the African continent. The crossings are all located on a desert/water interface and are randomly selected from varying scan line positions within the image. Like the coastline crossing in Figure 3, each water/land interface is perpendicular to the scan direction. For this comparison the PRF has been converted to pixel distances and the pixel values have been normalized. Each of these randomly selected crossings are plotted as thin black lines. It can be seen that our modelled coastline crossing does represent the actual AVHRR data very well.

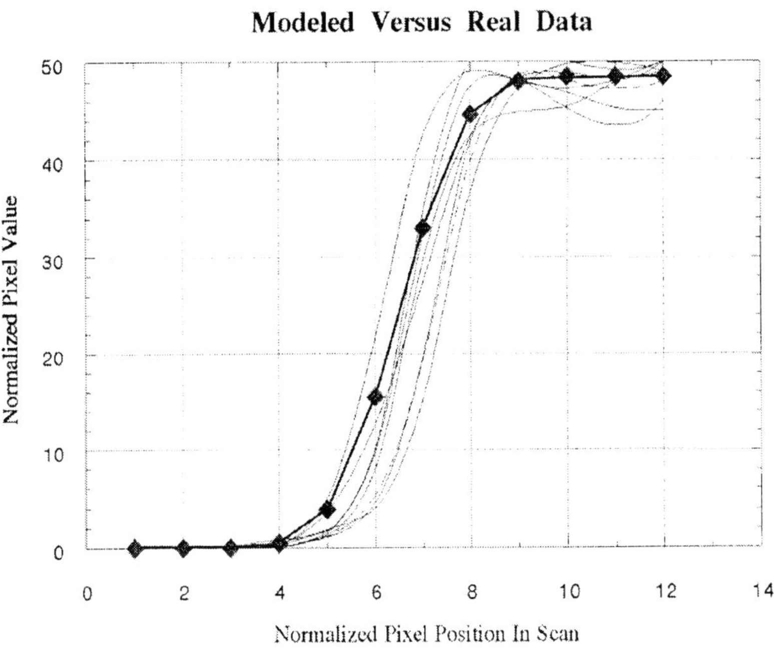

Figure 6. The AVHRR PRF has been used to model a coastline crossing (bold line)
and the individual modelled pixel values are plotted (diamonds).
Randomly selected coastline crossings from across the entire scan line
have been plotted for comparison to the modelled results.

There are additional complexities that have been avoided in this study. The PRF can have a more complex shape in the extremities (the low-energy regions) (Breaker, 1990). We have chosen to use a more refined version of the PRF than Breaker for presentation and modelling. The actual differences in values would be extremely small and would have only complicated this discussion. Additionally, the coastline modelling in Figure 6 was done with a pixel centred on the water/land interface boundary. In reality, the pixel centre can be offset from the interface boundary and this would skew the results slightly. Some of this skew is seen in the actual crossings in Figure 6 as a 1–2 pixel (left to right) shift that exists at any normalized pixel value. It should also be noted that the examples presented have focused on AVHRR. All imagers, including SPOT and Landsat, have similarly broad PRF's relative to their typically quoted resolutions.

4.   FIRE MODELLING

We have developed a physical fire model which allows us to understand the fire detection limitations of the AVHRR, the DMSP–OLS, and the MODIS instruments. The modelling begins on the ground with the derived PRF for each instrument in the appropriate fire detection channel. The model is run for each instrument with the optimum case of the fire fronts passing through the centre of the pixel. In much the same way in which the footprint has been examined, the behaviour of fire must be considered also. For present purposes, three different types of landscape fires are considered for evaluation in the model, with each type possessing overlapping fire characteristics. The atmosphere has been modelled using radiative transfer methods and the flame front energy has been transferred from the surface to the instrument using the appropriate solid angles for each instrument aperture. The differences in the transmission of energy within each instrument's optical system were also taken into consideration. The evaluations of fire detectability were based on selected criteria which will be defined in this section. In the end, a physically-based fire detection model has been developed in order to understand what the limitations are for fire detection using each instrument. Further, the results of the model will also provide insights into how to interpret the fire products that are being derived using various instruments and methodologies (e.g., Cahoon *et al.*, 1992a; Elvidge *et al.*, 1996; Flasse and Ceccato, 1996; Justice and Dowty, 1994; Kasischke *et al.*, 1993; Lee and Tag, 1990).

Fire behaves differently across the landscape depending on a variety of parameters (e.g., fuel moisture, winds, ...). But most importantly, when considering regions that are already prone to widespread fire development, the total energy release from a fire can largely be attributed to the structure of the fuels. Forest fires, with their complex fuel structure and high fuel loading, generally experience wide fire fronts with tremendous amounts of energy being released per unit area (Figure 7a). In contrast, savannas, with a simpler fuel structure and a lower fuel loading, have more narrow flame fronts and less energy being released per unit area (figure 7b). Field programmes (e.g., Alexander, 1998; Brass *et al.*, 1996; Kaufman and Justice,

1994; Sneeuwjagt and Frandsen, 1977; Stearns *et al.*, 1986; Stocks *et al.*, 1996; Stocks and Hartley, 1995; Stocks and Jin, 1988) have generated data that can be used to define fire characteristics in various landscapes. The characteristics for the three vegetation classes defined for this study are shown in Table 1. For the agricultural fires category, since there is little in the way of documented field measurements, personal observations of the flame front sizes by the authors have been included and the flame front temperatures are adopted from the savanna category.

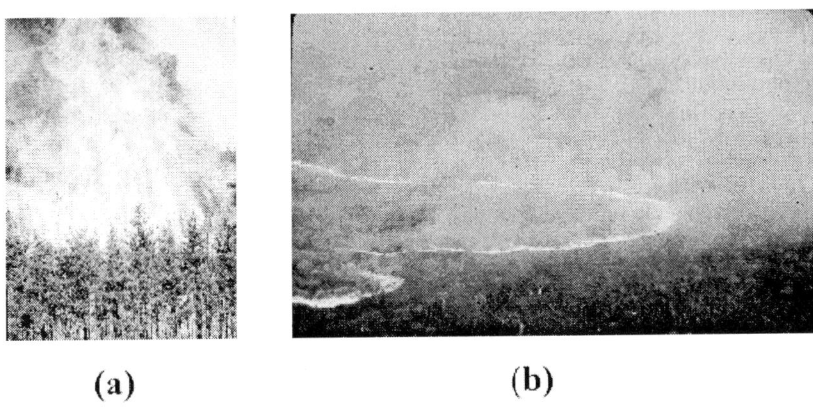

(a)   **(b)**

Figure 7. (a) Boreal forest fire demonstrating the intense energy released by forest fires relative to that of (b) savanna fires. Flames in the boreal fire are easily an order of magnitude higher than those of the grassland fires.

Table 1. Fire Characteristics

Vegetation Class	Flame Front			
	Residence Time (*s*)	Speed (*km/hr*)	Depth (*m*)	Temperature Range (*K*)
Forest Fires (Crown Fires)	30–60	1–8	8–133	800–1100
Savannas	5–16	1–3	1–13	900–1300
Agricultural	–	–	0.2–13	900–1300

For the modelling, the flame fronts are assumed to have had the time to grow since ignition and are spreading across the landscape at an equilibrium spread rate. Fires that have either just started to spread from their ignition point or are in a smouldering state are not considered. The flame fronts are assumed to be continuous and of even depth, with the depths ranging for each vegetation class as denoted in Table 1. The depth of a flame front is determined using field measurements of the fire's rate of spread and nominal flame front residence time (the length of time required for the flaming zone or fire front to pass a given point (Merrill and Alexander, 1987)). The horizontal lengths of the flaming fronts are varied in order to spread across the entire range of conceivable sizes. The limits of detection, either undetectable or instrument saturation, are identified for each vegetation system as the entire ranges of depths and lengths are exploited. Only single flame fronts within a footprint have been modelled and the flame fronts are considered to be uniform. The two primary variables that control the energy release in the model are the flame front depth and length. The flame fronts can be shifted off centre within a footprint, but by no more than half the distance between neighbouring footprint centres (only a few hundred meters). A sensitivity study determined that the differences were not significant enough to warrant additional modelling. This analysis will concentrate only on flame fronts that pass directly though the centre of the footprint.

The appropriate PRF weighting is applied to integrate the energy, both areas of flame front and background temperatures, emerging from the entire footprint area. The temperature of the flame fronts is nominally taken to be 1000 K and the background temperature is 305 K, that of a hot afternoon when fires are prevalent. The atmospheric transmission is modelled using the MODTRAN software package and the spectral response of each instrument channel that is used for fire detection has been inserted. The atmospheric modelling profiles of temperature and water vapour are those of a climatological mid-latitude summer and a very hazy environment. The atmospheric modelling for the DMSP–OLS instrument's visible wavelength channel is for nights with no moonlight and hazy skies. The threshold of detection for the thermal channels is selected based on observations of known fires found in the AVHRR imagery and knowledge of the AVHRR and MODIS fire-channel noise characteristics. To gain confidence in the classification of a fire pixel, the fire pixel needs to be significantly hotter than its surrounding background temperature. For the purpose of this analysis, the difference between the pixel that contains a fire and a pixel of only the background temperature must exceed 10 K in order for the fire to be detected. This 10 K criteria has been applied to both AVHRR and MODIS in this modelling effort. In reality, the background temperature would likely be higher than the 305 K used in this study and would push the threshold of detection upward since a larger fire would be required to meet the detection criteria of 10 K. For DMSP–OLS, using a night-time visible channel, the minimum detectable radiance has been chosen since all other pixel values will be that of the background.

The results of the modelling effort are intriguing. Figure 8 shows the relationship between fire detection for each instrument relative to the flame front area. Also, the typical range of flame front sizes is presented for each of the three vegetation classes. Figure 8 shows that MODIS will saturate only for the largest forest fires and that AVHRR will saturate for forest fires an order of magnitude smaller in flame front area than MODIS. To saturate any instrument channel, the energy from the input scene exceeds the designed measurement range of that instrument channel and only the maximum possible value is reported. It is interesting to note that a single savanna fire front would not saturate AVHRR, but saturation by savanna fires can be found in the imagery. In this case, as seen in Figure 9, multiple fire fronts can exist in close proximity and would be required to saturate the AVHRR instrument. For the minimum fire size that can be detected, MODIS (ca. 213 m^2) appears to be slightly more sensitive than AVHRR (ca. 435 m^2) to small fires. However, there is a very small difference in the PRF for each instrument and even the resolution of AVHRR and MODIS are typically quoted as being the same (~1 km). The advantage of MODIS over AVHRR can be attributed to the MODIS platform being 128 km lower in orbit than the nominal 833 km orbit of the AVHRR platform. It is also apparent from Figure 8 that the DMSP–OLS instrument can detect flame fronts an order of magnitude smaller in area (~45 m^2) then the smallest flame fronts that AVHRR and MODIS can detect. The ability of the DMSP–OLS instrument to detect very small fires is attributed to the unique low visible-light detection capability of that instrument (Croft, 1978).

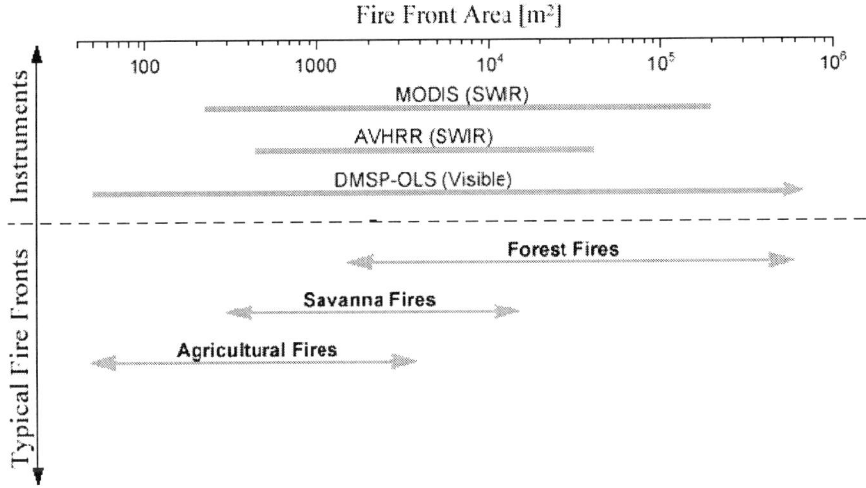

Figure 8. The modelled fire detection for AVHRR, MODIS, and DMSP–OLS has been drawn in as horizontal lines along the fire front area scale. The left side of each line represents the minimum detectable fire and the right side shows where each instrument will saturate. Coincidently, the typical range of fire front sizes for three vegetation classes is also drawn.

Figure 9. Aircraft picture of two savanna fire fronts in close proximity to one another. The larger fire front (on the right) is being fueled by the wind and is being pushed into the wind. The left-most front is advancing in the direction from which the wind is coming. The two fire fronts will grow into each other by which time the fuels will be burned in both directions and the fire will be suppressed.

Analyzed AVHRR imagery has been utilized to appraise the results of this modelling effort. Two AVHRR images were chosen, one covering a widespread fire event in the region of Angola, and the other covering a widespread fire event in Canada. A fire detection algorithm (Baum and Trepte, 1999) is used to classify and map the image pixels containing active fires. A cluster analysis is performed on the classified fire pixel maps that are derived from each image. The purpose of the cluster analysis is to determine the frequency of occurrence of contiguous fire pixels (clusters) in groups of number from 0 to n, where n is the largest number of contiguous fire pixels. The results of this analysis are shown in Figure 10. The forest fires cover an extensive amount of surface area and the results show that the clusters of contiguous fire pixels in the image are often large in count and that the large fire pixel clusters are very frequent in their occurrence. This means that the forest fires are not close to the limit of detection and are easy to identify. The results of the model have shown that the forest fires should be large enough to easily detect with the AVHRR instrument. For the savanna fires in Angola, the cluster analysis shows that single pixel fires are the most frequent in occurrence and the histogram does not show the lower half of the frequency distribution. This suggests that the savanna fires are much smaller in size and are on the edge of being detected by the AVHRR instrument. The model confirms this conclusion and shows that, for savanna fires, the typical fire front range stretches below the minimum detectable limit of the AVHRR instrument.

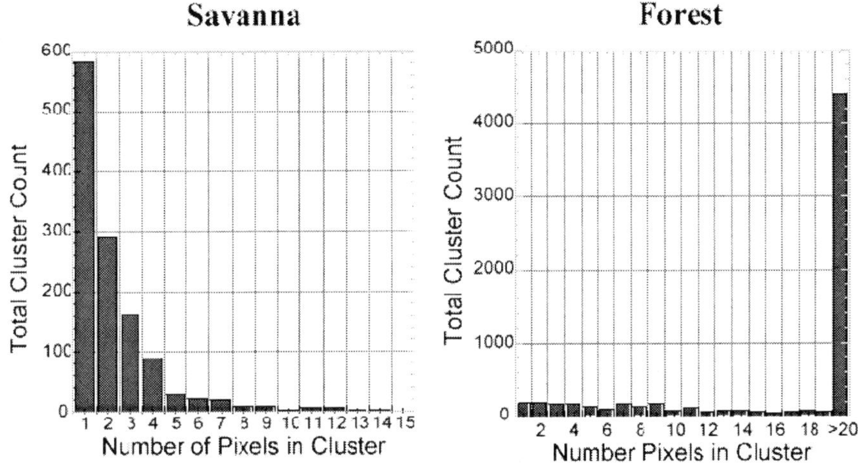

Figure 10. The fire pixels in two AVHRR scenes are detected using a fire detection algorithm (Baum and Trepte, 1999). The clusters of contiguous pixels are identified and the number of pixels in each cluster is counted. The skew of the cluster size distributions for the forest and savanna fires are completely different. The forest fire cluster size distribution is skewed toward a large number of pixels in each cluster and the savanna fire distribution is skewed toward single pixel fires.

There are a few additional caveats to consider in this analysis. It is possible, particularly for AVHRR and MODIS, to claim fire detection in some instances that are below the sizes that are shown in this paper. If the fire temperature is higher than 1000 K, then more energy is being released and slightly smaller fires can be detected. If the background temperature is lower than 305 K, then less energy is needed to exceed the defined detection criteria of 10 K and smaller fires will be detected. Of course, the converse is true for cooler fire temperatures and warmer backgrounds. It is believed that the model has been run to cover the nominal cases. The 10 K criterion, chosen for the purpose of providing a statistically significant increase in temperature of a fire pixel over that of the background temperature, could be relaxed if the location of a specific fire is known and there is confidence in the fire/no-fire judgement. It is also important to note that MODIS traded for more noise in the fire channel (2 K noise equivalent difference temperature (NE^{o}T)) to gain a higher saturation temperature and that AVHRR has a lower level of noise in its fire channel (1 K NE^{o}T). With less noise in the thermal channel being used for fire detection, the threshold of detection of 10 K could be relaxed slightly for the AVHRR modelling and in reality AVHRR could be slightly better or the same at detecting smaller fires as MODIS.

5. DISCUSSION AND CONCLUSIONS

For two decades it has been realized that fire detection was possible from space using meteorological satellite data. However, to date no instrument has been launched with the explicit design to monitor fires. To the knowledge of the authors, no study has evaluated the differences in fire detection limits of the instruments being used for fire monitoring. As various international projects are underway to map fires globally, having this very fundamental assessment will allow researchers to evaluate each of these developed products with a heightened awareness of their inherent limitations. More modelling is required to develop a similar understanding for each new instrument that will be used for fire detection so the derived products from each instrument can be appropriately compared. For example, MODIS provides an advantage over AVHRR in terms of having a higher saturation temperature, but as determined in this study MODIS does not have a significant advantage over AVHRR for detecting small fires. Hence, MODIS and AVHRR active fire detection products could be compared and even appropriately combined since their results should be very similar in mapping fire activity. In addition to understanding the capabilities of the exiting instruments for fire detection, this type of modelling can be used in the development of new instrumentation. One such result from this modelling effort is that a 250 m resolution (resolution in the conventional sense) thermal channel could provide a similar small fire detection capability as that of the DMSP–OLS instrument. This is a worthwhile goal given that many of the agricultural fires and fires in the early stages of growth in any ecosystem will escape detection with the current systems.

The modelling approach in this analysis is physically-based and begins with the basics. The behaviour of each instrument is determined and quantified in the form of the PRF. The PRF defines the actual footprint size and its weighting function. Fire behaviour was considered for developing the appropriate size fire fronts and temperatures to include in the model. Fire front sizes were developed for three classes of vegetation to simulate a wide range of fire behaviour characteristics that are attributed to their different fuel distributions. At each step in the process of developing the end-to-end model, the results were compared to satellite measurements. The results from the AVHRR analyses are presented in this paper, but the DMSP–OLS results are in equally good agreement with the modelling. Only after the MODIS instrument is launched can a similar comparative study be completed.

The PRF defines the radiometric footprint and provides the appropriate spatial weighting of the energy from within that footprint. The use of this weighting function prevents large fire fronts from inappropriately biasing the results. That is, if the front is further from the centre of the footprint it will have less influence on the overall pixel value. This is the reality of the measurements and this reality needs to be reflected in the models. Future work in the area of the PRF will provide a more rigorous mathematical examination, but for this paper the PRF has been presented in a more qualitative manner. Part of the reason for this is to present the idea of PRF itself, and hopefully generate thought about the appropriateness of imagery interpretations and analyses. Often, knowledge of the PRF would help to understand the results from studies like those of Belward and Lambin (1990) and Steyaert *et al.* (1997). Both of these studies have evaluated the ability of AVHRR to resolve spatial variations in the landscape relative to that of Landsat TM imagery and provide some insight into how large the AVHRR footprint might be. Another remote sensing research topic to think very carefully about is that of subpixel modelling. This is particularly true given how many image pixels fit within one PRF footprint (the element of a single pixel) as shown for AVHRR in Figure 5. And as noted, the PRF for each instrument, regardless of the quoted resolution, is similarly broad relative to a single pixel. The PRF is a mathematical description of the actual resolution of the instruments, and to some degree the PRF can provide a mathematical basis for sharpening the imagery (Gonzalez and Wintz, 1987).

6.   REFERENCES

Alexander, M. E., B.J. Stocks, B.M. Wotton, and R.A. Lanoville, 1998. An example of multifaceted wildland fire research: The International Crown Fire Modelling Experiment, Proc. The Joint III International Conference on Forest Fire Research/14th Conference on Fire and Forest Meteorology, November 16–20, Coimbra, Portugal, 83–112.

Baum, B. A., Q. Trepte, 1999. A grouped threshold approach for scene identification in AVHRR imagery, *J. Amos. Oceanic. Tech.*, 16, 793-800.

Belward, A. S., and E. Lambin, 1990. Limitations to the identification of spatial structures from AVHRR data, *Int. J. of Remote Sens.*, 11 (5), 921–927.

Brass, J. A., L. S. Guild, P. J. Riggan, V. G. Ambrosia, R. N. Lockwood, J. A. Pereira, and R. G. Higgins, 1996. Characterizing Brazilian fires and estimating areas burned by using the Airborne Infrared Disaster Assessment System, Biomass Burning and Global Change, MIT Press, Cambridge, MA, Vol. 2, 561–568.

Breaker, L. C., 1990. Estimating and removing sensor-induced correlation from Advanced Very High Resolution Radiometer satellite data, *J. Geophys. Res.*, 95 (C6), 9701–9711.

Cahoon, Jr., D. R., B. J. Stocks, J. S. Levine, W. R. Cofer III, and K. P. O'Neill, 1992a. Seasonal distribution of African savanna fires, *Nature*, 359 (29), 812–815.

Cahoon, Jr., D. R., B. J. Stocks, J. S. Levine, W. R. Cofer III, C. C. Chung, 1992b. Evaluation of a technique for satellite-derived estimation of biomass burning, *J. Geophys. Res.*, 97(D4), 3805–3814.

Croft, T. A., 1978. Nighttime Images of the Earth from Space, *Scientific American*, July, 86–98.

Elvidge, C. D., H. W. Kroehl, E. A. Kihn, K. E. Baugh, E. R. Davis, and W. Hao, 1996. Algorithm for the retrieval of fire pixels from DMSP Operational Linescan System data, Biomass Burning and Global Change, MIT Press, Cambridge, MA, Vol. 1, 73–85.

Fisher, P., 1997. The pixel: a snare and a delusion, *Int. J. Remote Sens.*, 18 (3), 679–685.

Flasse, S., and P. Ceccato, 1996. A contextual Algorithm for AVHRR fire detection, *Int J. of Remote Sens.*, Vol. 17, pp. 419–424.

Gonzalez, R. C., and Paul Wintz, 1987. Digital Image Processing, Second Edition, Addison-Wesley Publishing Company.

Lee, T. F., and P. M. Tag, 1990. Improved detection of hotspots using AVHRR 3.7 $\Box m$ channel, *Bull. Amer. Meteor. Soc.*, 71 (12), 1722–1730.

Justice, C.O., and P. Dowty, 1994. IGBP–DIS Satellite Fire Detection Algorithm Workshop Technical Report, IGBP–DIS Working Paper #9, Workshop held in Greenbelt, Maryland, USA, on 25– 26 February 1993, (NASA / GSFC).

Kasischke, E. S., N. H. F. French, P. Harrell, N. L. Christensen, S. L. Ustin, and D. Barry, 1993. Monitoring of wildfires in boreal forests using large area AVHRR NDVI composite data, *Remote Sens. Environ.*, 44, 61–71.

Kasischke E. S., N. L. Christensen, and B. J. Stocks, 1995. Fire, global warming, and the carbon balance of boreal forests, *Ecol. Appl.*, 5 (2), 437–451.

Kaufman, Y., and C. Justice, 1994. Fire Products, MODIS Algorithm Technical Background Document.

Kidwell, K. B., 1991. NOAA Polar Orbiter Data (TIROS–N, NOAA–6, NOAA–7, NOAA–8, NOAA–9, NOAA–10, NOAA–11) Users Guide, National Environmental Satellite Data and Information Service, Washington, D.C.

King, M. D., W. J. Kaufman, W. P. Menzel, and D. Tanré, 1992. Remote sensing of cloud, aerosol, and water vapor properties from the Moderate Resolution Imaging Spectrometer (MODIS). *IEEE Transactions on Geoscience and Remote Sensing*, 30, 2–27.

Kurz, W.A., M. J. APPS, B. J. Stocks, and W. J. A. Volney, 1995. Global climate change: disturbance regimes and biospheric feedbacks of temperate and boreal forests, Biotic Feedbacks in the Global Climate System: Will the Warming Speed the Warming?, G. Woodwell (ed.), Oxford Univ. Press, Oxford, UK., 119–133.

Matson, M., G. Stephens, and J. Robinson, 1987. Fire detection using data from the NOAA–N satellites, *Int. J. Remote Sens.*, 8 (7), 961–970.

Merrill, D.F. and M. E. Alexander, 1987. Glossary of Fire Management Terms (Fourth Edition), National Research Council No. 26516. Ottawa, Ontario.

Setzer, A. W., and J. P. Malingreau, 1996. AVHRR monitoring of vegetation fires in the tropics: Toward the development of a global product, Biomass Burning and Global Change, MIT Press, Cambridge, MA, Vol. 1, 25–39.

Sneeuwjagt, R. J., and W. H. Frandsen, 1977. Behavior of experimental grass fires vs. predictions based on Rothermel's fire model, *Can. J. For. Res.*, 7, 357–367.

Smith, W. J., 1966. Modern Optical Engineering, Mcgraw-Hill.

Stearns, J. R., M. S. Zahniser, C. E. Kolb, and B. P. Sandford, 1986. Airborne infrared observations and analyses of a large forest fire, *Applied Optics*, 25 (15), 2554–2562.

Steyaert, L. T., F. G. Hall, and T. R. Loveland, 1997. Land cover mapping, fire regeneration, and scaling studies in the Canadian boreal forest with 1 km AVHRR and Landsat TM data, *J. Geophys. Res.*, 102 (D24), 29581–29598.

Stocks, B. J. and J. Z. Jin, 1988. The Great China Fire of 1987: Extremes in fire weather and fire behavior, in Proc. 1988 Annual Meeting Northwest Fire Council: Fire Management in a Climate of Change, Victoria, British Columbia, 67–79 Nov. 14–15, 1988.

Stocks, B. J. and G. R. Hartley, 1995. Fire behavior in three jack pine fuel complexes, Canadian Forest Service, Great Lakes Stocks Forestry Centre, Sault Ste. Marie, ONT.

Stocks, B. J., B. W. Van Wilgen, W. S. W. Trollope, D. J. McRae, F. Weirich, and A.L.F. Potgieter, 1996. Fuels and fire behavior dynamics on large-scale savanna fires in Kruger National Park, South Africa, *J. Geophys. Res.*, 101 (D19), 23541–23550.

Stocks, B. J., M. A. Fosberg, T. J. Lynham, L. Mearns, B. M. Wotton, Q. Yang, J-Z Jin, K. Lawrence, G. R. Hartley, J. A. Mason, and D. W. McKenny, 1998. Climate change and forest fire potential in Russian and Canadian boreal forests, *Climatic Change*, 38 (1), 1–13.

Climate and Vegetation as Driving Factors in Global Fire Activity

EDWARD DWYER[1], JEAN-MARIE GRÉGOIRE[1], JOSÉ M.C. PEREIRA[2]
[1]Global Vegetation Monitoring Unit, Space Applications Institute, Joint Research Centre, European Commission, Ispra, Italy
[2]Laboratory for Remote Sensing and Geographical Analysis, Department of Forestry, Instituto Superior de Agronomia, Lisbon, Portugal

Abstract: Global active fire distributions have been determined for a 12 month period from daily acquired, low spatial resolution satellite imagery. These distributions have been grouped into a small number of classes based on the spatial and temporal characteristics of the data. A global climatology of monthly temperature and precipitation data was used to derive warmth and moisture indices. We show how different patterns of fire activity, as represented by the fire classes, can be related to particular climate conditions. Vegetation type is also shown to be important in determining fire activity, particularly in tropical regions. Our results support the premise that fire regimes will change under changed climate conditions and that the empirical approach to the investigation of the fire–climate relationship could provide a complementary tool to the physically-based climate change prediction models.

1. INTRODUCTION

The very extensive burning which occurred during 1997 and 1998 in many regions of the world, refocused world attention on the issue of vegetation fire. Many of these fires, such as those which occurred in Indonesia and Mexico, took place during extreme weather conditions related to the 1997–1998 El Niño event. The occurrence of fire is primarily dependent on climate, directly through weather conditions that enable fires to develop and indirectly through the supply of a sufficient vegetation fuel load to sustain fire. There is then the requirement for an ignition source that may be of natural or human origin.

Vegetation fires are found world-wide, the majority being set by man in connection with landuse. Satellite imagery shows that over 70% of global fire events in any one year are found within the tropics (Dwyer *et al.*, 1999a). Many areas of savanna are burned annually with the consequent emission of large quantities of carbon dioxide and other trace gases (Andreae *et al.*, 1996). Although the number and frequency of fires in temperate and boreal regions is much lower, large fire events, typical of the boreal region, can release huge amounts of carbon-based products, the atmospheric impacts of which are still not well understood (Cofer *et al.*, 1996).

There is a need to understand the complex relationship between fire, climate and vegetation as fire regimes are expected to change with alterations in temperature and precipitation conditions under a modified climate. Studies for boreal regions predict that more fire activity is a virtual certainty given more severe fire weather coupled with economic and infrastructure constraints concerning fire fighting. (Stocks *et al.*, 1998). In tropical forest and savanna ecosystems, climate change will augment fire risk, although in the long term there may be an overall decrease in fire activity because of a lower availability of fuels as a result of degradation, desertification and more intensive use of plant biomass for energy and food supply (Goldammer and Price, 1998). In a study of a Mediterranean region in Europe, fire hazard increased over the period 1941–1994, leading to the conclusion that climate warming will cause an increase in the number of wildfires in this region (Piñol *et al.*, 1998).

The construction of fire risk or danger indices relies on a number of meteorological parameters, such as temperature, relative humidity, wind speed and preceding rainfall, which are then combined with information on fuel characteristics such as curing and abundance (Tapper *et al.*, 1993). The indices tend to be empirical in nature, are often tuned to local conditions and undergo modifications based on observed fire occurrence. In their study of wildfire activity in southern Australia, Krusel *et al.* (1993) suggest that the mean maximum daily temperature and maximum relative humidity may be effective selectors of high fire activity days. The mean number of days since rain discriminated poorly between different levels of fire activity. Price and Rind (1994) identified water balance as one of the meteorological parameters best related to fire occurrence on a monthly scale and developed a model relating the number of fire events recorded in Arizona and New Mexico to effective precipitation, which is defined as the difference between precipitation and potential evapotranspiration. In their work modelling the long-term dynamics of forest fire in Siberia, Antonovsky *et al.* (1989) show that fire probability is strongly related to the mean air temperature, total precipitation and the maximum period between two successive rains, all calculated over the period of the fire season.

Most studies of the relationship between fire regimes and climate have been in the physical modelling field or have concentrated on observations over a limited geographical area. This paper proposes a complementary tool for empirically exploring the relationship between observed global active fire activity and climate and suggests how this may also have a role in predicting the spatial distribution of future fire regimes. The fire activity was determined from global satellite observations for the 12 month period from April 1992 to March 1993 and the climate parameters are based on a global temperature and precipitation climatology. Different patterns of fire activity are linked with particular temperature and moisture regimes. Different vegetation types under the same climatic conditions have a significant role in determining fire pattern characteristics and we suggest how this can be indicative of landuse.

2. DATA AND METHODS

2.1 Characterisation of fire activity

The geographical positions of those image pixels in which fires were detected (fire pixels) were determined from the global 1 km Advanced Very High Resolution Radiometer (AVHRR) data set for each day in the period from April 1992 to December 1993 (Stroppiana *et al.*, 1999). Although the observations are but a sample of the total fire activity that may take place in a day, the data set is globally consistent due to the use of the same processing algorithms for the entire data set. The overpass time of the satellite (early afternoon) is believed to coincide with the peak time for burning for tropical areas (Eva and Lambin, 1998). Observations in a Mediterranean environment by one of the authors indicate that the most severe fire weather conditions are in the early afternoon. In boreal regions, lightning is responsible for most of the area burned and fires can last for many days (Stocks *et al.*, 1996). For a 12 month period (April 1992 to March 1993) three parameters which describe the spatial and temporal distribution of fire pixels within cells of 0.5° by 0.5° (latitude/longitude) were calculated. No special processing was done for coastal cells containing both land and ocean. The parameters calculated were:

- fire number: For each grid cell the total number of fire pixels detected in the year was calculated;
- fire agglomeration size: This is a measure of the degree to which fire pixels are clustered within each grid cell; and

– fire season duration: The number of months from when 10 % of fire
  pixels were seen in a grid cell to when 90% of the fire pixels were
  detected.

Using an automatic clustering algorithm five classes of fire activity were
defined using these three parameters. The class characteristics can be
described as follows:

– Class 1: Low level of fire activity, there are few fire pixels in a cell. The
  fire season is short and the fire agglomerations are small.
– Class 2: Moderate level of fire activity with a moderate fire season
  duration, the fire agglomerations are small.
– Class 3: Moderate level of fire activity, long fire season duration. Fire
  events are unclustered.
– Class 4: Moderate to high level of fire activity with a moderate fire
  season duration. Fire agglomerations are large.
– Class 5: Very high level of fire activity, moderate to long fire season
  duration and moderate to large fire agglomerations.

The global geographical distribution of the 5 fire classes are shown in
Figure 1. The parameter extraction and the classification method is described
in detail in Dwyer *et al.* (1999b).

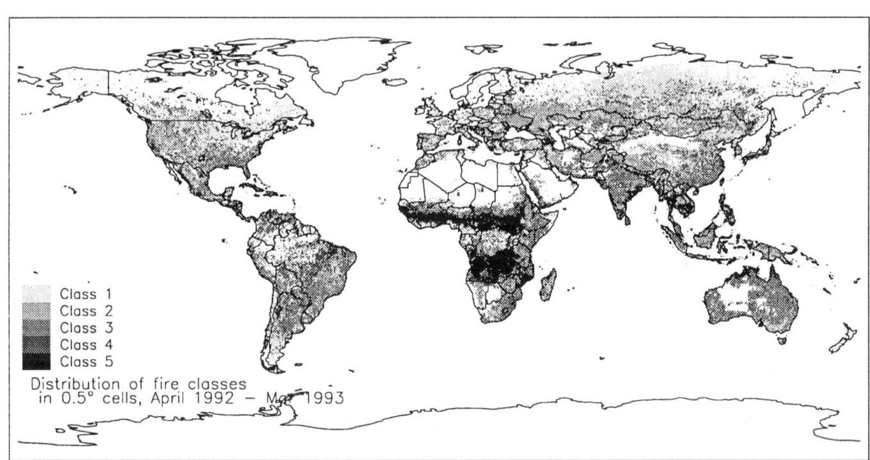

Figure 1. The spatial distribution in cells of 0.5° by 0.5° of the 5 fire classes defined by
clustering fire number, agglomeration size and fire season duration.

2.2 Climate data

The global network of weather stations has extensive coverage in the more developed regions of the world, but very large gaps exist in less populated and less technologically developed regions. Global measured climate parameters coinciding with the fire data were therefore unavailable. We used instead the gridded, global climate database of the International Institute for Applied Systems Analysis (IIASA). This has been constructed using weather station data of temperature, precipitation and other parameters taken from a minimum 5 year consecutive period from 1931 to 1960. Interpolation and surface fitting were used to create the final global product at 0.5° x 0.5° resolution, with monthly values for each of the cells (Leemans and Cramer, 1990).

For fire to occur, there must be an adequate fuel supply. The moisture content of the fuel must be such that ignition can take place and weather conditions such that the fire can be sustained. Both temperature and moisture balance have a determining role in vegetation growth and hence fuel load. The length and strength of any dry period will play a critical role in determining the potential for fires to develop and spread.

Vegetation growth slows with decreasing temperature and ceases around 0° C. Bio-temperature (T_b) was therefore selected as an appropriate warmth index, as it gives a measure of heat during the growing season which is likely to be more directly related to plant growth than mean temperature. It is calculated as the mean value of all daily temperatures above 0° C divided by 365. The daily temperature values are computed using smoothed pseudo-daily temperature values generated by fitting a curve through the monthly average temperature values. Bio-temperature was first used by Holdridge in the conception of his life zones and has been used subsequently in other vegetation classifications (Cramer and Leemans, 1993).

Total rainfall is not a complete measure of the moisture available for plant growth. "Combined evaporation from the soil surface and transpiration from plants, called evapotranspiration, represents the transport of water from the earth back to the atmosphere, the reverse of precipitation" (Thornthwaite, 1948). A measure of effective precipitation would therefore be the difference between precipitation and evapotranspiration. Actual evapotranspiration is rarely measured in meteorological stations, therefore Potential Evapotranspiration (PE) is used. PE is the evapotranspiration that would occur given unlimited water supply and it has a close relationship with mean monthly temperature. A number of empirical formulae have been developed to calculate PE. As Thornthwaite's (1948) formulation depends only on monthly mean temperature values it is the one used here. It is calculated as follows:

$$PE = 1.6(10t/I)^a \text{ where}$$

$$I = \sum_{j=1}^{12} \ldots (t_j/5)^1$$

and

$$a = 0.000000675 1I^3 - 0.0000771 1I^2 + 0.01792I + 0.49239$$

PE = monthly potential evapotranspiration (cm)
t = mean monthly temperature ($^\circ$C)
I = heat index summed over the 12 monthly values
j = month number

The PE values are modified by a latitude factor related to the mean possible duration of sunlight for each month of the year. The complete development of the formula is described in Thornthwaite (1948). The effective precipitation can then be readily calculated from the precipitation (P) and PE values.

Vegetation is more susceptible to fire under conditions of moisture stress, when the plants suffer a water deficit. The annual duration of this period, which can be considered a measure of the dry season length was calculated as the number of months when the water balance was negative i.e. PE greater than P.

2.3   Vegetation data

A global map of vegetation cover was required to investigate if vegetation types can discriminate between different classes of fire activity. We chose the recently produced global land cover map at 1 km resolution based on a 12 month period of satellite observations (Loveland & Belward, 1997). Although not completely validated, this map represents one of the more recent attempts to determine objectively actual vegetation cover. For this study we have aggregated the cells to 0.5° by 0.5° by assigning the vegetation type which occurs most frequently in the cell. We used is the 24 class United States Geological Survey (USGS) Land Use/Land Cover System legend. The characteristics of the data set are described at http://edcwww.cr.usgs.gov/landdaac/glcc/globdoc1_2.html, while the formulation of the USGS legend is described in Anderson *et al.* (1976).

3. RESULTS

3.1 Fire numbers and climate

The four bioclimatic variables used in the analysis are :

– biotemperature;
– total annual precipitation;
– effective precipitation, i.e. the difference between total annual precipitation and potential evapotranspiration; and
– months of negative effective precipitation or period of moisture deficit (dry period).

Figure 2 shows the average number of fire pixels detected in a $0.5°$ by $0.5°$ cell as a function of these variables. At a biotemperature of $19°$ C, there is a steep increase in the number of fire pixels detected. Above $20°C$, the fire number increases slowly then flattens off and the trend is for a decrease as the biotemperature increases towards $30°$ C. The peak at $27°$ C corresponds to a peak in the total number of land cells, irrespective of fire presence, with this biotemperature.

There is a rapid increase in the number of fire events seen above 700 mm of precipitation per annum, with peak activity at 1100 mm, above which the number of fires seen falls off. There is a weak relationship between the annual effective precipitation (P-PE) and the number of fires detected. The highest fire numbers are found in those cells with an effective precipitation from –200 mm to –800 mm. When there is a strong moisture deficit (very negative values) or increasing moisture supply (positive values) fire number decreases. The former is probably linked to lack of fuel load, while the latter is associated with the difficulty of igniting fuel with a high moisture content. Price and Rind (1994) show a significant relationship between the monthly number of fire events and the monthly effective precipitation for the south-western United States. In their study the number of fire events increased with increasing water stress. Here what appears more important is the period of moisture stress, rather than the overall annual moisture balance. Figure 2 (d) shows the relationship between the length of the period of moisture stress (defined as a month with negative effective precipitation) and the number of fires detected. Between four and eight months of moisture deficit there is an increase in the average fire number observed, above which fire number decreases again.

Although a clear relationship can be seen between the number of fire pixels detected in a cell and the bioclimatic parameters, there is a very large variability.

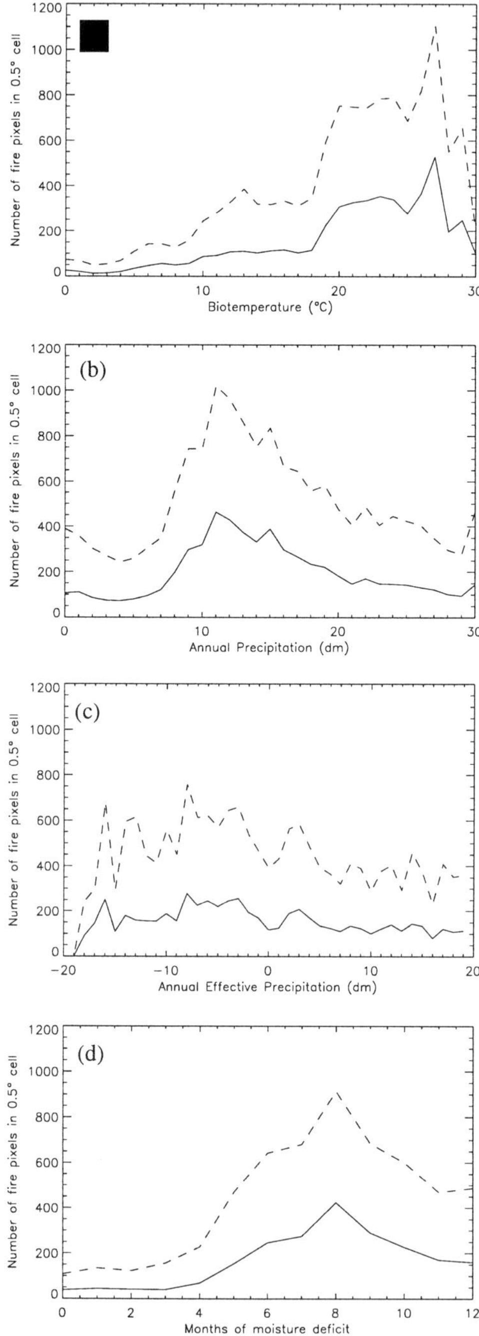

Figure 2. The mean and one standard deviation of the mean (dashed) number of fire pixels detected in each 0.5° x 0.5° cell is shown as a function of 4 bioclimatic variables.

Fire presence is related to the combined effect of warmth and moisture, not only during the period of burning itself but also in the preceding period, in that climatic conditions influence vegetation growth and hence potential fuel load. As a climatology is being used here, it cannot reflect the actual meteorological conditions experienced prior to and during the fire period.

3.2   Fire classes and climate

The number of fire pixels detected in a cell is only one aspect of the fire distribution characteristics. In addition to this, the classification shown in Figure 1 incorporates information on the spatial features of the fire distribution in a cell in terms of clustering, and one of its temporal aspects: the fire season duration. The dependency of the fire classes on climate was investigated using the above bioclimatic variables, excluding effective precipitation which was seen to have a weak link with the number of fire pixels detected.

Figure 3 shows the distribution of the fire classes for each of the three bioclimatic variables. The most important observations on these distributions are:

1. The fire classes cannot be separated using any one variable alone. An analysis of variance confirms the poor separation between the classes. Biotemperature is the best discriminator, with 21% of the variance in the data set explained by the five classes. The period of moisture deficit is the next best discriminator. An analysis of variance excluding all fire class cells with either 0 or 12 moisture deficit months indicates that 18% of the variance can be explained, while total annual rainfall is the poorest, only explaining 5% of the variance.
2. Class 5, which accounts for 45% of all fire events seen, shows a very distinctive curve on all three bioclimatic parameters, with respect to the other classes.
3. The other four classes show strong overlap on all three bioclimatic parameters. However, because of the requirement of appropriate warmth and moisture conditions for vegetation growth and favourable burning conditions, it is necessary to look at the fire classes in relation to a combination of the three variables.

  Edward Dwyer et al.

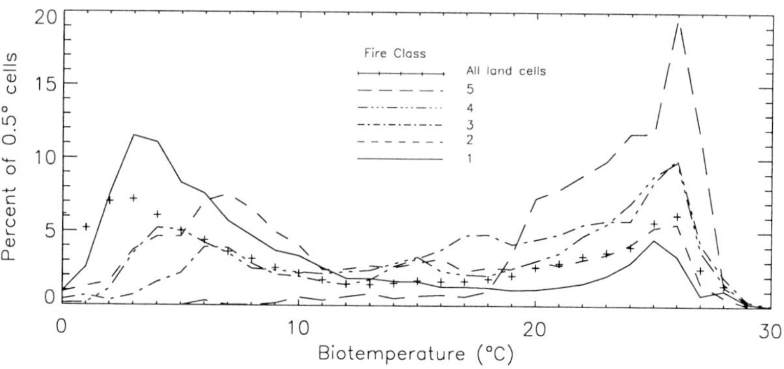

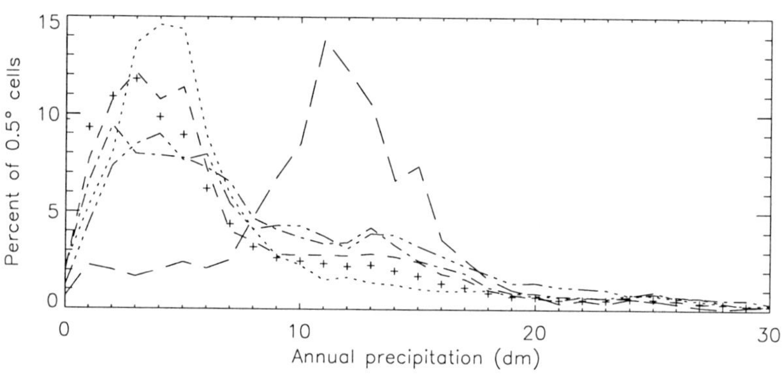

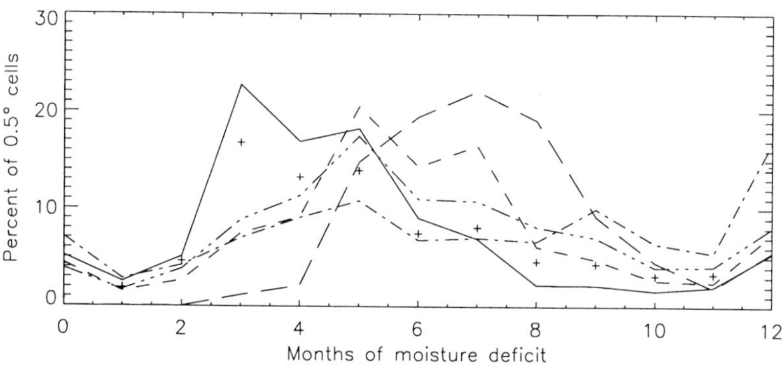

Figure 3. The distribution of the 5 fire classes against each of the 3 bioclimatic variables.

Figure 4 shows the distribution of the fire classes in the climate space defined by combinations of the three variables. The characteristics of the distribution of each fire class is different and in general the peak frequency of fire occurrence is in a different region of the climate space for each fire class.

- Class 1. The centre of activity of this class occurs in regions of low biotemperature ($< 5°$ C) , with annual rainfall around 500 mm and with a dry period from two to four months. The main characteristic of class 1 fires is a very low fire pixel count in each cell.
- Class 2. Here the centre of activity is in regions with biotemperatures in the range $5°$ C to $10°$ C, with less than 500 mm rain per year and a dry period from four to eight months. Class 2 fires are defined as a moderate number of fire pixels in a cell and with fire activity concentrated in a few months.
- Class 3. This is quite dispersed in terms of temperature and rainfall amounts. The strongest peak is found in areas with a biotemperature from $15°$ C to $25°$ C with low rainfall amounts – there is a small peak around 200 mm. The dry period is long, being between eight and 12 months. Most of these cells are found in the Australian interior. Another peak occurs in climate regions similar to those of class 2 but with slightly more rainfall. Class 3 is characterised by a moderate number of fires – on average less than class 2 – occurring over a long period of time.
- Class 4 : This class also has a large spread with multiple peaks. The main centre of activity is in regions of low biotemperature and annual rainfall amounts from 400 mm to 800 mm, similar to class 1. The length of the dry period is more similar to class 2 being centred between four and six months. Another peak is found in regions with biotemperatures $> 20°$ C , rainfall amounts between 500 mm and 1500 mm and dry periods from four to ten months. Class 4 is defined with a high number of fire pixels spatially close, to give large fire pixel clusters.
- Class 5: The peak is found in a region with biotemperatures between $20°$ C and $30°$ C, annual precipitation from 1000 to 1500 mm and a dry period from six to eight months. Class 5 is defined by a very large number of fire pixels occurring in a restricted time period and with variable fire agglomeration sizes. The climate conditions coinciding with this class are favourable for vegetation growth and for drying of fuel. This coupled with the wide range of landuses in these regions can explain the high number of fires detected in this class.

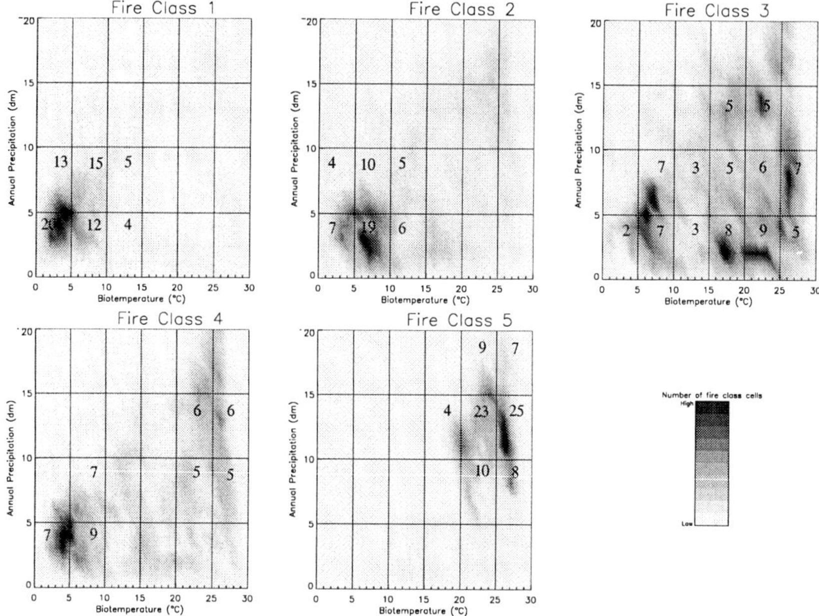

(a) Annual precipitation vs biotemperature

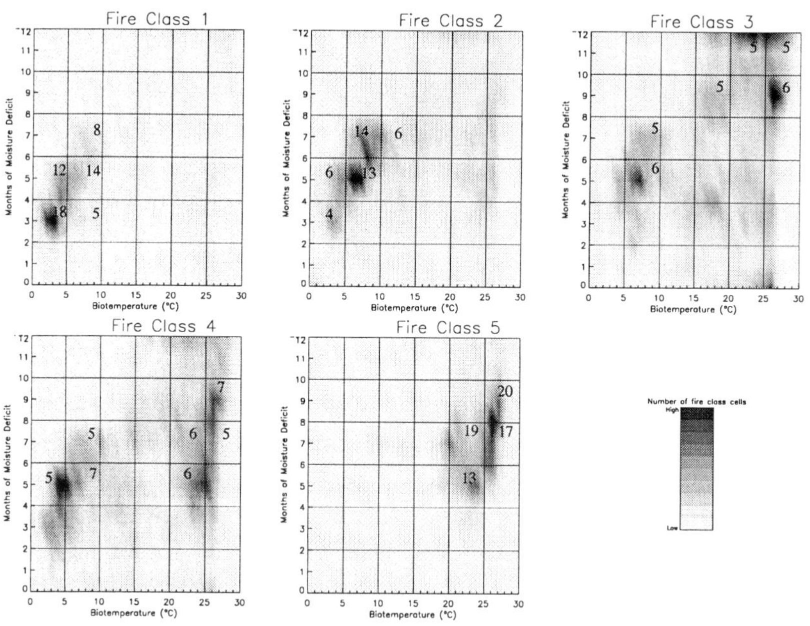

(b) Months of moisture deficit vs biotemperature

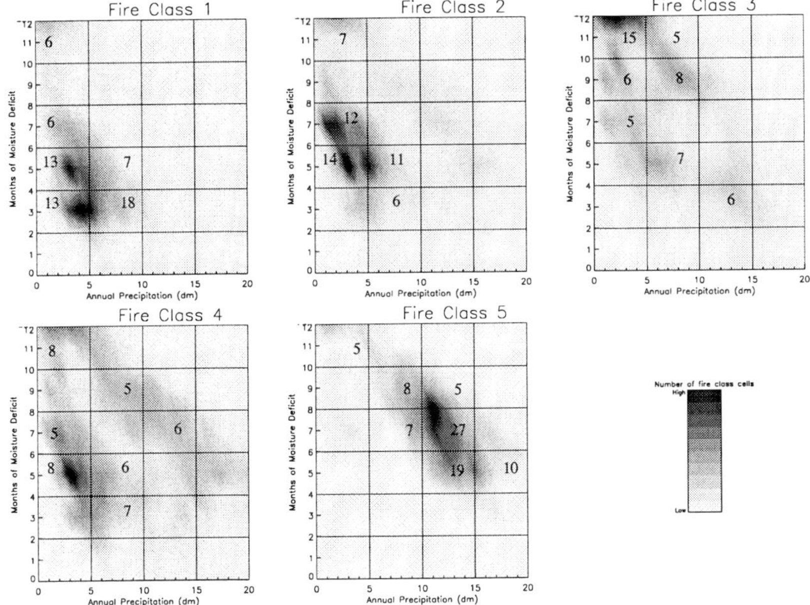

(c) Months of moisture deficit vs annual precipitation

Figure 4a,b,c. The distribution of the cells in each fire class against all combinations
of the 3 climate variables is shown. The percentage of cells, for a given fire class,
in each of the climate regions delimited by the horizontal and vertical lines is
overlayed in areas of peak fire class occurrence.

As the fire classes in the climate space are not normally distributed, it is
difficult to quantify how successfully the bioclimatic variables discriminate
the fire classes. A simple linear discriminant analysis shows that a
discriminant function drawn on the three combined variables accounts for
22% of the variance in the data set. This value together with the preceding
results indicate that climate discriminates between different types of fire
activity as represented by the fire classes defined here. However, there is still
significant overlap of the classes in the climate space. As an average
climatology was used it will not reflect actual meteorological conditions
immediately prior to and during the year in which fire activity is studied.
The fire classification itself is quite simple and does not explicitly
incorporate the timing of the actual period in which burning occurred. The
work of Price and Rind (1994) and Antonovsky *et al.* (1989) indicates that
the number of lightning-induced fires, in the southwestern United States and
boreal Siberia respectively, is more directly related to moisture conditions
during the period of fire occurrence than to any annual average moisture
measure. This may not be true, however, of human-induced fires under other
climate conditions and in other vegetation types.

3.3   Vegetation distributions

Vegetation distributions have been used in the past in the compilation of large-scale climate maps. These mapping techniques assumed that patterns of vegetation distribution and those of climate were in equilibrium and that climate was the single determining factor for vegetation (Cramer & Leemans, 1993). However, partly because of large-scale man-made modifications of land cover, different cover types are found today under similar climate conditions. For example, we have seen that class 1 and class 4 fire patterns overlap in cool regions, with a short dry period. Some of these regions contain forest, pasture and croplands. In any given fire season, a small number of very large fires is typical of boreal forest, whereas a low number of small, controlled fires are associated with pasture and crop management in these regions.

To determine if fire class distributions vary significantly as a function of vegetation type under similar climatic conditions, we have used the global land cover map already discussed above (Loveland and Belward, 1997). All $0.5°$ x $0.5°$ vegetation cells were identified for each of the climate regions, defined by the horizontal and vertical lines in Figure 4. Those vegetation types which represented at least 10% of the total cells in the region were retained. The percentage of cells in each vegetation type containing 1 or more fires was calculated and only those vegetation types which had a minimum of 10% of the cells affected by fire were retained. For all cells containing fires within these vegetation types, the percentage of cells in each of the five fire classes was calculated. As an example, a complete set of results is shown in Table 1 for one of the climate regions.

All vegetation types, as defined in the USGS Land Use/Land Cover System legend (Anderson *et al.*, 1976). show the occurrence of a number of fire classes. However, it is the relative distribution that is of interest. Taking the example of the region shown in Table 1, both the *savanna* and *deciduous broadleaf forest* types are dominated by class 5. In these climatic conditions both types occur mostly in southern Africa. In this region the *deciduous broadleaf forest* of the USGS legend is also called *miombo* woodland and is known to undergo frequent burning. The *savanna* class includes a range of subtypes from grass to bush savanna which may explain the presence of different fire classes. Almost all the *cropland/woodland mosaic* in this climate region occurs in Brazil, where it is dominated by class 3 fires, indicative of moderate levels of burning spread over a very long dry period. Finally, the *evergreen forest* has an almost equal distribution among four classes: class 5 is found almost exclusively in southeast Africa, again in an area that is more likely open forest, while classes 1 and 2 dominate in South America.

Table 1. The fire class distributions for one climate region delineated by the horizontal and vertical lines in Figure 4. For those vegetation types which occupied a minimum of 10% of the total cells and had at least 10% of cells occupied by fires, the distribution of the five fire classes was calculated. The similarity /dissimilarity of the distributions was determined using rank correlations.

Climate Region: Biotemperature (°C):   20 – 25
                   Annual Precipitation (mm):   1000 – 1500
                   Dry Period (months):   4 – 5

Total vegetation cells in region: 553
Total with 1 or more fires: 547

Vegetation Type (USGS legend)	% of cells in region with this vegetation	% of vegetation with fire cells	Fire Class				
			1	2	3	4	5
1. Cropland/ Woodland Mosaic	28.39	100.00	7.6	28.0	45.2	16.6	2.5
2. Deciduous Broadleaf Forest	22.24	100.00	0.0	5.7	4.1	8.9	81.3
3. Evergreen Forest	18.63	94.17	22.7	22.7	8.2	16.5	29.9
4. Savanna	14.47	100.00	7.5	15.0	7.5	12.5	57.5

Rank Correlation Coefficients for each vegetation combination

1 – 2	1 – 3	1 – 4	2 – 3	2 – 4	3 – 4
-0.40	-0.82	-0.46	0.41	0.87	0.71

The rank correlation coefficients between each vegetation pair were calculated to express the similarity or difference of fire distributions in these vegetation types. When the coefficients are low or negative, it means that the fire distributions are different in the two vegetation types being considered; when the coefficient approaches 1 it means that the distributions are similar. For the example, four of the combinations give a low correlation coefficient while two give a high value, therefore in this climate region the presence of different vegetation types can partly explain the overlap seen between the fire classes.

The horizontal and vertical lines shown in Figure 4 divide the climate space into 144 regions. Sixty-four of these are occupied by vegetation with a sufficient number of fire cells to fulfil the criteria already described. The following rule was used to indicate if the presence of different vegetation types affected the fire class distributions in these climate regions:

If at least one of the rank correlation coefficients is less than 0.5 then the fire class distributions are considered DIFFERENT in that climate region, otherwise they are considered to be SIMILAR.

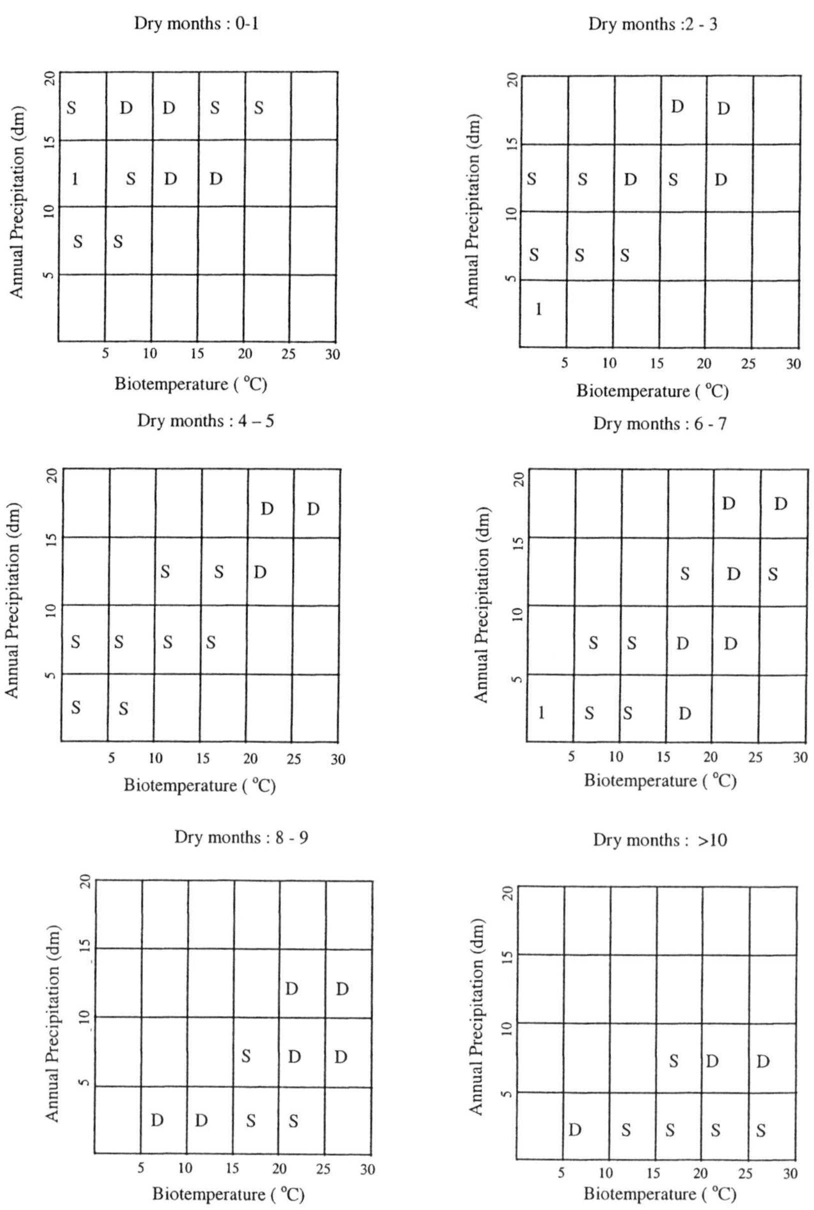

Figure 5. An indicator of difference or similarity in fire class distribution in vegetation types is shown for each climate region delimited by the horizontal and vertical lines. (D – fire class distributions different in different vegetation types, S – fire class distributions similar, 1 – only one vegetation type affected by fire in this region, blank squares means that there were few vegetation cells or few fire events).

The results are shown in Figure 5. Three different scenarios can be identified:

1. In cool regions ($T_b < 10°$ C), with less than eight dry months and irrespective of rainfall amounts, fire class distributions tend to be the same, independent of vegetation type. The fire class distributions, not shown in the figure, indicate that class 1 fires dominate when the period of moisture stress is short, while the percentage of class 2 fires increase as the moisture stress period increases above four months.
2. In warm and hot areas ($T_b > 15°$ C) with low rainfall amounts (< 500 mm) and long periods of moisture stress (> 7 months), fire class distributions tend to be the same, independent of vegetation type. Class 3 fires predominate, which are those where moderate levels of fire activity are seen over very long periods of time. Fuel supply is expected to be low under these conditions.
3. In hot areas ($T_b > 20°$ C) with more than 500 mm of rain annually and two or more months of moisture stress, the fire distributions vary in different vegetation types. Class 5 fires dominate grasslands/savanna but few occur in closed forest and croplands.

4. DISCUSSION AND CONCLUSIONS

Bioclimatic conditions as described by mean annual biotemperature, length of dry period and mean annual precipitation discriminate between different types of fire activity. These results are in keeping with those of Krusel *et al.* (1993) and Antonovsky *et al.* (1989). Working on wildfire activity in south-eastern Australia, Krusel *et al.* (1993) identified, in descending order of merit, maximum daily temperature, days since rain and a drought index as the best discriminators between different levels of fire activity. Antonovsky *et al.* (1989) linked fire probabilities with mean seasonal air temperature, maximum seasonal period between two successive rains and the seasonal sum of precipitation. Unlike these studies, for which coincident and localised meteorological data were available, we used indices derived from a global climatology of temperature and precipitation. From a number of possible measures of warmth and moisture, the three bioclimatic parameters used were chosen as the best combination of discriminators between the different types of fire activity represented by the five classes.

Fire activity fluctuates from year to year in terms of number of fires, area burned and timing of the fire season. This has been illustrated for Africa using a multi-annual time series of remotely sensed data (Koffi *et al.*, 1996; Barbosa *et al.*, 1998). Fire statistics collected over many years show large inter-annual variations in the number of fires and the area burned in boreal regions (Stocks, 1992). These fluctuations are related to both seasonal and inter-annual variations in fuel supply and weather. In addition, the characteristics of individual fires are controlled by a range of meteorological factors including wind speed and relative humidity. More comprehensive meteorological data coincident with the fire period studied here should show a better correlation than the average climatology that we used.

The fire classification is based on numbers of fires detected in $0.5° \times 0.5°$ cells, their position in relation to one another and the length of the period of time in which they occur. The onset of the fire season is linked to fuel moisture content (Ceccato and Flasse, 1998), which in turn is related to meteorological conditions. A more complex fire classification could incorporate the onset and decline in the fire season but coincident meteorological data would be required to investigate the link with the seasonal development of the fire activity.

Here, class 5 fire activity, which accounts for 45% of all fire pixels detected, is quite well separated in the climate space from the other classes. Class 4 is the least well discriminated in terms of the bioclimatic variables. Its main characteristic is that it identifies those cells in which fire events are strongly clustered. Such cells are distributed throughout all climate regimes and may depend more on the continuity of the fuel bed, landuse and degree of control of the fire than on climatic conditions.

We also show that vegetation type is significant in determining the class of fire activity, especially in hot, moist regions, with a well developed dry season. In such conditions, most fires are man-made and are used for a variety of purposes. Numerous and often large fires are seen in grass and bush savanna, while infrequent and small fires are associated with forest slash and burn activity and smaller more frequent but controlled fires are associated with burning of crop residues and preparation of seed beds. Fire practices vary with vegetation type, but within the same vegetation type regional differences are also apparent. Part of this variation may be due to the restricted vegetation legend used, which masks real variation within a vegetation type.

We have shown that currently observed fire patterns can be partially explained in terms of climate factors and vegetation type. This opens up the exciting possibility of using such observations to help to predict fire regimes under changed climate conditions. Current modelling techniques using global and regional General Circulation Models (GCMs) explore how fire weather and hence fire potential will change (Stocks *et al.*, 1998). Here, the spatial extent of the data set has allowed us to examine and derive relationships for a wide range of climate–vegetation–fire conditions. An extension of the work could involve extraction of at least the equilibrium fire regime change vectors as a function of climate change. For example, in sub-Saharan northern Africa, there is a progressive change in fire pattern, from north to south, from class 1 to class 3 to class 5, as the climate becomes less arid and vegetation becomes more abundant. Future fire regimes may see a reduction in the extent of class 5 activity and an increase in class 3, if there is an extension of the Sahel region toward the south, with encroaching desertification, as has been predicted. One caveat, however, is that the fire–climate–vegetation relationships found in this work are based on ecosystems existing in a more-or-less steady state. Future fire regimes may diverge from these relationships if changes in vegetation type do not keep pace with a changing climate or if the human use of fire responds to changes in a more complex manner.

Given that this empirical approach to the investigation of the fire–climate–vegetation relationship can provide a complementary tool to the physically-based climate change prediction models, it is worth mentioning how the analysis could be improved to provide more quantitative results. A significant part of the ambiguity in the derived relationships is probably because of a number of shortcomings in the input datasets and the low spatial resolution of the grid over which the analysis was done. The analysis could be improved by increasing the number of years of fire observations to give a fire 'climatology'. Over the next few years, continuous and improved information on global fire distributions should become available with the launch of new satellite sensors, such as MODIS, and the development of appropriate algorithms for the calculation of burned area (Pereira *et al.*, this book). The fire classification used here could be improved by including information on the start and end of the fire season and further climate parameters related to the onset and end of significant dry periods could be incorporated in the climate parameterisation. Analysis at a finer spatial resolution may be appropriate, as the scale of $0.5°$ x $0.5°$ masks local variability in both fire pattern and vegetation distribution. Finally, given that most fire activity is either human-induced or controlled, modelling of future fire regimes should include a socio-economic dimension, otherwise there is a risk of incorrect predictions and the consequent adoption of costly and ineffective fire management policies.

5. ACKNOWLEDGEMENTS

E. Dwyer was funded by the TMR doctoral programme of the European Commission. Thanks to Paulo Barbosa and Simon Pinnock for their comments on the text and to Michel Verstraete for some fruitful discussions.

6.   REFERENCES

Anderson, J.R., Hardy, E.E., Roach, J.T. and Witmer, R.E., (1976) A land use and land cover classification system for use with remote sensor data. U.S. Geological Survey, Professional paper 964, 28p.

Andreae, M.O., Atlas, E., Cachier, H., Cofer III, W.R., Harris, G.W., Helas, G., Koppmann, R., Lacaux, J.-P. and Ward, D.E., (1996) Trace gas and aerosol emissions from savanna fires. In *Biomass Burning and Global Change* vol. 1, edited by J. Levine (MIT Press, Cambridge, Mass.), pp.278–295.

Antonovsky, M.Ya., Ter-Mikhaelian, M.T. and Furyaev,V.V., (1989) A spatial model of long-term forest fire dynamics and its applications to forests in western Siberia. WP–89–109, International Institute for Applied Systems Analysis, Laxenburg, Austria.

Barbosa, P.M., Stroppiana, D., Grégoire, J.-M., and Pereira, J.M.C., (1999) An assessment of vegetation fire in Africa (1981–1991): burned areas, burned biomass and atmospheric emissions. *Global Biogeochemical Cycles* (submitted).

Ceccato, P. and Flasse, S.P., (1998) Assessing vegetation fuel moisture from satellite NOAA–AVHRR data in the context of EXPRESSO experiment. Proceedings of the 24th annual conference and exhibition of the remote sensing society, The University of Greenwich, 9-11 September 1998 (Compiled by Burt, P.J.A., Power, C.H. and Zukowskyj, P.M.) pp 28–34.

Cofer III, W.R., Winstead, E.L., Stocks, B.J., Overbay, L.W., Goldammer, J.G., Cahoon, D.R. and Levine, J.S., (1996) Emissions from Boreal forest fires: are the atmospheric impacts underestimated? In *Biomass Burning and Global Change* vol. 2, edited by J. Levine (MIT Press, Cambridge, Mass.), pp.834–839.

Cramer, W.P. and Leemans, R., (1993) Assessing impacts of climate change on vegetation using climate classification systems. *Vegetation dynamics and global change* (eds. Solomon, A.M. & Shugart, H.H.) pp 190 – 210 (Chapman and Hall, New York).

Dwyer, E., Pinnock, S., Grégoire, J.-M. and Pereira, J.M.C., (1999a) Global spatial and temporal distribution of vegetation fire as determined from satellite observations, *International Journal of Remote Sensing* (in print).

Dwyer, E., Pereira, J.M.C., Grégoire, J.-M. and DaCamara, C.C., (1999b) Characterization of the spatio-temporal patterns of global fire activity. *Journal of Biogeography* (accepted).

Eva, H.D. and Lambin, E.F., (1998) Remote Sensing of Biomass burning in tropical regions: sampling issues and multisensor approach. *Remote Sensing of the Environment*, 64, pp. 292–315.

Goldammer, J.G., and Price, C., (1998) Potential impacts of climate change on fire regimes in the tropics based on MAGICC and a GISS GCM-derived lightning model. *Climatic Change*, 39, pp 273–296.

Koffi, B, Grégoire, J.-M., and Eva, H.D., (1996) Satellite monitoring of vegetation fires on a multiannual basis at continental scale in Africa. In *Biomass Burning and Global Change* vol. 1, edited by J. Levine (MIT Press, Cambridge, Mass.), pp.225–235.

Krusel, N., Packham, D. and Tapper, N.J. (1993) Wildfire activity in the malee shrubland of Victoria, Australia. *International Journal of Wildland Fire*, 3 (4). pp 217–227.

Loveland, T.R. and Belward, A.S. (1997) The IGBP–DIS global 1km land cover data set, DISCover: first results. *International Journal of Remote Sensing* 18 (15), pp 3289–3296 .

Piñol J, Terradas, J. and Lloret, F., (1998) Climate warming, wildfire hazard, and wildfire occurrence in coastal eastern Spain. *Climatic Change*, 38, pp 345–357.

Price, C. and Rind, D. (1994) The impact of a 2 x CO_2 climate on lightning-caused fires. *Journal of Climate*, 7, pp 1484–1494.

Leemans, R., and Cramer, W.P., (1990) The IIASA database for mean monthly values of temperature, precipitation and cloudiness of a global terrestrial grid. IIASA, Laxenburg, Austria. WP–90–41, 64p.

Pereira, J.M.C., Vasconcelos, M.J.P., and Sousa, A.M. (1999) A rule-based system for burnt area mapping in temperate and tropical regions using NOAA/AVHRR imagery (this publication).

Stocks, B.J., Cahoon, D.R., Cofer III, W.R., and Levine, J.S. 1996. Monitoring large-scale forest-fire behaviour in northeastern Siberia using NOAA–AVHRR satellite imagery. In: Biomass burning and global change, Vol. 2, edited by J.S. Levine. MIT Press, Cambridge, Mass. 802–807.

Stocks, B.J., Fosberg, M,A., Lynham, T.J., Mearns, L., Wotton, B.M., Yang, Q., Jin, J.-Z., Lawrence, K., Hartley, G.R., Mason, J.A. and McKenny, D.W., (1998) Climate Change and forest fire potential in Russian and Canadian boreal forests *Climatic Change*, 38, pp 1–13.

Stocks, B.J., (1992) The extent and impact of forest fires in northern circumpolar countries. In *Global Biomass Burning, Atmospheric, Climatic, and Biospheric Implications,* edited by J. Levine (MIT Press, Cambridge, Mass.), pp. 197–202.

Stroppiana, D., Pinnock, S. and Grégoire, J.-M., (1999) The Global Fire Product: daily fire occurrence, from April 1992 to December 1993, derived from NOAA–AVHRR data.. *International Journal of Remote Sensing* (in print).

Tapper, N.J., Garden,G., Gill, J., and Fernon, J., (1993) The climatology and meteorology of high fire danger in the northern territory. *Rangeland Journal*, 15 (2) pp 339–351.

Thornthwaite, C.W. (1948) An approach toward a rational classification of climate. *The Geographical Review*, 38, pp 55–94.

Modelling the Impact of Vegetation Fires, Detected from NOAA–AVHRR Data, on Tropospheric Chemistry in Tropical Africa

DANIELA STROPPIANA[1], PIETRO ALESSANDRO BRIVIO[2] and JEAN-MARIE GRÉGOIRE[1]

[1] *Global Vegetation Monitoring Unit, Space Applications Institute, JRC, Ispra (VA), Italy*
[2] *Remote Sensing Dept.-IRRS, National Research Council, Milan, Italy*

Abstract: Burned biomass, gas and aerosol emissions for the African continent were estimated using information on active fires whose spatial distribution was derived from NOAA–AVHRR HRPT (High Resolution Picture Transmission, 1.1 km resolution) and GAC (Global Area Coverage, 4.5 km resolution) imagery. AVHRR–GAC images covering a four year period (November 1984–October 1988) provided information on seasonality and inter-annual variability of fire activity while AVHRR–HRPT were used for quantitative assessment purposes for the period November 1992–October 1993. Above-ground biomass loading, burning efficiency and emission factors for each vegetation type were retrieved from the literature. We estimated that 1360.45 Tg of biomass burned in 1992–1993. The ecosystem mostly affected by vegetation fires was the Guinean Savanna located in the northern hemisphere of the African continent. Results derived using the multi-annual data set of AVHRR–GAC images showed a high inter-annual variability for the estimates produced.

1.   INTRODUCTION

Biospheric processes on the Earth's land masses have profound effects on the composition of the entire global atmosphere. In this context a key role is played by the tropical belt which accounts for about 40% of the planet's biomass and 60% of net primary production per year. Biomass burning is one of the factors which influences these processes and is mainly related to human landuse and land management (deforestation, agricultural residue burning, hunting, fire wood use, etc.). Within the tropical belt, the African continent, with large areas of savanna, contributes significantly to the total global amount of biomass burned. In Africa, there are two main periods of burning. In the northern hemisphere, the fire season generally runs from November to March, while in the southern hemisphere, it runs from May to October (Koffi *et al.*, 1996).

Several studies have developed different methodologies to assess the quantity of burned biomass since the problem was first approached by Seiler and Crutzen (1980). Estimates of burned area in the Tropics based on statistical surveys have been used to produce a quantitative assessment of the phenomenon for the different sources contributing to the total amount of biomass burned, and gases and aerosols emitted (Seiler and Crutzen, 1980; Hao *et al.*, 1990; Crutzen and Andreae, 1990; Andreae, 1991; Hao and Liu, 1994). Remote sensing techniques developed in the past decades have enabled a more detailed knowledge of the spatial and temporal distribution of fire occurring in the biomes of the world. Vegetation fire monitoring using satellite images has been performed with two different methodologies: active fires and burned-area detection. The first relies on measuring the thermal emission of the combustion process, hence enabling the mapping of fires occurring when the satellite passes overhead, while the second is based on fire-induced changes in the spectral characteristics of the land cover. Several algorithms have been developed using the available sensors in order to detect both active fires (Dwyer *et al.*, 1998; Stroppiana *et al.*, 1998; Koffi *et al.*, 1996; Eva and Flasse, 1996; Elvidge *et al.*, 1996; Prins and Menzel, 1992) and burned areas (Barbosa *et al.*, 1998c; Eva and Lambin, 1998; Eva *et al.*, 1998; Kasischke and French, 1995; Kasischke *et al.*, 1993) from regional to global levels.

In this work, spatial and temporal distributions of active fires for the African continent were derived from the AVHRR (Advanced Very High Resolution Radiometer) sensors on the NOAA (National Oceanic and Atmospheric Administration) satellite series. A quantitative assessment of burned biomass and gas and aerosol emissions was produced for the period November 1992–October 1993 on a monthly basis using a temporal extract of the GFP (Global Fire Product) daily fire maps (Dwyer *et al.*, 1998; Stroppiana *et al.*, 1998). This data set, which covers a 21 month period from April 1992 to December 1993 was derived from AVHRR–HRPT 1.1 km (High Resolution Picture Transmission) imagery, in the frame work of the IGBP–DIS (International Geosphere–Biosphere Programme Data and Information System) project.

Seasonal and inter-annual variability of the same parameters were analysed for the four year period November 1984–October 1988 using monthly active fire maps derived from AVHRR–GAC (Global Area Coverage) 4.5 km images (Koffi *et al.* 1996).

2. METHODOLOGY DEFINITION

Burned biomass, and gas and aerosol emissions for the African continent were estimated by applying the following equation pixel by pixel and summing up over the continent (Scholes *et al.* 1996):

$$M = A \times BD_i \times BE_i \times EF_i \tag{1}$$

M: quantity of emitted gas and aerosol [kg]
A: burned area [ha]
BD_i: biomass loading [t ha^{-1}]
BE_i: burning efficiency [g g^{-1}]
EF_i: emission factor [g kg^{-1}]
i: vegetation class

Burned area, the factor with the highest uncertainty, was estimated using active fire maps derived from both the GFP and the AVHRR–GAC data sets. Daily fire pixel positions for the period November 1992–October 1993 for the African continent were extracted from the GFP and synthesised into monthly maps with a spatial resolution of 4.5 km by 4.5 km in order to produce an input data set with the same spatial resolution as the one from AVHRR–GAC. Koffi *et al.* (1996) assumed negligible fire activity between March and May and therefore the AVHRR–GAC images for this period were not processed. This three month period will be subtracted from the results derived from the GFP when compared to those derived from GAC images.

AVHRR–GAC images are obtained onboard the satellite by applying a spatial sampling procedure to the full resolution LAC (Local Area Coverage) 1.1 km images. This spatial sampling leads to an underestimate of the fire activity and makes AVHRR–GAC images unreliable for any quantitative estimation purpose, such as number of fire events and burned area (Belward *et al.*, 1994). Nevertheless, as the data set covers a four year period it provides useful information on the seasonal and the inter-annual variability of burning activity. The GFP data set, although it is a temporal sample of the total daily fire activity, with a resolution of 1.1 km instead of 4.5 km, can provide more detailed information on the number of fire events and performs better for burned-area estimation purposes.

Burned-area estimates were derived from active fire maps by multiplying the number of detected fire pixels by the pixel area, assuming that a pixel is completely burned when it is classified as fire. An active fire is an indicator of the presence of a burned area, but the extent to which the area covered by the pixel is burned is unknown. Therefore this assumption leads to an overestimation of the burned area.

In order to estimate the actual extent of the area burned, information on fire size, at a sub-pixel scale, is necessary. Active fires are identified mainly by the enhancement of the signal in the middle infrared domain: most of the detection algorithms are based on thresholds applied to the AVHRR channel 3 (3.7 μm). In savanna ecosystems where, during the dry season, the background temperature is relatively high, the low saturation value of channel 3 prevents any attempt to estimate sub-pixel fire size. In fact, with high background temperatures, a fire occupying a small fraction of the pixel may saturate channel 3 (Matson and Dozier, 1981). Nevertheless, the overestimation may be negligible in the case of savanna fires where the area burned can be very large; it is likely that a detected fire burns the whole pixel. However, the error induced by the temporal sampling, caused by the single daily pass of the satellite, is still present in the final estimates of burned area. This error might be balanced by the overestimation produced by assuming a pixel completely burned. Eva and Lambin (1998) addressed in detail the issue of the spatial and temporal sampling associated with the detection of burned areas and active fires for a study area in the Central African Republic.

As the parameters in equation 1 other than burned area vary spatially and with vegetation type, the African continent was stratified into the major ecosystems (Figure 1). These were defined using a functional vegetation cover map derived from AVHRR–GAC images (Janodet 1995) in order to assign each class the corresponding value of biomass loading (BD), burning efficiency (BE) and emission factors (EF) retrieved from the available literature and field experiment measurements. This vegetation map is not very detailed in the southern part of the continent. However, the number of active fires detected in this area is very low and so the contribution to the total amount of burned area, gas and aerosol emissions at a continental level is small. In addition, a more detailed map would require an equally detailed data set of biomass loading values.

Biomass loading for Dense Forest was derived from a spatial distributed density map (Brown and Gaston, 1996) while a constant value of 2 t ha^{-1} was assigned to Desert and Semi-Desert ecosystems and to the Sudanian–Sahelian Savanna. A minimum–maximum range was defined using the highest and the lowest values found for Guinean (Oak Ridge National Laboratory–Distributed Active Archive Center), Sudanian (Menaut *et al.*, 1991) and South Africa (Oak Ridge National Laboratory–Distributed Active Archive Center) Savannas in order to represent the inter-annual variability of the quantity of biomass through the years. Dry and Humid Miombos are characterized by a constant value of 5 t ha^{-1} (Shea *et al.*, 1996) as it was not possible to use a range due to the lack of published data. For the Somalian Bush Thickets class a constant value of 4 t ha^{-1} was used (Table 1).

Figure 1. Functional vegetation cover map derived from AVHRR-GAC images used in this work to define the main ecosystems in the African continent: (1) North Africa Desert, (2) North Africa Semi-Desert, (3) Sudanian-Sahelian Savanna, (4) Sudanian Savanna, (5) Guinean Savanna, (6) Dense Forest, (7) Somalian Bush-Thickets, (8) Humid Miombo, (9) Dry Miombo, (10) South Africa Bush Savanna, and (11) South Africa Deserts and Semi-Deserts.

Biomass loadings used in this work were kept constant over time, even though the vegetation is characterized by a dynamic which influences the quantity of biomass during the year. As fires are mainly concentrated in the dry season of each ecosystem when the biomass loading has reached its maximum value after the rainy season this kind of approximation may be reasonable.

Burning efficiency is the fraction of biomass that actually burns relative to the quantity before the fire occurs and its value is mainly related to the water content of the vegetation. Early fires burn less biomass than fires that occur later in the season when the vegetation is dry: burning efficiency can vary in the range 0.2 to 1. Here, a mean value of 0.80 was assigned to all the savanna vegetation types in accordance with the literature (Delmas *et al.*, 1991); for the Dense Forest ecosystem a value of 0.25 (Seiler and Crutzen, 1980) was chosen because of the high relative humidity, the short dry season and the presence of a large size biomass with a woody component that hardly burns during a fire.

Table 1. Biomass density for each vegetation class retrieved from the literature and field measurements

Vegetation classes	Biomass Density (min-max) [t ha^{-1}]
Dense Forest	Spatially distributed [1]
Guinean Savanna	4-8 [2]
Humid and Dry Miombo	5 [3]
Sudanian Savanna	2-4 [4]
South Africa Bush Savanna	1-4 [2]
Somalian Bush Thickets	4 [5]
Sudanian-Sahelian Savanna	2 [5]
Desert and Semi-Desert	2 [5]

[1] Brown and Gaston, 1996
[2] ORNL-DACC
[3] Shea et al. 1996
[4] Menaut et al. 1991
[5] Best guess

Emission factor is defined as the quantity of emitted gas or aerosol per unit of burned biomass. The proportion and the quantity of emitted gas depend on the stage of the combustion process and on the type of fire: flaming fires mainly produce CO_2, H_2O and NO_x, while smouldering fires are related to the emission of secondary combustion products such as CO (Andreae, 1996).

Mean emission factors for CO, CH_4, NO_x and aerosols were used for forest and savanna ecosystems (Scholes *et al.*, 1996), while a constant value of 1640 g kg^{-1} (Andreae, 1996) was used for emitted CO_2 in all ecosystems (Table 2). For the gaseous emissions the quantity of each compound, expressed as Tg of emitted gas, was converted in Teragrams of emitted carbon (TgC) for CO_2, CO and CH_4 and Teragrams of emitted nitrogen (TgN) for NO_x by using the corresponding factors. Nitrogen compounds, here referred to as NO_x, are the sum of NO and NO_2; for the conversion from $TgNO_x$ to TgN it was assumed that NO almost exclusively predominates (Lobert *et al.*, 1991).

Table 2. Emission factors for main gases and aerosols with a particle diameter less than 2.5 μm (Scholes et al. 1996, Andreae 1996)

Gases and aerosols	Emission factors [g kg⁻¹] for the forest ecosystem	Emission factors [g kg⁻¹] for the savanna ecosystems
CO_2	1640	1640
CO	230	68
CH_4	9.47	2.05
NO_x	16.1	4.75
Aerosols (PM < 2.5 μm)	12.3	5.45

3.   BURNED AREA ASSESSMENT

Using the information on active fires detected in AVHRR–HRPT imagery (GFP data set), we estimated that 304.39 million ha burned in the African continent between November 1992 and October 1993, of which 51% burned in the four month period November–February. This result agrees with other studies that have shown that the savanna ecosystems (Guinean and Sudanian) in the northern hemisphere contribute most to vegetation burning in Africa (Barbosa *et al.*, 1998a; Hao and Liu, 1994; Menaut *et al.*, 1991).

Fire activity decreases between March and May, with less than 38 million hectares burned in three months (13% of the annual total), but increases again from June with a monthly contribution ranging between 5 and 9% (Table 3). During the period March–May, the monthly quantity of burned area is considerably lower than the previous months but not negligible as assumed in the AVHRR–GAC data set.

Table 3. Monthly burned area [10^6 ha] and percent contribution in 1992-93 estimated using the Global Fire Product derived from AVHRR-HRPT 1.1 km

	Burned Area [10^6 ha]	Monthly contribution [%]
Nov-92	23.68	8
Dec-92	57.52	19
Jan-93	48.73	16
Feb-93	23.30	8
Mar-93	14.07	5
Apr-93	11.10	4
May-93	12.77	4
Jun-93	19.43	6
Jul-93	26.80	9
Aug-93	25.08	8
Sep-93	25.70	8
Oct-93	16.22	5
Total	304.39	100

The four year data set shows how annual burned area estimates vary from a minimum of 15.28 10^6 ha in 1984–1985 to a maximum of 43.51 10^6 ha in

                                                      Daniela Stroppiana et al.

1985–1986, which is almost three times greater (Table 4). In the same Table, the total amount of area burned in 1992–1993 is calculated for the nine month period November–February and June–October. All amounts and percentages for 1992–1993 that we refer to when comparing results derived from the two data sets are calculated excluding the March–May estimates.

Table 4. Burned area estimates derived from active fire maps.
All the amounts are expressed in million of burned hectares [10^6 ha]

	AVHRR-GAC					GFP
Burned	**1984-85**	**1985-86**	**1986-87**	**1987-88**	**Mean**	**1992-93**
Area	15.28	43.51	36.37	16.48	27.91	266.45

Looking at the contribution of individual months to the annual burning activity for the 1984–1988 period (Figure 2), a high monthly variability between one year and another is evident in addition to the variability in total annual amounts. Burned area estimates depict a strong fire activity at the beginning of the season (November–February) when vegetation fires are mainly located in the northern hemisphere. In the period 1985–1988, more than 65% of the area totally burned is concentrated in those four months. This percentage decreases to 16% in 1984–1985, but we believe that these values do not reflect what actually happened and that they are more likely related to a lack of images or poor data rather than to a sudden change in fire activity.

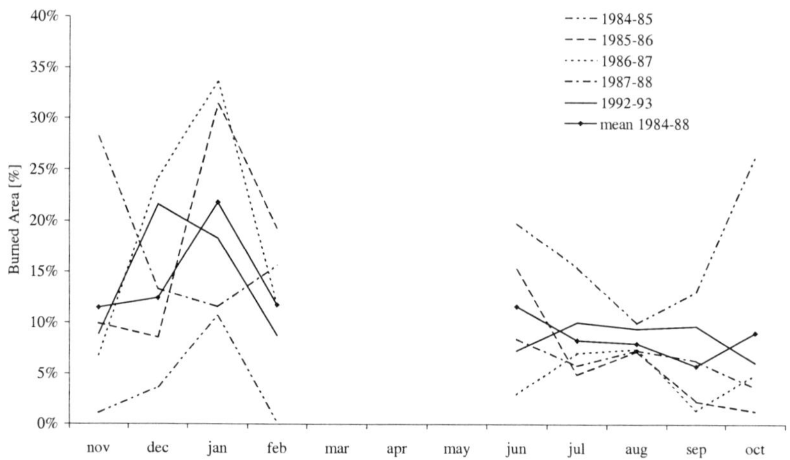

Figure 2. Monthly contributions to the total amount of area annually burned estimated from active fire maps derived from either AVHRR-GAC (1984-88) and AVHRR-HRPT (1992-93) data sets. All the values are expressed in per cent.

In 1985–1986 and 1986–1987, most of the area burned in January (31% and 34%) while in 1987–1988 28% of the total burned in November. Between 1985 and 1988, July shows a narrow range of variability (5–7%) while August shows a constant value of 7% which increases to 9% in GFP-derived estimates. In 1992–1993, the contribution of the northern hemisphere burning season (November–February) is 58% while that of the southern hemisphere burning season (June–October) is 42% (Figure 2). Mean values from the 1984–1988 data set are 57% and 43%, respectively. The two data sets show quite good overall agreement.

4. ESTIMATES OF BURNED BIOMASS AND EMISSIONS

Monthly burned biomass, and gas and aerosol emissions were assessed using both data sets and by applying Equation 1. Annual mean estimates for November 1992–October 1993 based on the GFP data set amount to 1360.45 Tg of burned biomass, 608 TgC as CO_2, 60 TgC as CO, 3.7 TgC as CH_4, 4.5 TgN as NO_x and 9.36 Tg of aerosols (PM <2.5μm). The availability of a range of variability for biomass density enabled assignment of a minimum–maximum range combined with the above average values (Table 5).

Table 5. Annual estimates of burned biomass, gas and aerosol emissions as derived using AVHRR-HRPT imagery between November 1992 and October 1993. The minimum and maximum values are derived from a variability range defined for biomass density

	Annual estimates		
	Min	**Max**	**Mean**
Burned Biomass [Tg]	1140.04	1580.88	1360.45
Aerosols [Tg]	8.15	10.58	9.36
CO_2 [TgC]	509.44	706.44	607.93
CO [TgC]	53.4	66.25	59.81
CH_4 [TgC]	3.34	4	3.68
NO_x [TgN]	4	4.94	4.47

In 1992–1993, 52% of the total amount of burned biomass was concentrated between November and February while 34% occurred from June to October. Almost 180 Tg of burned biomass was calculated for the months of March, April and May (13%). Based on these amounts of burned biomass, monthly quantities of emitted gases and aerosols were assessed. Figure 3 represents monthly estimates of emitted CO_2; its spatial distribution is shown in Figure 4, where monthly maps are grouped in three month periods. The main characteristics of the spatial and temporal distribution of fire activity are evident.

                                                                                    Daniela Stroppiana et al.

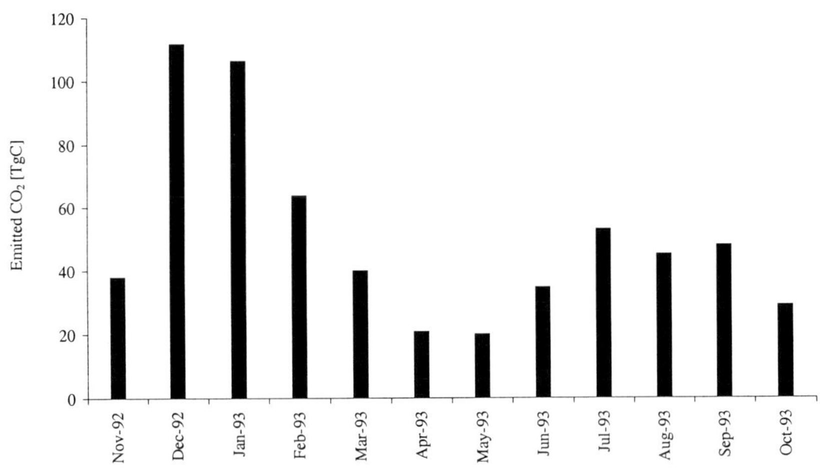

Figure 3. Monthly estimates of emitted CO_2 for the period November 1992-October 1993 derived using the AVHRR-HRPT data set. All the amounts are expressed in Teragrams of carbon released ($1TgC = 10^{12}$ gC).

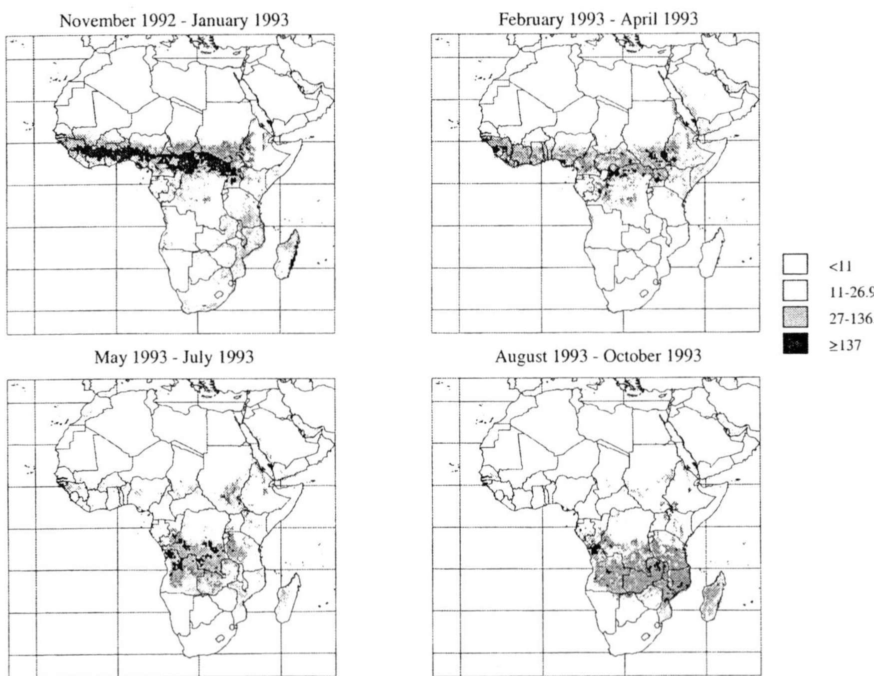

Figure 4. Spatial distributions of the monthly quantity of CO_2 emitted between November 1992 and October 1993 grouped in three month periods. The amounts are expressed in Gigagrams of carbon released ($1GgC = 10^9$ gC) and each grid cell has a resolution of $0.4°$ by $0.4°$.

Active fire maps from AVHRR–GAC images showed the annual variability of burned biomass. Annual estimates vary from a minimum value of 62.24 Tg in 1984–1985 to a maximum, more than four times higher, of 275.07 Tg in 1985–1986 (Table 6). Burning activity may also vary annually in terms of spatial distribution (Figure 5). Gas and aerosol emissions, being calculated from burned biomass by applying constant factors, follow the same pattern of variability. A constant distribution of burned biomass within a year is not definable: every year is characterized by a specific monthly distribution (Table 7). Several studies have attempted to identify a seasonal distribution to burning activity (Hao *et al.*, 1990; Hao and Liu, 1994) but these results indicate that this approach may be an over simplification of the process. The very high inter-annual variability of the monthly distribution is related to the regional climate and short-term weather conditions. Climate determines the availability of fire fuel, fire frequency and fire season. Short-term weather conditions (temperature, precipitation, and wind) influence the occurrence of vegetation fires as well as their behaviour (Lobert and Warnatz, 1992).

Table 6. Quantitative estimate of burned biomass, gas and aerosol emissions as derived from AVHRR-GAC imagery

	AVHRR-GAC				
	1984-85	**1985-86**	**1986-87**	**1987-88**	**Mean**
Burned biomass [Tg]	62.24	275.07	234.30	63.92	158.89
Aerosols [Tg]	0.36	1.66	1.42	0.37	0.95
CO_2 [TgC]	27.70	122.84	103.47	28.42	70.62
CO [TgC]	2.04	9.6	8.26	2.1	5.5
CH_4 [TgC]	0.11	0.54	0.47	0.11	0.31
NO_x [TgN]	0.15	0.71	0.61	0.15	0.41

Table 7. Monthly percentage distribution of burned biomass. Values for the period 1992-93 have been recalculated excluding March-May estimates

	AVHRR-GAC					AVHRR-HRPT
Month	**1984-85**	**1985-86**	**1986-87**	**1987-88**	**Mean**	**1992-93**
Nov	1	6	6	23	9	7
Dec	4	6	17	15	10	21
Jan	12	51	48	14	31	20
Feb	0	17	15	18	13	12
Mar-May		Not processed				
Jun	20	10	2	9	10	7
Jul	15	3	4	6	7	10
Aug	9	4	4	7	6	9
Sep	12	2	1	6	5	9
Oct	26	1	2	3	8	5

Koffi *et al.* (1995) analysed bush fire dynamics in central Africa, derived from AVHRR–GAC imagery for the period 1984–1989, in relation to pluviometric patterns. They found that fire activity, burning season length and fire calendar are related to the annual amount of rainfall, the length of the dry season and the pluviometric calendar, respectively. However, the monthly pluviometric patterns do not explain the inter-annual spatio-temporal variability of fire activity and fire calendar, suggesting that non-climatic factors may be involved. Dwyer *et al.* (this book), using the GFP data set, analysed the relation between vegetation fires and climate at a global level. Their conclusions confirm that in addition to climatic parameters, information on vegetation type and landuse is necessary to explain the characteristics of vegetation fires in each ecosystem.

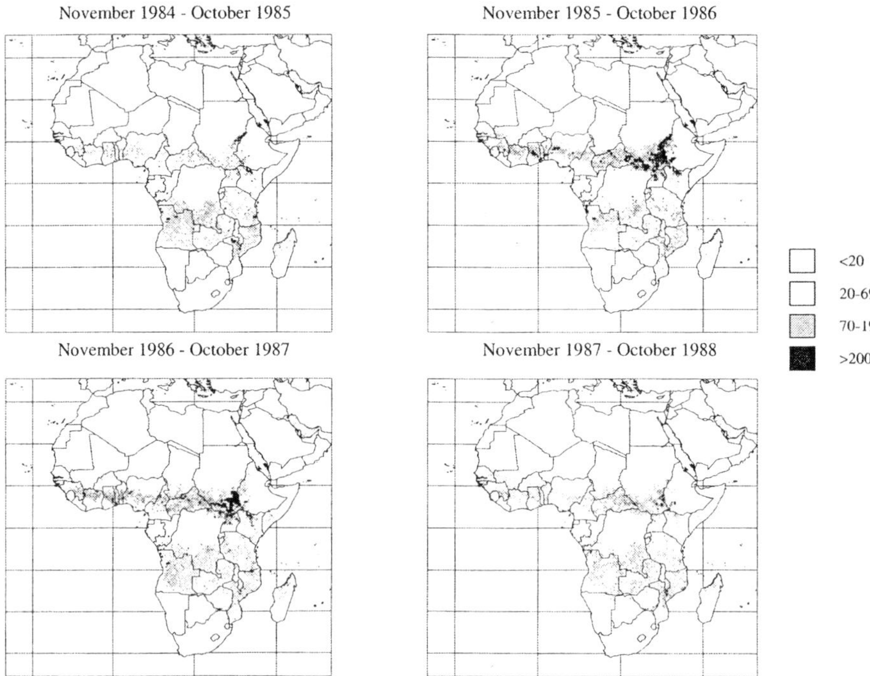

Figure 5. Spatial distribution of annually burned biomass between November 1984 and October 1988. The amounts are expressed in Gigagrams (1Gg = 10^9 g) of biomass burned and each grid cell has a resolution of 0.4° by 0.4°.

The monthly percentage distribution of burned biomass differs from burned area mainly because of the location of vegetation fires. Biomass loading plays a key role in this difference: ecosystems with higher vegetation density could have a stronger effect than others with a comparable extent of burned area. For instance, in August, a stable contribution to the total burned area of 7% in the period 1985–1988 (Figure 2) becomes 4% in 1985–1987 and 7% in 1987–1988 in terms of burned biomass (Table 7) because of a change in fire location within the vegetation cover classes. The spatial distribution of fire activity, which at a continental level could be provided only by satellites, is still an important element in biomass burning effects assessment.

5. BURNED AREA AND BURNED BIOMASS BY VEGETATION TYPES

Koffi *et al.* (1996) identified the seasonality of burning activity in the African continent using active fire maps derived from AVHRR–GAC imagery. Assuming the onset of the burning season to be in November, when the dry period begins in the northern hemisphere, a season is considered to run from November to October of the following year. In view of the spatial distribution of vegetation fires and the ecosystems affected by them, three main periods can be identified (see also Figure 4): November–February, March–May and June–October. From November to February, vegetation fires are located in the northern hemisphere mainly in the Guinean and Sudanian Savannas and move during that period from the northern and drier savannas towards the south. In March–May, fire activity decreases while in June–October vegetation fires are located in the southern hemisphere (Humid and Dry Miombo, South Africa Bush Savanna) where they move eastward (Somalian Bush Thickets). November, which has a strong fire activity in the northern hemisphere, could be considered a transition period because of the fires still affecting the south-east part of Africa.

Most of the biomass annually burned is concentrated in the first period when fire activity is mainly located in the Guinean and Sudanian Savanna ecosystems. Between November 1992 and October 1993 1360.45 Tg of biomass burned in the African continent of which 37% (510.16 Tg) was in the Guinean Savanna and 21% (289.28 Tg) was in Dense Forest; these two classes contributed 58% to the total quantity (Table 8).

A comparison between burned area extent and burned biomass (Table 8) by vegetation type confirms how biomass loading influences the relative contribution of each biome. For instance, in the Dense Forest ecosystem, a relatively low contribution to the total area burned in 1992–1993 (5%) is transformed into 21% of burned biomass. The opposite happens in the case of Sudanian–Sahelian Savanna, where low values of biomass loading reduce the contribution in terms of burned biomass.

Monthly contributions for the main ecosystems have been grouped in the following two classes: Northern savannas (Guinean, Sudanian and Sudanian–Sahelian Savannas) and southern savannas (Dry and Humid Miombo, South Africa Bush Savanna) (Table 9). The predominance of vegetation fires in the northern hemisphere is confirmed again by these results; the highest contributions come from December 1992 (86%) and

 Daniela Stroppiana et al.

January 1993 (80%). In those months the Guinean Savanna contributes 74% and 73% respectively (184.10 Tg and 174.10 Tg). In March 1993, a low contribution of savanna ecosystems from both hemispheres increases the forest percentage up to 50%. In February and March 1993, 57.25 Tg and 44.92 Tg of biomass respectively burned in the Dense Forest ecosystem. This corresponds to 2.66 and 2.40 millions ha of burned forest. The high percentages might be explained by a real increase of fire activity in the forest domain but also by a lower number of fire events in the other biomes, which enhances the forest contribution. In any case, fire activity in March is still considerable with a total amount of burned biomass close to 90 Tg that decreases to 45.9 Tg in April and 44.19 Tg in May. Between June and October 1993 the southern savanna contributions range between 67% (June 1993) and 74% (August 1993) with a lower monthly variation than in the northern hemisphere. Monthly quantitative assessments of burned biomass in the main ecosystems are shown in Figure 6.

Table 8. Total amount of burned area and biomass in each biome in 1992-93 together with the percentage showing each contribution to the total. Deserts and Semi-Deserts have been excluded because of their negligible contribution

Vegetation classes	Burned area		Burned biomass	
	[10^6 ha]	%	[Tg]	%
Dense Forest	11.55	5	289.28	21
Guinean Savanna	106.28	26	510.16	37
Dry Miombo	40.41	15	161.62	12
Humid Miombo	47.79	18	191.18	14
Sudanian Savanna	44.76	15	107.42	8
South Africa Bush Savanna	12.08	4	24.17	2
Somalian Bush Thickets	7.55	3	22.28	2
Sudanian-Sahelian Savanna	11.02	5	17.64	1

Table 9. Percentage of monthly burned biomass in each biome. The last two rows show grouped values for the main vegetation classes of the Northern and Southern hemisphere

	N	D	J	F	M	A	M	J	J	A	S	O
Dense Forest	14	8	16	40	50	29	20	19	24	18	21	18
Guinean Savanna	30	74	73	48	34	37	17	1	0	0	0	1
Dry Miombo	1	0	0	0	1	2	24	55	50	33	11	3
Humid Miombo	19	2	0	0	0	0	4	11	18	37	55	63
Sudanian Savanna	29	12	6	5	5	14	19	7	1	0	0	7
South Africa Bush Savanna	3	1	1	0	0	0	1	1	1	5	7	4
Somalian Bush Thickets	2	1	1	2	4	6	3	1	1	2	2	1
Sudanian-Sahelian Savanna	1	1	0	1	1	3	5	2	2	3	1	1
Northern Savannas	*61*	*86*	*80*	*54*	*41*	*55*	*40*	*10*	*3*	*3*	*1*	*9*
Southern Savannas	*22*	*3*	*1*	*1*	*1*	*3*	*29*	*67*	*70*	*74*	*73*	*70*

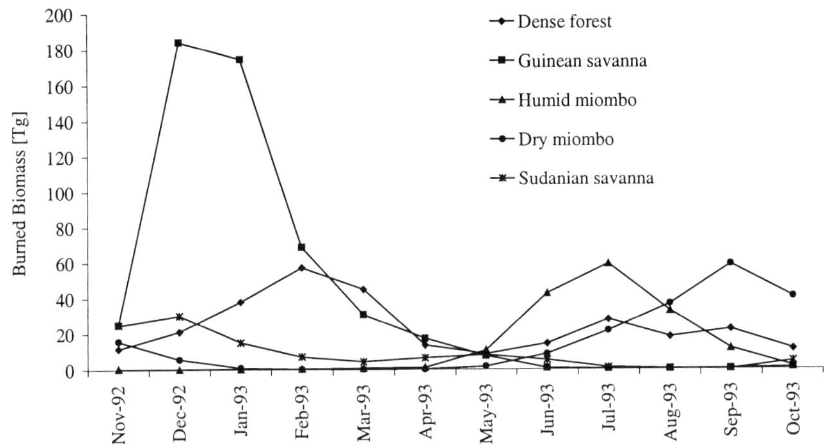

Figure 6. Monthly quantity of biomass burned between November 1992 and October 1993, as derived using the AVHRR-HRPT data set, for those ecosystems that mainly contribute to the total amount. All the amounts are in Teragrams ($1Gg = 10^{12}$ g).

Estimates derived from the AVHRR–GAC data set define how each ecosystem assumes a different weight in the total amounts through the years (1984–1988). The annual percent contribution of each biome to the total amount of biomass burned during the period November 1984 and October 1988 showed a high annual variability (Table 10). However, for most of the vegetation ecosystems the mean value calculated over the four year period agrees well with estimates derived from the GFP data set (Table 10). In the Dense Forest ecosystem a mean value of 7% in 1984–1988 becomes 19% in 1992–1993: AVHRR–HRPT images with a higher resolution can detect small fires in the forest domain and this could justify a contribution of the forest class in 1992–1993 which is almost three times higher than in the period 1984–1988. In AVHRR–GAC images small fires cannot be detected, thus limiting the utility of this data set for fire monitoring in forest ecosystems (Belward *et al.*, 1994).

Monthly percent contributions of the main ecosystems from the northern and southern hemispheres depict the main patterns of vegetation burning seasonality between 1985 and 1988 (Figure 7 and Figure 8). In the northern hemisphere, fire activity is concentrated between November and March even though the peak changes its position within those months. In the Sudanian Savanna, activity is still detectable in June 1985 and 1986 while a one month advancement in the onset of the burning season is evident in 1987 and 1988. For the southern hemisphere, the burning activity runs from June to October with no residual burning in the other months. The peak's position is more variable between 1985 and 1988. With the exception of 1984–1985, November is when fire activity is detected in that part of the Dry Miombo located in the south.

Table 10. Burned biomass by vegetation type expressed as percent contribution to the total annual amount

[%]	1984-85	1985-86	1986-87	1987-88	Mean	1992-93
Dense Forest	5	8	9	5	7	19
Guinean Savanna	16	59	51	37	41	38
Dry Miombo	29	5	8	22	16	16
Humid Miombo	26	6	6	12	13	13
Sudanian Savanna	12	15	13	14	13	7
South Africa Bush Savanna	3	1	2	6	3	2
Somalian Bush Thickets Savanna	6	5	10	3	6	1
Sudanian-Sahelian Savanna	3	1	2	1	2	1

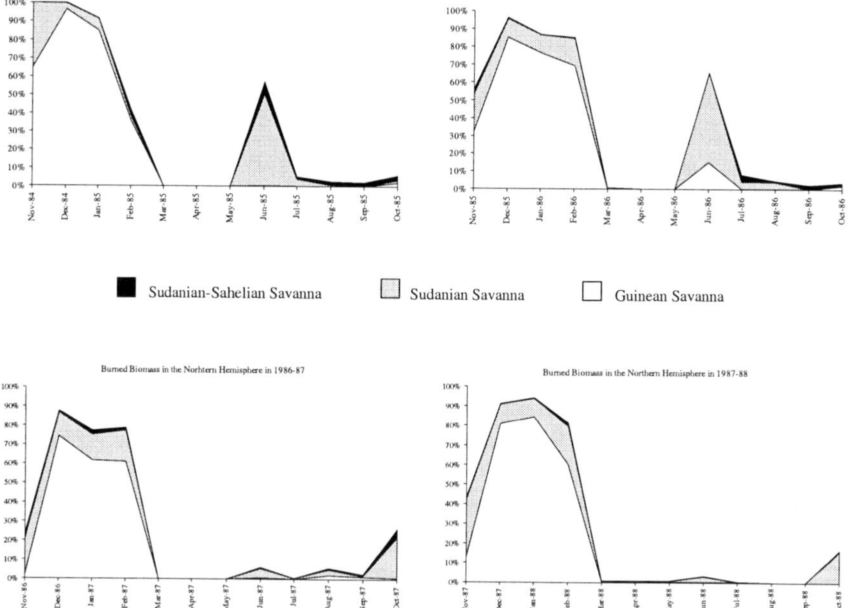

Figure 7. Monthly percentages of burned biomass in the main vegetation classes of the Northern hemisphere. Each graph represents one of the four burning seasons covered by the multi-annual data set of AVHRR-GAC images between November 1984 and October 1988.

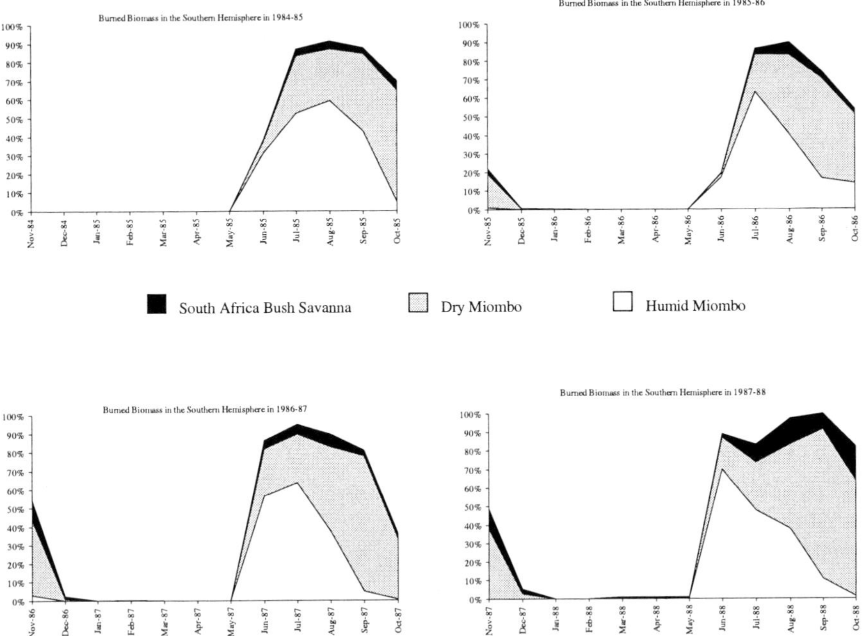

Figure 8. Monthly percentages of burned biomass in the main vegetation classes of the Southern hemisphere. Each graph represents one of the four burning seasons covered by the multi-annual data set of AVHRR-GAC images between November 1984 and October 1988.

6.   CONCLUSIONS AND PERSPECTIVES

In this work, a quantitative assessment of monthly burned biomass and gas and aerosol emissions from vegetation fires for the African continent was derived from active fire maps provided by the GFP data set for the period November 1992–October 1993. Seasonal and inter-annual variations for the same parameters were analysed using a four year data set (1984–1988) of active fires derived from AVHRR–GAC images.

Using the GFP data set we estimated that 304.39 million ha in the African continent burned between November 1992 and October 1993, leading to a loss of biomass of 1360.45 Tg and an emission of 608 TgC as CO_2.

Hao and Liu (1994) estimated burned biomass in tropical Africa using statistical data based on a FAO survey on the changes of landuse in tropical countries (1975–1980) to be 2320 Tg per year. This estimate was obtained by summing the contributions from each source (deforestation, savanna fires and agricultural residues burning) considered by the authors. A higher value might be expected because of the reduction in burned area estimates because of the use of satellite images. The only work that used remotely sensed information on burning activity was developed by Scholes *et al.* (1996). They calibrated active fire maps to assess the quantity of burned area and estimated that 177 Tg of biomass burned from April to October 1989 in the Southern Hemisphere of the African continent. Our corresponding value of burned biomass for the period April–October 1993 is 557 Tg, three times higher than the estimates produced by Scholes *et al.*. We derived from their published data the estimate of burned area to be 168.4 million ha; a comparison with the value obtained in this work (137 million hectares) showed that the main discrepancy occurs between biomass loading and burning efficiency data sets.

Seasonal and inter-annual variability of burned area and burned biomass, which at a continental level can be provided only by remotely sensed data, is an important input to the models studying atmosphere dynamics and changes. Monthly maps of the spatial distribution of burned area and burned biomass could present the required information at the right spatial resolution for those models rather than total annual estimates.

As a complementary result, the AVHRR–GAC active fire maps data set showed that the seasonal distribution of burning activity is unstable and characterized by inter-annual variability. The peak month of burning activity is not definable *a priori*, nor is the contribution to the total amounts. These vary from a minimum of 12% in January 1985 to a maximum of 51% in January 1986. Nevertheless, for the period 1985–1988, it is always between November and February, confirming that savanna ecosystems from the northern hemisphere contribute most to the total burning activity. This kind of dynamic and multi-annual information should improve knowledge regarding the relationship between biomass burning and other phenomenon acting at a continental or global scale such as ENSO (El Niño – Southern Oscillation) events.

Comparing results derived using AVHRR–GAC images to those from AVHRR–HRPT imagery, this study confirms that AVHRR–GAC images are unsuitable for the quantitative assessment of burned area if estimated from active fires. The two data sets are complementary because they provide information regarding different aspects of the phenomenon of burning activity.

The analysis by vegetation type reveals the quantitative contribution to the burning activity together with the seasonal behaviour of each biome. We estimated burned-area extent from active fires but for burned area, burned biomass and gas and aerosol emissions estimation purposes an algorithm capable of directly detecting burned areas would perform better. A burned-area detection algorithm has been developed for AVHRR–GAC images

(Barbosa *et al.*, 1998b, 1998c). Research studies are developing methods, based on remotely sensed data, to estimate the dynamic variation of biomass loading (Barbosa *et al.*, 1998a) and the assessment of the vegetation water content (Ceccato and Flasse, 1998), the latter being the main factor influencing burning efficiency. Emission factors are mainly related to the stage of the combustion process and a more detailed definition of each stage (smouldering and flaming) might improve gas emission estimates.

In the near future the improved thermal imaging capabilities of new sensors, such as the Moderate Resolution Imaging Spectrometer (MODIS) of the NASA Earth Observing System programme and the Advanced Along Track Scanning Radiometer (AATSR) of the ESA ENVISAT mission, will offer the opportunity to estimate the rate of emission of radiative energy from fire as well as the fraction of area that it covers.

7.   ACKNOWLEDGEMENTS

This work contributed to the degree thesis of D.S. submitted to "Politecnico di Milano". We acknowledge Sandra Brown for having kindly provided some data regarding vegetation.The authors would also like to thank Ned Dwyer for his comments on the manuscript.

8.   REFERENCES

Andreae, M. O., 1996 Emissions of trace gases and aerosols from southern African savanna fires, in *FIRE in Southern African Savannas. Ecological and Atmospheric Perspectives*, edited by B.W. van Wilgen *et al.*, pp. 161–183, Witwatersrand University Press, Johannesburg.

Andreae, M. O., 1991 Biomass burning: Its history, use, and distribution and its impact on environmental quality and global climate, in *Global Biomass Burning, atmospheric, climatic and biospheric implications,* edited by J.S. Levine, pp. 3–21, MIT Press, Massachusetts.

Barbosa, P.M., D. Stroppiana, J.-M. Grégoire, and J.M.C. Pereira, 1998a An assessment of vegetation fire in Africa (1981–1991): burned areas, burned biomass and atmospheric emissions, *Global Biogeochemical cycles*, (submitted).

Barbosa, P.M., J.M.C. Pereira, and J.-M. Grégoire, 1998b. Compositing Criteria For Burned Area Assessment Using Multitemporal Low Resolution Satellite Data, *Remote Sensing of Environment*, 65, 38–49.

Barbosa, P.M., J.-M. Grégoire, and J.M.C. Pereira, 1998c. An Algorithm for Extracting Burned Areas from Time Series of AVHRR GAC Data Applied at a Continental Scale, *Remote Sensing of Environment* , (in press).

Belward, A.S., P.J. Kennedy, and J.-M. Grégoire, 1994. The limitations and potential of AVHRR–GAC data for continental scale fire studies, *International Journal of Remote Sensing*, 15(11), pp. 2215–2234.

 Daniela Stroppiana et al.

Brown, S., and G. Gaston, 1996. Estimates of biomass density for tropical forests, in *Biomass burning and global change. Remote sensing, modeling and inventory development, and biomass burning in Africa,* edited by J.S. Levine, Vol. 1, pp. 133–139.

Ceccato, P., and S. Flasse, 1998. Assessing vegetation fuel moisture content from satellite NOAA–AVHRR data in the context of EXPRESSO experiment, Proceedings of the 24th Conference of the Remote Sensing Society, Chatham, U.K..

Crutzen, P.J., and M.O. Andreae, 1990. Biomass burning in the tropics: impact on atmospheric chemistry and biogeochemical cycles, *Science*, 250, pp. 1669–1678.

Delmas, R.A., P. Loudjani, A. Podaire, and J-C. Menaut, 1991. Biomass burning in Africa: an assessment of annually burned biomass, in *Global Biomass Burning, atmospheric, climatic and biospheric implications,* edited by J.S. Levine, pp.126–132, MIT Press, Massachusetts.

Dwyer, E., S. Pinnock, J.-M. Grégoire, and J.M.C. Pereira, 1999. Global vegetation fire distribution as determined from satellite observations, *International Journal of Remote Sensing* (in press).

Dwyer, E., Grégoire, J.-M. and Malingreau, J.-P. 1998. A globala analysis of vegetation fires using satellite images: spatial and temporal dynamics. Ambio 27 (3), 175–181.

Dwyer, E., J.-M. Grégoire, and J.M.C. Pereira, Climate and vegetation as driving factors in global fire activity, *this book.*

Elvidge, C.D., H.W. Kroehl, E.A. Kihn, K.E. Baugh, E.R. Davis, and W.M. Hao, 1996. Algorithm for retrieval of fire pixels from DMSP operational linescan system data, in *Biomass burning and global change. Remote sensing, modeling and inventory development, and biomass burning in Africa,* edited by J.S. Levine, Vol. 1, pp. 73–85.

Eva, H.D., and S. Flasse, 1996. Contextual and Multiple-threshold algorithms for Regional active fire detection with AVHRR Data, *Remote Sensing Reviews*, 14, pp. 333–351.

Eva, H.D., and E.F. Lambin, 1999. Burnt area mapping in Central Africa using ATSR data, *International Journal of Remote Sensing*, 19, pp. 3473–3497.

Eva, H.D., and E.F. Lambin, 1998. Remote Sensing of Biomass Burning in Tropical Regions: Sampling Issues and Multisensor Approach, *Remote Sensing of Environment*, 64, pp. 292–315.

Eva, H.D., J.P. Malingreau, J.-M. Grégoire, A.S. Belward, and C.T. Mutlow, 1998. The advance of burnt areas in Central Africa as detected by ERS–1 ATSR–1, *International Journal of Remote Sensing*, 19, pp. 1635–1637.

Flasse, S., and P. Ceccato, 1996. A contextual algorithm for AVHRR fire detection, *International Journal of Remote Sensing*, 17, pp. 419–424.

Hao, W.M., and M.-H. Liu, 1994. Spatial and temporal distribution of tropical biomass burning. *Global Biogeochemical Cycles*, 8(4), pp. 495–503.

Hao, W.M., M.-H. Liu, and P.J. Crutzen, 1990. Estimates of annual and regional releases of CO_2 and other trace gases to the atmosphere from fires in the tropics, based on the FAO statistics for the period 1975–1980, in *Fire in the tropical biota. Ecosystem processes and global challenges*, edited by J.G. Goldammer, pp. 440–462, Springer Verlag, Berlin.

Janodet E., 1995. Stratification saisonnière de l'Afrique continentale et Cartographie fonctionnelle des écosystemes forestiers africains. TREES project–MTV–IRSA–JRC Ispra.

Kasischke, E.S., N.H.F. French, P. Harrell, N.L. Christensen, S.L. Ustin, and D. Barry, 1993. Monitoring of wildfires in boreal forests using large area AVHRR NDVI composite image data, *Remote Sensing of Environment*, 45, pp. 61–71.

Kasischke, E.S., and N.H.F. French, 1995. Locating and estimating the areal extent of wildland fires in Alaskan boreal forests using multiple-season AVHRR NDVI composite data, *Remote Sensing of Environment* , 51, pp. 263–265.

Koffi, B., J.-M. Grégoire, and H.D. Eva, 1996. Satellite Monitoring of Vegetation Fires on a Multiannual Basis at Continental Scale in Africa, in *Biomass burning and global change. Remote sensing, modeling and inventory development, and biomass burning in Africa,* edited by J.S. Levine, Vol. 1, pp. 225–235.

Koffi, B., J.-M. Grégoire, G. Mahé, and J.-P. Lacaux, 1995. Remote sensing of bush fire dynamics in Central Africa from 1984 to 1989: Analysis in relation to regional vegetation and pluviometric patterns, *Atmospheric Research,* 39, pp. 179–200.

Lobert, J.M., D.H. Scharffe, W.M. Hao, T.A. Kuhlbusch, R. Seuwen, P. Warneck, P.J. Crutzen, 1991. Experimental evaluation of Biomass Burning Emissions: Nitrogen and Carbon Compounds, in *Global Biomass Burning, atmospheric, climatic and biospheric implications,* edited by J.S. Levine , pp.289–304, MIT Press, Massachusetts.

Lobert, J.M., and J. Warnatz, 1992. Emissions from the combustion process in vegetation, in *Fire in the Environment: the ecological, climatic importance of vegetation fires,* edited by P.J. Crutzen and J.G. Goldammer, pp. 15–37, John Wiley & Sons.

Matson, M., and J. Dozier, 1981. Identification of Subresolution High Temperature Sources Using a Thermal IR Sensor, in *Photogrammetric Engineering and Remote Sensing,* 47, 9, pp.1311–1318.

Menaut, J.C., Abbadie, L., Lavenu, F., Loudjani, P., A. Podaire, 1991. Biomass burning in west African savannas, in *Global Biomass Burning, atmospheric, climatic and biospheric implications,* edited by J.S. Levine, pp. 132–142, MIT Press, Massachusetts.

Oak Ridge National Laboratory–Distributed Active Archive Center, 1998 *Net Primary Production database,* http://www-eosdis.ornl.gov/npp/npp_home.html.

Prins, E.M., and W.P. Menzel, 1992. Geostationary satellite detection of biomass burning in South America, *International Journal of Remote Sensing,* 13, pp. 2783–2799.

Scholes, R.J., J.D. Kendall, and C.O. Justice, 1996. The quantity of biomass burned in southern Africa, *Journal of Geophysical Research,* 101(D19), pp. 23667–23676.

Seiler, W., and P.J. Crutzen, 1980. Estimates of gross and net fluxes of carbon between the biosphere and the atmosphere from biomass burning, *Climatic Change,* 2, pp. 207–247.

Shea, R.W., B.W. Shea, J. B. Kauffman, D.E. Ward, I.C. Haskins, and M.C. Scholes, 1996. Fuel biomass and combustion factors associated with fires in savanna ecosystems of South Africa and Zambia, *Journal of Geophysical Research,* 101(D19), 23551–23568.

Stroppiana, D., S. Pinnock, and J.-M. Grégoire, 1998. The Global Fire Product: daily fire occurrence, from April 1992 to December 1993, derived from NOAA–AVHRR data, *International Journal of Remote Sensing* (in press).

A Rule-Based System for Burned Area Mapping in Temperate and Tropical Regions Using NOAA/AVHRR Imagery

JOSÉ M.C. PEREIRA[1,2], MARIA J.P. VASCONCELOS[2,3], and ADÉLIA M. SOUSA[1]

[1]*Laboratory for Remote Sensing and Geographical Analysis, Department of Forestry, Instituto Superior de Agronomia, Lisbon, Portugal*
[2]*Centre for Forest Studies, Department of Forestry, Instituto Superior de Agronomia, Lisbon, Portugal*
[3]*National Centre for Geographical Information (CNIG), Lisbon, Portugal*

Abstract: The feasibility of deriving a single classifier capable of mapping burned areas in Iberia and central Africa, using NOAA/AVHRR satellite imagery was investigated. A supervised classification approach based on the Classification and Regression Trees (CART) algorithm was used to classify a single date image from Africa and a multi-temporal composite from Iberia into three classes: burned, unburned, and cloud. A classification tree with 22 terminal nodes constructed was constructed with CART, using albedo, GEMI, and channel 4 brightness temperature as independent variables. The accuracy of this classifier was assessed on a set of independent data and found to be higher than 98% for each class. All burned area pixels in the test data set were correctly classified. Visual comparison with high resolution data in the case of Iberia, and with active fire data in the case of Africa, confirm this good performance. Advantageous characteristics of rule induction approaches for the classification of satellite imagery are discussed from the perspective of deriving a system for global burned area mapping.

1.   INTRODUCTION

The issue of biomass burning as a global phenomenon, with potentially significant impacts on the biogeochemical cycles of major elements, and particularly on the atmospheric concentration of CO_2, was raised by Seiler and Crutzen (1980). Based on data available at the time, they estimated that biomass burning contributed an annual gross CO_2 flux of 2000–4000 teragrams (Tg) of carbon to the atmosphere. Research conducted during the 1980s confirmed that biomass burning is a significant source of atmospheric CO_2, and more recent estimates suggest that an average 8680 Tg of biomass dry matter are burned each year, resulting in the production of about 3500 Tg of carbon per year in the form of CO_2. If these estimates are correct, then biomass burning is responsible for approximately 40% of the world's annual production of CO_2 (Levine, 1991).

The estimation of burned biomass, and consequent atmospheric emissions, is usually performed with a formula of the type (Seiler and Crutzen, 1980; Hao *et al.*, 1990):

$$M = A \times B \times \alpha \times \beta \tag{1}$$

where
M = amount of dry biomass burned (t.ha^{-1})
A = area burned (ha)
B = biomass density (t.ha^{-1})
α = fraction of above-ground biomass (dimensionless)
β = combustion efficiency (dimensionless)

For a given biome or region, the area burned, A, is probably the most variable of all factors in the above equation, and it is therefore essential to estimate it as accurately as possible. In fact, Levine (1996) considers that the greatest single challenge to the scientific community studying biomass burning is the accurate assessment of the spatial and temporal distribution of burning, over periods of time ranging from a few weeks to a year.

Some recent studies have demonstrated the feasibility of burned area estimation with remotely sensed imagery at spatial scales ranging from regional (Kasischke, 1993, 1995; Cahoon, 1994; Eva and Lambin, 1998; Pereira, 1998) to continental (Barbosa *et al.*, 1998), and over time periods of a few days to multiple years. However, the only global overview of fire activity over an extended period of time was based on the analysis of active fires, i.e. the remote sensing of the thermal signal (Dwyer *et al.*, 1998a,b). This approach has drawbacks because the short persistence and strong diurnal cycle of the thermal signal leads to temporal sampling problems, which tend to underestimate the areas actually burned (Pereira, 1990; Pereira *et al.*, 1998). Given the importance of the problem of burned area mapping at a global scale, we decided to analyse the adequacy of a modern statistical supervised classification methodology. For this purpose we developed a classification tree to segment NOAA/AVHRR satellite imagery into unburned surfaces, burned surfaces, and clouds. Data from two different biomes were pooled together to test the feasibility of developing a unified classifier capable of handling data from very distinct ecological zones and, ultimately, from the whole Earth.

2.   METHODS

2.1   Study areas

In order to assess the feasibility of deriving a single classifier capable of mapping burned surfaces under a broad range of environmental conditions, two study areas were selected, one in Western Europe, and another in Central Africa. The first study area is the Iberian Peninsula, i.e. Portugal and Spain, and has a total area of 590,000 km². The area is remarkably diverse from climatic and biotic standpoints, ranging from very hot and dry areas in the southeastern part of Spain, almost completely devoid of vegetation, to much cooler and more humid areas, with temperate forest and diverse agricultural systems in the northwestern part of the Peninsula, to high mountain forest and areas of permanent snow in the Pyrenean Mountains of northeastern Spain. Throughout most of this area the climate can be described as predominantly Mediterranean, warm to hot and dry during the summer, and cool and wet in the winter time. The fire season is concentrated in the summer months, with the vast majority of vegetation fires occurring between June and September.

The second study site is located in the border region between the Central African Republic and the Sudan, covering an area of approximately 300,000 km². The dominant vegetation types are Sudanian savanna to the North, Guinean savanna in the central part, and dense tropical forest in the south of the area. The area is sparsely populated, and very large areas are burned usually between the months of November and January, progressing from the northeast to the southwest of the region, driven by hot and dry Harmattan winds.

The spatial pattern of fire occurrence is drastically different between the two study sites. During a typical Iberian summer season, fires burn a very small percentage of the vegetated area (less than 1%), and display a patchy pattern, due to the abundance of natural and man-made fuelbreaks. On the contrary, in the Central African site, almost unbroken flame fronts often 10 to more than 100km wide (as seen in 1km satellite imagery) burn freely through vast extensions of dry savanna grasslands, creating some of the largest seasonal fire scars observable on Earth. Post-fire dynamics, and particularly the persistence of the fire scar signal, also differ between the two areas. Iberian vegetation fires primarily affect temperate forests and Mediterranean type shrublands. These fires burn large quantities of biomass, and produce an abundant charcoal residue, a good part of which is made up of relatively large fuel particles. The resulting fire scars are quite persistent, and are often still detectable four or five years after the fire. In areas where dry farming of cereals is practised, the burning of stubble causes less persistent fire scars. In Central Africa the dominant fuel type is herbaceous, and the charcoal and ash residue from this plant material has small dimensions and is easily blown away. Additionally, regeneration of the grasses on the burned surfaces is very rapid, reducing the persistence of the fire scars to periods that range between two weeks and two or three months (Barbosa *et al.*, 1998) .This diversity in the spatial and temporal pattern of the burned area signal was considered a desirable feature to test the performance of the proposed methodology for burned area mapping.

2.2   Data set

Our analysis was based on satellite imagery from the NOAA Advanced Very High Resolution Radiometer (AVHRR), in High Resolution Picture Transmission (HRPT) format, with 1.1km pixel size. For the Central Africa study area we used a single NOAA–14 image, from the early afternoon overpass on November 29, 1996. The Iberian image is a multi-temporal composite produced using the early afternoon NOAA–11 overpasses of September 1–10, 1991. All images were navigated using an orbital model and then geometrically corrected with ground control points to an accuracy of about one pixel. The visible channels were radiometrically calibrated to top-of-the-atmosphere reflectance, taking into account the drift of the calibration coefficients since the day of launch (Teillet, 1994; Rao and Chen, 1996). The thermal channels were calibrated to brightness temperature, corrected for the non-linear response of the sensor (Kidwell, 1997).

Multi-temporal compositing of the Iberian dataset was performed using a new formula, which combines channel 4 temperature and the normalized difference vegetation index (NDVI). It has been shown that the conventional NDVI maximum-value compositing procedure tends to degrade the burned vegetation signal, and that minimum channel 2 (Cahoon *et al.*, 1994), minimum albedo or maximum channel 4 temperature (Barbosa *et al.*, 1998) are better alternatives. However, none of these alternatives deals adequately with the problem of cloud shadows, which is the reason why we proposed a new two-channel compositing criterion (Sousa, 1998):

$$\text{Max} \left[T4^{1.5} + \left(10 \, |NDVI| \right)^{3} \right] \qquad (2)$$

The rationale for this provisional formula, the parameters of which appear to be satisfactory but are certainly not optimal, is as follows: exponentiation of channel 4 temperature (T4) with an exponent that is positive and larger than one tends to increase the separation between recent burns, which are typically hot, and most other land cover types. There is still confusion between, for example, dense dark vegetation and cloud shadows. As we also aim at preserving the vegetation signal over that of shadows, we then use the NDVI to raise the value of pixels containing dense dark vegetation over that of shadowed pixels, which typically have an NDVI value close to zero. The exponents and multiplicative constant in the formula are chosen to produce a stratification of values such that burned pixels are preferred over vegetated pixels, and these are selected over cloudy or cloud-shadowed pixels. We are currently experimenting with two-step compositing criteria, where the second criterion is applied only to a subset of the better pixels identified by the first criterion (e.g. selecting the hottest pixel of the three pixels with the lowest albedo, in a 10–day set of images).

Assessment of classifier performance was performed visually, using different auxiliary data at the two study areas. In Iberia, the burn scars detected by the rule-based AVHRR image classifier were overlaid with fire scar perimeter obtained from visual interpretation of 30m resolution Landsat 5 TM satellite imagery, available only for Portugal. High resolution satellite data were unavailable for the Central African site, and so we extracted the active fire pixels present in the image using the IGBP–DIS contextual algorithm for active fire detection (Flasse and Cecatto, 1996). These were overlaid on the classified image to check for spatial consistency between the position of the active fire fronts and the edges of burned surfaces.

2.3 Induction of rules with classification and regression trees (CART)

Most studies of remote sensing of fire, including those dealing with the thermal signal of active fires and those based on the surface albedo and vegetation signal of burned surfaces, have relied on multiple threshold approaches. These thresholds may refer to ranges of values of spectral channels or indices in single-date images (Kaufman *et al.*, 1990; Kennedy *et al.*, 1994; Flasse and Cecatto, 1996; Kasischke and French, 1993; Kasischke *et al.*, 1995), or to the magnitude of changes in multi-temporal change detection and time series analyses (Barbosa *et al.*, 1998; Eva and Lambin, 1998). Multiple threshold methods are, essentially, rule-based systems since thresholds specify conditions for acceptance or rejection of a pixel as containing a fire-related signal. Definition of a threshold is equivalent to specifying an IF–THEN rule, whose antecedent is a range of values of one or more spectral indices or channels, and whose consequent is a YES–NO decision on whether the pixel analysed displays a fire-induced signal.

This approach facilitates the incorporation of knowledge concerning the physical properties of the targets, and is relatively easy to implement. However, the current implementation of this type of classifier also has some significant disadvantages, namely:

- Structure identification for the classifier system, i.e. selection of the variables to include in the system and the definition of the number and nature of the rules are essentially deductive and subjective, dependent on each expert's knowledge and interpretation of the problem.
- System parameter identification, i.e. definition of threshold values for each variable in each rule is based on visual analysis of very large numbers of images, a process which is very time-consuming, tedious, error-prone and also subjective.

There is not an objective way to assess the completeness of the rule set, which can only be improved by trial and error.

In order to overcome these limitations, we decided to use a formal statistical methodology for inducing classification rules from data, and selected one that is widely used and considered to have a rigorous mathematical foundation. Rules were extracted from a decision tree automatically induced from training data using the classification and regression trees (CART) algorithm (Breiman *et al.*, 1984). CART fits models by binary recursive partitioning, successively splitting the data set into increasingly homogeneous subsets. Results are presented in the form of an inverted tree that begins with a root node and generates descendent nodes through series of YES–NO questions. Some nodes are terminal, meaning that a final classification is reached, while other nodes need additional splitting until a terminal node is reached. Each split separates parent node into exactly two child nodes, and is usually based on a question relating to a single variable. The key methodological questions in CART are (Breiman *et al.*, 1984):

- how to split each node;
- how to decide when the tree is complete; and
- how to assign each terminal node to a class outcome.

Node splitting is based on YES–NO questions of the kind:

$$\text{Is Temperature} <= 300\text{K} \ ?$$

CART performs an exhaustive search of all possible splits for all variables included in the analysis (a brute force search). Then, splitting rules are rank ordered based on some figure of merit criterion for the split. One of the most commonly used node impurity criteria is the Gini diversity index, which attempts to minimise the diversity (i.e. maximise the purity) in each node. It is especially appropriate for problems with a small number of classes, and was chosen for this study.

The decision concerning tree completeness is not solved in CART through trying to stop splitting at the right set of terminal nodes. Instead, node splitting and consequent tree growing is continued until all terminal nodes are very small, yielding a very large tree. This maximal tree is then selectively pruned upward, generating a series of sequentially smaller sub-trees. Independent test samples or cross-validation techniques are used to pick out the tree with the lowest estimated classification error rate. A complexity parameter is linearly combined with the error measure, in order to penalise large trees, such that the tree pruning strategy is guided by a cost-complexity measure. As a result, the minimum misclassification cost tree is not necessarily the best one, and the optimal tree is usually smaller, to a degree determined by the value of the complexity parameter (Breiman *et al.*, 1984).

Terminal node classification is a relatively simple task. If the node is completely pure, it is obviously allocated to the only class it contains. Otherwise, a majority rule is followed and the node is allocated to its most abundant class. Sometimes this rule is modified to allow for different classification error costs for the various classes, or to compensate for over- or under-representation of classes.

The main outputs of a CART analysis are:
- the tree-structured rule set;
- an estimate of the importance of the various variables used to grow the tree;
- a cost-complexity assessment of the various trees generated in the analysis process; and
- matrices for misclassification rates and predictive accuracy, based on learning and test samples.

The classification tree structure is easily translated into a series of rules, one per terminal node. These are IF–THEN rules whose antecedent conditions are a conjunction of sets of ranges in the predictor variables, and the consequent is a class of the response variable. For example:

IF Temperature > 310°K AND Albedo < 0.08 AND
VegIndex <= 0.00 THEN Class = Burned

A series of rules of this kind is applied to the data set, thus segmenting the images into the desired classes. The estimates of variable importance take into account potential masking effects, so that a variable not used in the tree may have a relatively high importance rating. This indicates that the predictor variable is not unrelated to the response variable, but that this relationship is hidden by the effect of another variable with better predictive ability.

2.4   Rule-based image classification

In order to classify the images previously described, we extracted training and testing data from the two study areas through the conventional procedure of on-screen digitising over AVHRR colour composites containing channel 4 in the red colour gun, channel 2 in the green, and channel 1 in the blue. The data extracted from the two study areas were pooled together in order to assess CART's ability to induce a classification tree capable of generating rules applicable to both sites. We considered three classes, namely unburned surfaces, burned surfaces, and clouds. Clouds were included as a separate class in order to distinguish between the situation where the surface is observed and it is not burned and the situation where it is impossible to observe the surface due to cloud cover.

As predictor variables we selected channel 4 temperature, albedo (defined as the arithmetic mean of the TOA reflectances of channels 1 and 2; Saunders, 1990), and the GEMI vegetation index (Pinty and Verstraete, 1992). This choice was dictated by our concern with physical interpretability of the classification rules and also by the results of exploratory analysis of variable importance. There were of course alternative choices, such as calculating a split-window surface temperature, using a more sophisticated method to determine albedo, or selecting another vegetation index. However, we wanted to incorporate in the classification system spectral information clearly related to the physical conditions that prevail in recently burned areas. The most important of these are a charred, dark surface, partially or completely devoid of green vegetation, and hot as a result of high absorption of solar radiation as a consequence of the lowered albedo and evapotranspiration (Whelan, 1995).

We have previously demonstrated the superior capability of a modified GEMI, defined in the near- infrared – mid-infrared bi-spectral space, to detect burns in comparison with the most commonly used NDVI (Pereira, 1998; Pereira *et al.*, 1998). This index, GEMI3, uses channel 3 reflectance data instead of channel 1 data. However, with the data sets used in this study we had problems in applying GEMI3 due to channel 3 saturation, which prevents the calculation of channel 3 reflectance. Also, the fact that clouds were included as a desired class, and were not previously screened, reduces the attractiveness of channel 3 reflectance data for our purposes. The great advantage of these data for fire detection in cloud-screened AVHRR images is that recent burns are almost systematically the most reflective surfaces in the landscape, but this is not the case when clouds are present. For these reasons, and also because GEMI has consistently shown to outperform the NDVI (Pereira, 1998; Martín, 1998; Sousa, 1998), and other vegetation indices (Sousa, 1998) such as the Soil Adjusted Vegetation Index (SAVI) (Huete, 1988) and the modified SAVI (Qi *et al.*, 1994), it was chosen as the preferred vegetation index for this study.

3.   RESULTS AND DISCUSSION

The classification tree was constructed with image data extracted from the AVHRR scenes according to the procedure described above. The numbers of observations (pixels) used from each site and for each class are given in Table 1. Few burned area pixels were taken from the Iberian Peninsula, because the extent of area burned in any season, in this region, is much smaller than that burned in Central Africa. On the contrary, most cloud pixels are from the Iberian composite, because the Central Africa image had only small, scattered clouds in the southeastern part. Out of a total of 12028 pixels, 75% (9037) were used to train the classifier and 25% (2991) were left out for testing purposes. The proportions of each class in the total dataset were kept equal in the learning and testing subsets.

Table 1. Number of pixels used to develop and test the classification tree, by study site and class.

	Unburned surface	Burned surface	Clouds
Central Africa	4656	2459	736
Iberia	3848	89	236

The process of tree construction generates a maximally sized tree, which is successively pruned to yield smaller sub-trees, according to the minimum cost-complexity criterion. Figure 1 shows the relationship between the number of terminal nodes in each tree produced in the analysis, and its cost-complexity score, relative to that of a single-node "tree", which has unit cost-complexity. The decrease in cost-complexity is initially very fast but quickly stabilises, fluctuating slightly. The maximal tree in our CART analysis had 38 nodes and a test set relative cost-complexity score of 0.0092. The minimum cost tree had 24 nodes and a cost-complexity of 0.0078. The optimal tree, i.e. the one presenting the best trade-off between misclassification rate and tree size was identified as a tree with 22 terminal nodes, and a cost-complexity score of 0.0089.

The structure of this optimal tree is presented in Figure 2, where root and intermediate nodes are represented as blue hexagons, and terminal nodes as red squares. The class assignment, splitting variable and splitting threshold, and number of cases contained, are given for all root and intermediate nodes.

For terminal nodes, class assignment and number of cases are reported. Of the 22 paths down the tree (i.e., rules), 10 lead to terminal nodes classifying for unburned surfaces (class 0), eight classify clouds (class 2), and four classify burned surfaces (class 1). This illustrates CART's ability to handle non-homogeneous data by inducing various rules for each class. If, for example, tropical savanna fires and temperate forest fires have distinct spectral responses in the measurement space, they will be classified by various rules. These can invoke different predictor variables and use different thresholds rather than having to rely on a global set of coefficients applied to all instances of the same class.

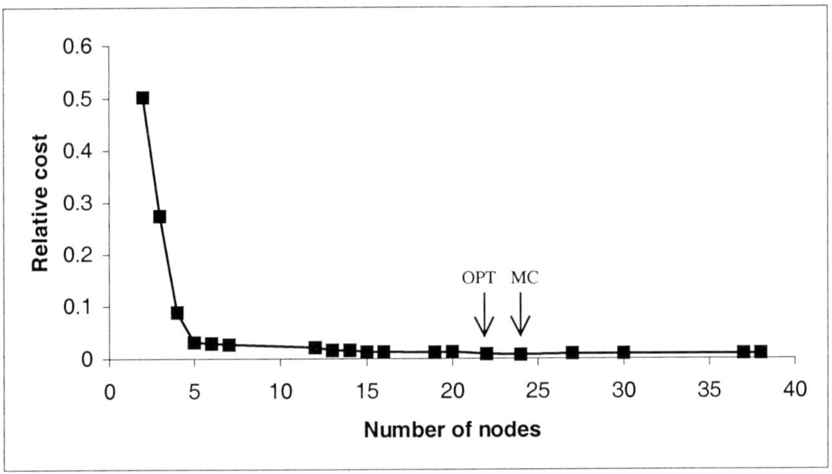

Figure 1. Cost-complexity of the trees generated in the analysis. OPT and MC indicate, respectively, the optimal tree and the minimum cost tree.

Some of the terminal nodes contain a very small number of cases (down to the absolute minimum of one, for nodes 15 and 18), with six of the 22 terminal nodes containing fewer than 10 cases. It is feasible to eliminate these rules or better to merge them into neighbouring rules, thus significantly reducing tree complexity, at a very small cost in terms of classification error. However, we kept them in order to illustrate a CART feature that is considered important for outlier detection.

Figure 3 shows the relative importance of the variables used to construct the classifier. The most important variable is always assigned a score of 100, and the other variables are rated relative to the best one. The ratings, and even the ranking are contingent on the specific classification tree they refer to, and should not be considered as absolute. Although all variables are considered very important, thus supporting their initial selection, it is interesting that the vegetation index is considered the least important of all variables when it is known that the loss of greenness is the strongest fire-induced spectral signal (Pereira *et al.*, 1998). This is because, simultaneously with burned area mapping, we are also detecting clouds, and obviously albedo and temperature are crucial variables to discriminate between land surface and clouds. The major distinction appears to take place between these two types of pixels, and the discrimination between burned and unburned land surfaces is performed at a lower hierarchical level. This interpretation is corroborated by the splitting variable for the tree root node being channel 4 temperature (identified in Figure 2 as C4) at a brightness temperature of 305.95°K, uncorrected for atmospheric effects.

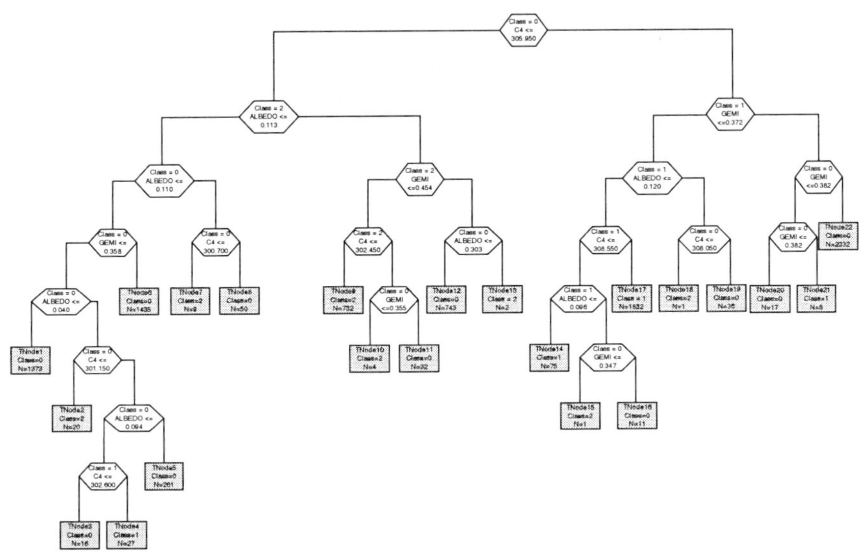

Figure 2. Optimal tree generated by the CART analysis, using channel 4 temperature, albedo and GEMI. The tree has 22 terminal nodes / rules.

Optimal tree misclassification statistics and test sample prediction success are reported in Tables 2 and 3, respectively. Table 2 shows that there is little difference in misclassification between learning and test samples, with slightly higher values for the test samples, as expected. Nevertheless, the classifier performs extremely well, mis-classifying a total of only 53 pixels out of 9037 in the learning sample, and a total of 30 pixels out of 2991 in the test sample. Even better, no burned surface pixels were mis-classified in either sample.

Table 3 is a confusion matrix and reveals the magnitude and types of confusions made by the classifier. The unburned class has a slightly higher error than the other two classes, possibly due to it being spectrally much more heterogeneous than either clouds or recent burns. Unburned surfaces range from ocean water to densely vegetated surfaces to bare soils, while the other two classes are much more spectrally homogeneous. The remarkable feature of both tables is the excellent performance of the classification tree. However, one must keep in mind that the data used for validation, although independent from those used for model fitting, were sufficiently prototypical of their respective classes to have been selected as potential training candidates, and this is likely to inflate accuracy assessment figures (Stehman and Czaplewski, 1998).

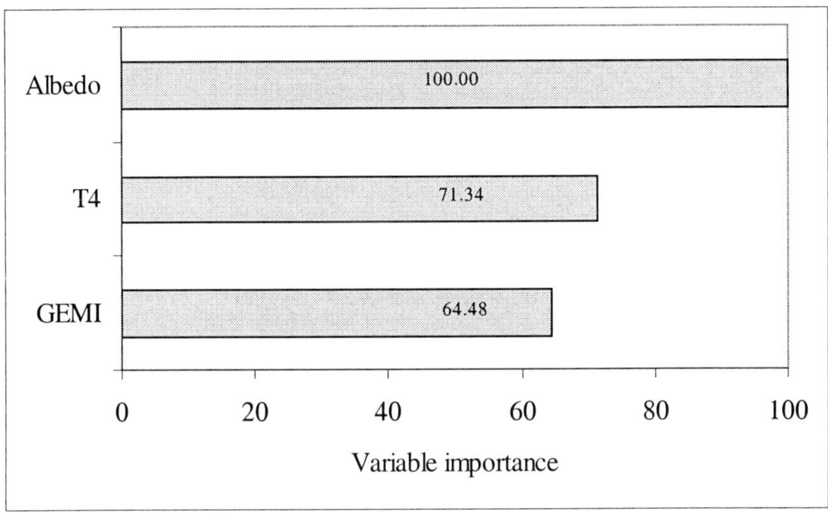

Figure 3. Importance of the variables used to derive the classification tree. The scores are relative, and contingent on the specific tree to which they refer, in this case the optimal tree.

Table 2. Optimal tree misclassification statistics

	Learning sample			Test sample		
Class	# cases	# mis-classified	% error	# cases	# mis-classified	% error
unburned	6379	53	0.83	2125	29	1.36
burned	1920	0	0.00	630	0	0.00
cloud	738	0	0.00	236	1	0.42
TOTAL	9037	53		2991	30	

Table 3. Optimal tree test sample prediction success table (%)

	Predicted class			
Actual class	unburned	burned	cloud	TOTAL
unburned	98.63	0.85	0.52	100.00
burned	0.00	100.00	0.00	100.00
cloud	0.42	0.00	99.58	100.00

Each path down the optimal tree was converted into a rule, and the set of 22 rules was applied to the Central African single date image and to the Iberian multi-temporal composite. Figures 4 and 5 are GEMI greyscale images of the Central African study area and of Iberia, respectively. As the Iberian fire scars are quite small, at the scale of the whole region, Figure 5 includes a zoom-in on an area in central Portugal showing a more detailed view of the burns with an overlay of fire perimeters obtained from Landsat 5 TM imagery. In the Sudan/Central African Republic region, the recently burned surface is very visible in dark grey and black, and active fires detected with the Flasse–Cecatto algorithm are shown in white.

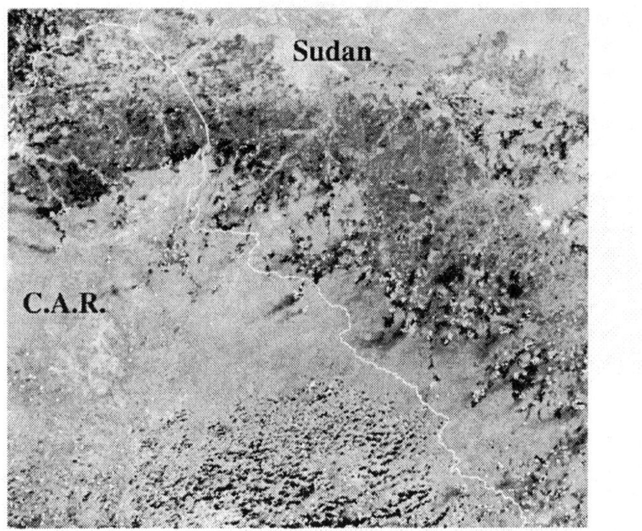

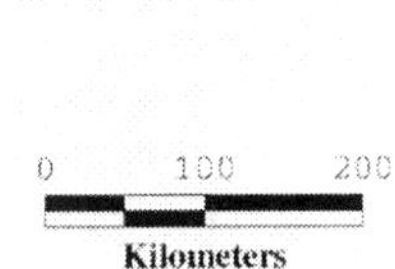

Figure 4. GEMI greyscale image of the Central African study area. Recently burned areas are dark grey and black, forming a crescent-shaped region extending from the north-west to the southeast corner of the image. The lighter-coloured areas at the northeast corner are bare or sparsely vegetated soils, and the lighter grey region located southwest of the fire scars is unburned vegetation. Scattered clouds (black) can be seen at the bottom of the scene. Active fires are represented as white polygons.

Figures 6 and 7 show the classified scenes. In Figure 6 the active fires detected on the Central African scene were overlaid on the classification, and display a spatially coherent position, at the edge of the burned area. Many active fires were also detected by the Flasse–Cecatto algorithm in the southern part of the image, interspersed with a broken cloud cover. While some of these can be confirmed visually on the computer screen as active fires, others are believed to be false alarms generated by cloud edge effects. A comparable type of visual assessment of the classification results is

                                                          José M.C. Pereira et al.

impossible to do for Iberia, at the same scale, because of the small size of the fire scars. Therefore, we show the same window as in Figure 5, now with the high resolution fire perimeters overlaid on the classification results.

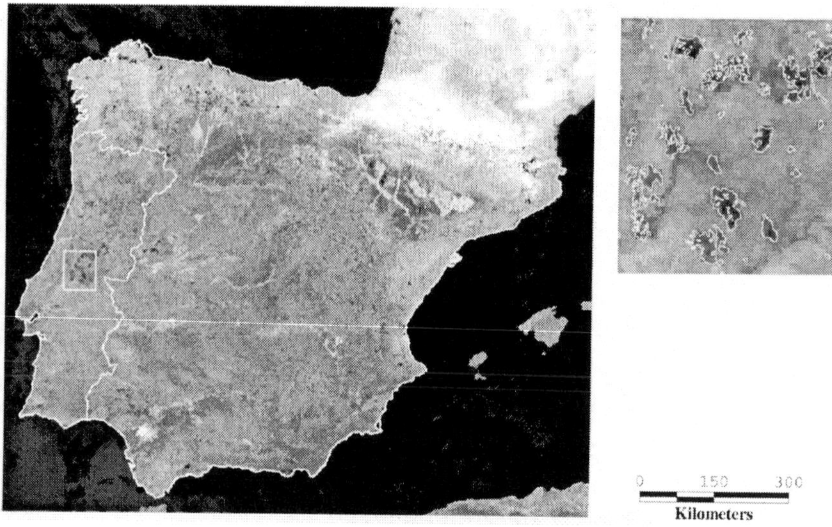

Figure 5. GEMI greyscale image of the Iberian study area. Lighter-coloured areas are irrigated agriculture and dense natural vegetation, and darker areas are fire scars, water bodies, and bare soils. The inset is a zoom-in on central Portugal, showing fire perimeters obtained from Landsat–TM interpretation outlining the fire scars.

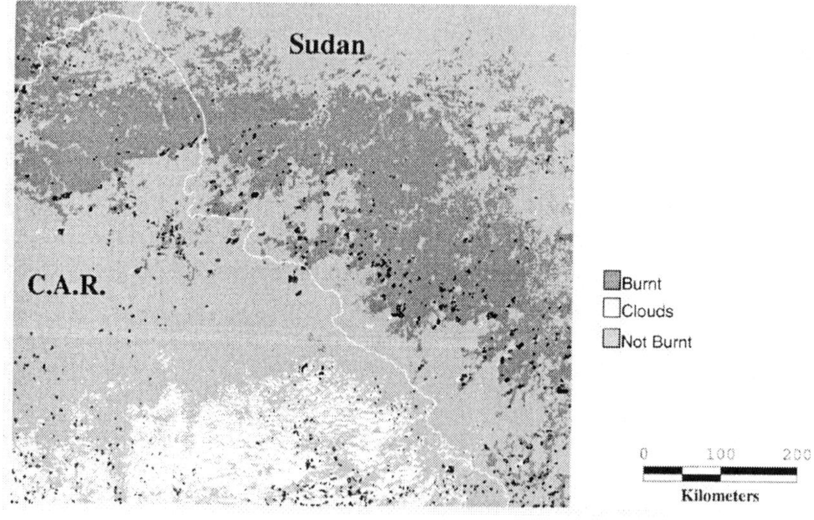

Figure 6. Classification of the Central African study area, identifying burns, unburned areas and clouds. Active fires are represented as black polygons.

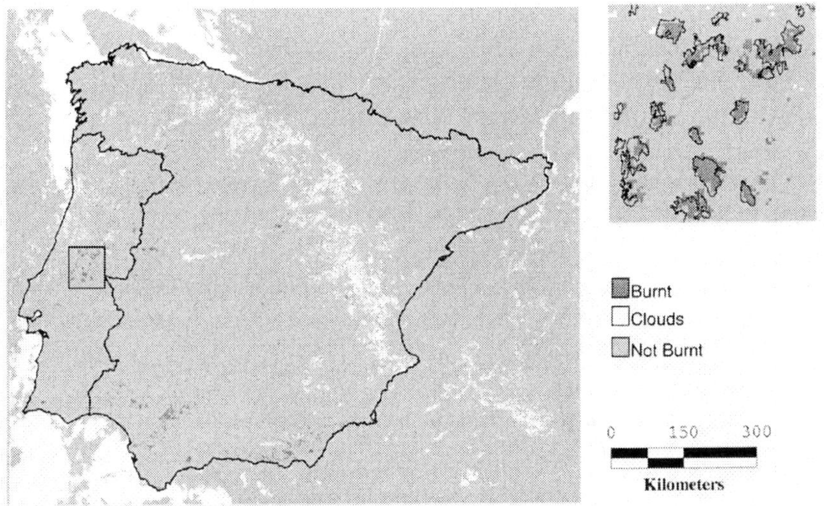

Figure 7. Classification of the Iberian study area, identifying burns, unburned areas and clouds. On the zoom window, white fire perimeters over burns were obtained from interpretation of Landsat5 TM imagery.

This is an area where the classifier performed very well, but in other parts of Iberia, we believe it made commission and omission errors. In the southern part of Spain, in the Andalusian autonomous region, some of the burns detected may actually be hot, bare soil surfaces that were mistakenly taken for burns, while due east of Valencia, in the central part of the Mediterranean coast of Spain, some fires with areas larger than 10,000 ha were missed. However, visual inspection of the multi-temporal image composite has shown that the fire scars were barely visible, probably because of cloud cover and poor atmospheric conditions during the compositing period. These are however relatively insignificant errors when compared to the extent of areas burned at the regional scale, especially in tropical and boreal regions, and burned area mapping at this level of accuracy is expected to represent a big improvement over current estimates based on active fire counts (Pereira *et al.*, 1998; Barbosa *et al.*, 1998).

4.   CONCLUSIONS

We have shown that tree-based classifiers are an adequate tool for performing burned area mapping simultaneously with cloud screening. Data from two regions with very different climatic types, land cover patterns, and fire regimes were pooled together and classified using a unified set of rules

extracted from a classification tree generated with the CART algorithm. Classification accuracy assessed in an independent test sample was very high, and transposition of the classification rules to the two study areas also produced very encouraging results. The methodology is easy to implement, has a solid statistical foundation, and includes rigorous procedures for accuracy assessment. Tree- based classifiers have several attractive features for global burned area mapping, as they:
– can handle massive amounts of data;
– are trained by a procedure that is standard in supervised image classification;
– are a non-parametric tool that does not require stringent assumptions regarding distributional properties of the data;
– are extremely robust to the effects of outliers;
– can handle a mixture of data types, and can be stratified with categorical variables; and
– are non-homogeneous, dealing easily with the fact that different relationships hold between variables in different regions of the measurement space.

We are currently investigating an improvement of these classifiers, aimed at replacing the crisp node splitting thresholds by fuzzy thresholds (Jang, 1997; Janikow, 1996), thus increasing robustness of the system relatively to problems such as variable image calibration, illumination and observation geometry, and atmospheric conditions. We are also preparing data from other continents and biomes, to test inductive rule-based classifiers over a broader range of environmental conditions and fire regimes.

5.   ACKNOWLEDGEMENTS

We thank Ana Sá (DEF/ISA) for assistance in preparation of the figures. The data used in this study were acquired in the framework of projects MegaFiReS (ENV–CT96–0256), funded by CEC–DGXII, E. Chuvieco, Project Co-ordinator, and EXPRESSO (BiBEx/ IGAC), J.M. Grégoire (GVMU/SAI/JRC, Principal Investigator).

6.   REFERENCES

Barbosa, P.M., J.-M. Grégoire, and J.M.C. Pereira (1998) An algorithm for extracting burned areas from a time series of AVHRR GAC data, applied at a continental scale. *Remote Sensing of Environment*, in press.

Breiman, L., J.H. Friedman, R.A. Olshen, and C.J. Stone (1984) *Classification and Regression Trees*. Wadsworth, Pacific Grove, CA.

Cahoon Jr, D.R., B.J. Stocks, J.S. Levine, W.R. Cofer III, and J.M. Pierson (1994) Satellite analysis of the severe 1987 forest fires in northern China and southeastern Siberia. *Journal of Geophysical Research* 99 (D9): 18627–18638.

Dwyer, E., J.-M. Grégoire, and J.-P. Malingreau (1998a) A global analysis of vegetation fires using satellite images: spatial and temporal dynamics. *Ambio* 27 (3): 175–181.

Dwyer, E., S. Pinnock, J.-M. Grégoire, and J.M.C. Pereira (1998b) Global spatial and temporal distribution of vegetation fire as determined from satellite observations. *International Journal of Remote Sensing*, in press.

Eva, H., and E.F. Lambin (1998) Burnt area mapping in Central Africa using ATSR data. *International Journal of Remote Sensing*, in press.

Flasse, S.P. and P. Ceccato (1996) A contextual algorithm for AVHRR fire detection. *International Journal of Remote Sensing* 17(2) :419–424.

Hao, W.M., M.H. Liu, and P.J. Crutzen (1990) Estimates of annual and regional releases of CO_2 and other trace gases to the atmosphere from fires in the tropics, based on the FAO statistics for the period 1975–1980. In *Fire in the Tropical Biota* (J.G. Goldammer, ed.), pp.440–462. Springer-Verlag.

Huete, A.R. 1988. A soil adjusted vegetation index (SAVI). *Remote Sensing of the Environment* 25, 295–309.

Jang, J.-S.R., C.-T. Sun, and E. Mizutani (1997) *Neuro-Fuzzy and Soft Computing*. Prentice Hall, Upper Saddle River, NJ.

Janikow, C.Z. (1998) Fuzzy decision trees: issues and methods. *IEEE Transactions on Systems, Man, and Cybernetics* 28(1): 1–14.

Kasischke, E.S. and N.H.F. French (1995) Locating and estimating the areal extent of wildfires in Alaskan boreal forests using multiple-season AVHRR NDVI composite data. *Remote Sensing of Environment* 51: 263–275.

Kasischke, E.S., N.H.F. French, P. Harrel, N.L. Christensen, Jr., S.L. Ustin, and D. Barry (1993) Monitoring wildfires in boreal forests using large area AVHRR NDVI composite image data. *Remote Sensing of Environment* 45: 61–71.

Kaufman, Y.J., Tucker, C.J., and Fung, I. 1990. Remote Sensing of Biomass Burning in the Tropics. *Journal of Geophysical Research* 95 (D7):9927–9939.

Kennedy, P.J., A.S. Belward, and J.-M. Grégoire (1994) An improved approach to fire monitoring in West Africa using AVHRR data. *International Journal of Remote Sensing* 15 (11):2235–2255.

Kidwell, K. (Ed.) (1997) NOAA Polar Orbiter Data User's Guide (TIROS–N, NOAA–6, NOAA–7, NOAA–8, NOAA–9, NOAA–10, NOAA–11, NOAA–12, NOAA–13 AND NOAA–14). August 1997 Revision. U.S. Department of Commerce, National Oceanic and Atmospheric Administration.

Levine, J.S. (1991) Global biomass burning: atmospheric, climatic and biospheric implications. In *Global Biomass Burning* (J.S. Levine, ed.), pp.xxv–xxx. MIT Press.

Levine, J.S. (1996) Introduction. In: *Biomass Burning and Global Change. Remote Sensing, Modeling and Inventory Development, and Biomass Burning in Africa*, Ed. J.S. Levine. MIT Press, Cambridge, Massachusetts, Vol. 1, xxxv–xliii.

Martín Isabel, M.P. (1998) Cartografía e inventario de incendios forestales en la Península Ibérica a partir de imágenes NOAA–AVHRR. Unpublished Doctoral Thesis, Universidad de Alcalá, Facultad de Filosofia y Letras, Departamento de Geografia. Alcalá de Henares, Spain.

Pereira, J.M.C. (1998) A comparative evaluation of NOAA/AVHRR vegetation indices for burned surface detection and mapping. Accepted for publication, *IEEE Transaction on Geosciences and Remote Sensing*.

Pereira, J.M.C., B.S. Pereira, P.M. Barbosa, and D. Stroppiana (1998) Quantitative assessment of fire occurrence in the EXPRESSO study area during the 1996 dry season experiment: active fires, burnt area and atmospheric emissions. In preparation, for the *Journal of Geophysical Research* EXPRESSO Special Issue.

Pereira, M.C., S. Amaral, N.J. Zerbini, and A.W. Setzer (1990) Estimativa da área total queimada no Parque Nacional das Emas com o uso de imagens da banda 3 do AVHRR: comparação com estimativas do TM–Landsat. Anais VI Simpósio Brasileiro de Sensoriamento Remoto, Vol.2, pp.302–310. Manaus, 21–29 Junho 1990.

Pinty, B. and Verstraete, M.M. (1992) GEMI: a non-linear index to monitor global vegetation from satellites. *Vegetatio* 101:15–20.

Qi, J., A. Chehbouni, A.R. Huete, Y.H. Kerr, and S. Sorooshian (1994) A Modified Soil Adjusted Vegetation Index. *Remote Sensing of Environment* 48: 119–126.

Rao, C.R.N. and Chen, J. (1996) Post-launch calibration of the visible and near-infrared channels of the Advanced Very High Resolution Radiometer on the NOAA–14 spacecraft. *International Journal of Remote Sensing* 17 (14):2743–2747.

Saunders R. W. (1990) The determination of broad band surface albedo from AVHRR visible and near infrared radiances. *International Journal of Remote Sensing*, Vol. 11, pp. 59 – 67.

Seiler, W. and Crutzen, P.J. (1980) Estimates of gross and net fluxes of carbon between the biosphere and the atmosphere from biomass burning. *Climatic Change* 2:207–247.

Sousa, A.M. (1998) Cartografia de áreas ardidas na Península Ibérica com imagens NOAA/AVHRR. Master Thesis, Universidade Técnica de Lisboa, Instituto Superior Técnico, Departamento de Engenharia Civil. Lisboa, Portugal.

Stehman, S.V. and Czaplewski, R.L. (1998) Design and analysis for thematic map accuracy assessment: fundamental principles. *Remote Sensing of Environment* 64 (3): 331–344

Teillet, P.M. and B.N. Holben (1994) Towards operational radiometric calibration of NOAA AVHRR imagery in the visible and near-infrared channels. *Canadian Journal of Remote Sensing*, 20 (1): 1–10.

Whelan, R.J. (1995) *The Ecology of Fire*. Cambridge University Press, UK.

Fire Regime Sensitivity to Global Climate Change: An Australian Perspective

GEOFFREY J. CARY and JOHN C. G. BANKS
Department of Forestry, Australian National University, Canberra, Australia

Abstract: The Australian eucalypt forests are highly adapted to fire, and their component species possess well-developed response mechanisms that ensure post-fire recovery of these ecosystems. Fire regimes, which may alter forest floristics and structure, have changed since pre-European times because of management practices and may again change because of a changing climate. Two complimentary approaches are used to determine spatial and temporal patterns of fire regimes, a) dendrochronology to determine pre- and post-European fire histories for specific sites and b) fire–climate–landscape modelling to predict spatial patterns in fire regimes for topographically complex landscapes. This paper brings together these two approaches which have been applied independently to the same forest in the Southern Tablelands of New South Wales. The model predictions of spatial patterns in fire regimes under the present climate provide reasonable results when compared with observed site fire histories. Also, model results indicate that around half of the landscape is likely to experience a significant increase in fire frequency as a result of climate change. These findings, which have implications for fire-prone forest environments world-wide, are discussed in relation to the effects that anthropogenic ignition have had on the fire frequency in the study area over the last century.

1.   INTRODUCTION

A fire regime is a description of the nature of the periodic fires which burn a site in a fire-prone landscape. It has the characteristics of fire frequency, fire intensity and season of fire occurrence (Gill, 1975) and affects the occurrence and abundance of plant species in many fire-prone plant communities (Keeley & Zedler, 1978; Fox & Fox, 1986; Taylor, 1993; Cary & Morrison, 1995). Therefore, a good understanding of the nature of fire regimes, including the potential effects of climate change, are fundamental for the optimal management of natural systems.

This paper is concerned with the effect that a changed climate may have on the fire regime of forests and woodlands in south eastern Australia. The objectives are to explore how two different approaches to obtain insights into the nature of long-term fire regimes, namely dendrochronological studies and computer simulation modelling, can be brought together in a complimentary fashion, and to make some predictions about likely changes in fire regimes that may be expected under a simple scenario for climate change.

Dendrochronological approaches involve determining the timing between fires, which may be recorded as fire scars or other aberrations of regular growth (Brown & Swetnam, 1994) in the chronosequence described by the annual growth rings of trees. This method provides information on temporal patterns of real fires but has a number of limitations including inter- and intra-specific variation in the ability to record all fires (McBride & Lewis, 1984) and the considerable time and effort required to construct long-term fire histories that encompass substantial spatial extents.

Simulation modelling provides a relatively recent alternative approach to studying patterns of fire regimes at the level of the landscape. These include FIRE–BGC (Keane *et al.*, 1996), a process-based theoretical model of forest dynamics in topographically complex landscapes, and the landscape implementation of EMBYR (Gardner *et al.*, 1996). Apart from other limitations, both are based on the fire spread algorithms developed by Rothermel (1972) which are not widely used in Australia, making the models difficult to implement in Australian forest systems.

FIRESCAPE (Cary, 1997a; Cary, 1997b; Cary, 1998) is a new landscape-level fire regime simulator that has sought to address some of the deficiencies in the existing models, particularly the reliance on the US fire spread algorithms. Originally, the model was conceived for generating fire regimes under current climate scenarios in the absence of other suitable fire history records. However, given the process-based nature of the model, it is ideal for investigating the effects of changed climate on fire regimes.

In this paper we compare predictions of fire frequency from an implementation of the FIRESCAPE model in the forests and woodlands of the Australian Capital Territory (ACT) region with fire frequencies determined by dendrochronology in the same system (Banks, 1982). Then, predictions about changes in fire frequency for the region under a simple scenario of climate change for Australia are produced using the model. The importance of these changes in relation to the ability of humans to manipulate fire regimes are then discussed from the point of view of temporal trends in the dendrochronological data set.

The study area of approximately 85 x 110 km, centred on the Australian Capital Territory in south eastern Australia (Figure 1), is representative of landscapes and landuse on the Southern Tablelands of New South Wales. The topography of the area includes the undulating plains and low hills of the tablelands between 500–700 m elevation in the northeast of the study region, with dissected mountainous terrain extending to 1700 m toward the southwest. The natural vegetation on the tablelands is dominated by the *Eucalyptus melliodora–E. blakelyi* Alliance, and the mountains, from the footslopes to the subalpine zone, by forests comprising of the alliances *E. macrorhyncha–E. rossii*, *E. fastigata–E. viminalis*, *E. delegatensis–E. dalrympleana*, *E. pauciflora–E. stellulata* and *E. niphophila*. Landuses vary across the study area, with sheep and cattle grazing and hobby farms on the tableland and with most of the forest areas under conservation and management for water catchment. Some 30,000 ha of the mountain footslopes are under commercial conifer plantations while some 300,000 people reside in the national capital, Canberra, located to the north of the centre of the study region.

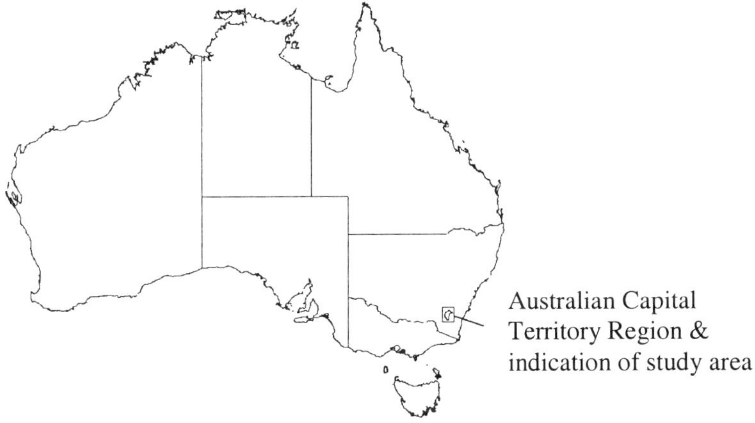

Figure 1. Location of the Australian Capital Territory Region, Australia.

2. DESCRIPTION OF THE DENDROCHRONOLOGICAL DATA

Banks (1982) determined the fire history for five stands of *Eucalyptus pauciflora* (snow gum) along a 40 km transect on the Brindabella Range which lies along the western border of the ACT. Composite chronologies were developed using 4 to 14 trees per stand, so as to capture as complete a site fire record as possible. Snow gum, a subalpine species, was an ideal candidate for this study as it has a relatively thin, fire-sensitive bark, and readily accumulates distinctive fire scars. The tree rings are clear and distinct and this ensured an accurately-dated, point-source fire record.

A few old-growth trees provided a diminishing fire record back to the 1750's. This record was severely limited as anthropogenic burning in this forest during the grazing era caused major losses of old-growth stands which have been progressively replaced by even-aged stands of the same species over the period from 1860's to 1939.

3.   DESCRIPTION OF THE FIRESCAPE MODEL

FIRESCAPE is a landscape-level, fire-regime simulation model. It was developed and implemented for the ACT region, Australia (Cary 1997b; Cary, 1998). The simulation landscape is approximately 900,000 hectares and is represented by an array of square 1 hectare pixels. Fire regimes are simulated by igniting and spreading individual fires which, through time, result in the development of a landscape-level fire regime.

In the model, fire ignitions result from cloud-to-ground lightning strikes and their spatial locations are generated by an empirical model developed from data used to construct a similar model by McRae (1992). In the revised ignition model used in FIRESCAPE, the probability of lightning ignition is positively associated with the macro-scale elevation at the broad spatial scale, primarily reflecting the orographic effect of mountain ranges on storm occurrence, and also positively associated with the magnitude of the meso-scale elevation residual at finer spatial scales. These patterns reflect those found in similar studies in Yosemite and Sequoia National Parks, California, USA (Vankat, 1983). The probability of lightning occurrence on any particular day is related to the maximum temperature and whether the day is wet or dry, and is described by empirical functions generated from eleven years of meteorological observations in the ACT.

A modified version of the Richardson-type stochastic climate generator (Richardson, 1981, Cary & Gallant, 1997) is used to generate daily weather data, based on the underlying stochastic structure of the meteorological process (Richardson, 1981). The original model was modified to include the variables required for modelling fire spread. Daily rainfall amounts are generated using the truncated power normal model (Hutchinson, 1995) and the weather generator was parameterised using the meteorological data described above.

The rate of spread from a burning cell toward each of its eight neighbours is determined by the elliptical fire spread model (Van Wagner, 1969), which has been modified by Anderson *et al.* (1982) and Wallace (1993) to predict fire spread under non-uniform meteorological, fuel and topographic conditions. Head fire rate of spread is determined for a cell, as it burns, using the equation form of McArthur's Forest fire-danger meters (McArthur, 1967) and derived hourly meteorological information, and drought factor and fuel load derived using the relationships published in Mount (1972) and Olson (1963), respectively. The parameters of the litter accumulation model were determined from published data on 234 litter quadrats in the study region. Fireline intensity (I) (kW m^{-1}) (Byram, 1959) is calculated for the spread of fire from one pixel to the next for the purpose of determining this component of the fire regime and for determining whether a fire spreads, is extinguished, or persists in the same pixel.

Like the rate of spread of fires, fire extinction is an important process for determining the pattern of fire regimes in a landscape. In FIRESCAPE, the spread of a fire in a particular direction only occurs if the intensity of the fire is above a critical threshold of 80 to 100 kW m^{-1}.

Model output consists of, amongst other fire regime parameters, maps which detail the predicted average inter-fire interval and the predicted standard deviation of the inter-fire interval for each cell. The fire frequency (average inter-fire interval) map for the ACT implementation of FIRESCAPE is presented in Figure 2. It is worth noting that, for a given set of parameters, the outcome of the model is repeatable in the sense that the coefficient of variation of the majority of the output parameters for each individual cell from five replicate simulations are below 20% in 84% of the cells. This outcome arises from the importance of the structure of the landscape in the ignition neighbourhood (Cary, 1997b) of a particular cell for determining its fire regime and means that few simulation runs are required for a particular set of parameters.

4. TESTING THE MODEL AGAINST MEASUREMENTS OF FIRE FREQUENCY IN THE STUDY AREA

The fire frequency outputs of the ACT implementation of the model were compared with the dendrochronological data compiled for *Eucalyptus pauciflora* stands in the Brindabella Ranges (Banks, 1982). The comparison was limited to the dendrochronological data for the period 1840 to 1880 because of the scarcity of fire history before 1840 and the increased influence of Europeans after 1880. For each stand, the proportion of trees which recorded each fire was summed over the 40 year period and the fire frequency was determined by dividing the 40 year period by this number.

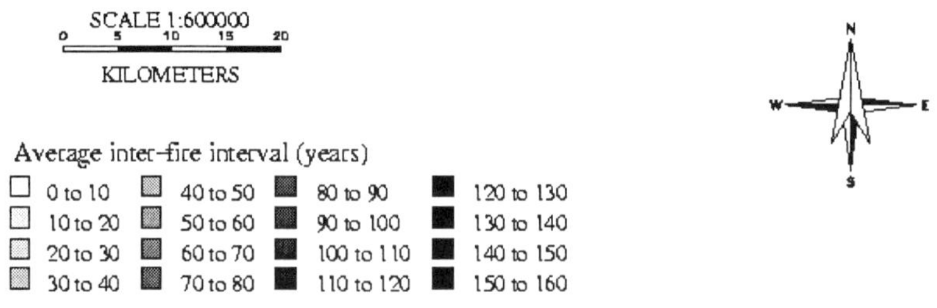

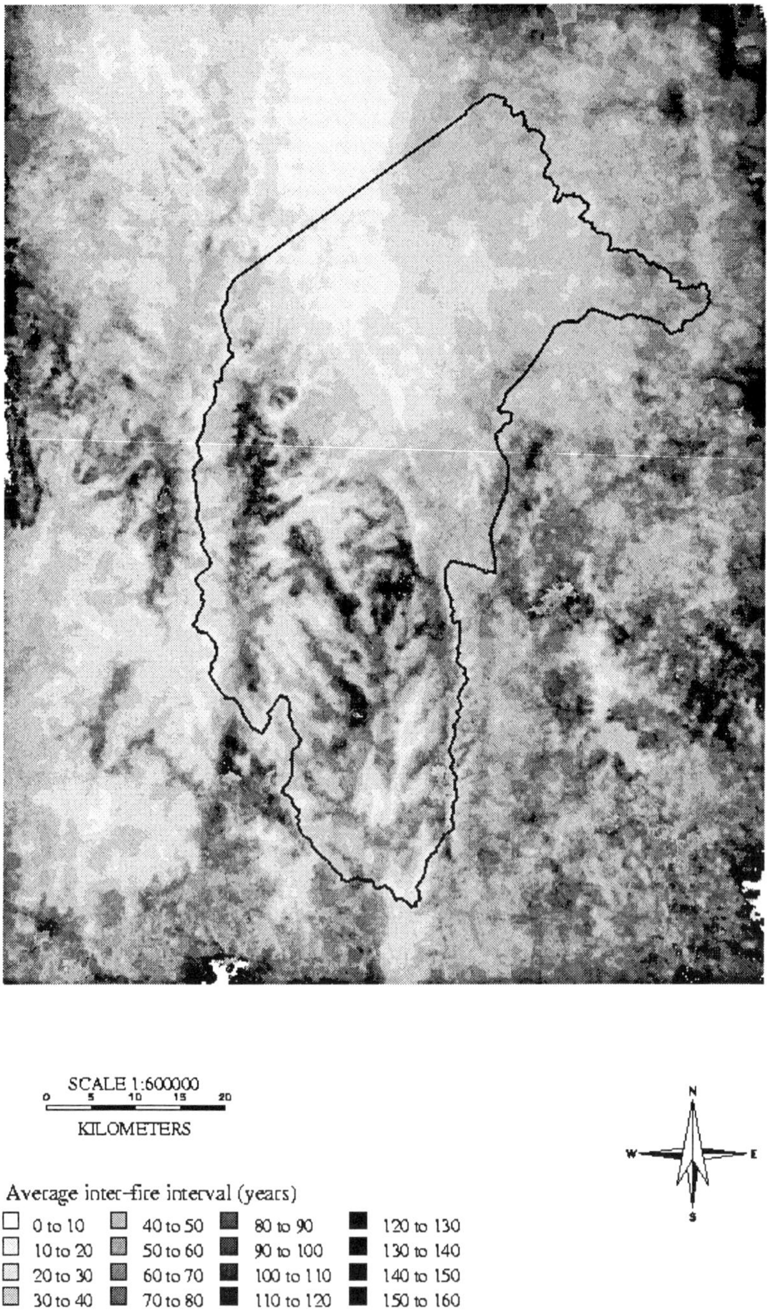

Figure 2. Spatial pattern in the average inter-fire interval (years) from a FIRESCAPE simulation of the Australian Capital Territory (ACT) region, Australia. ACT border indicated.

This assumes that each tree will record each fire and that when only a subset of trees recorded a fire it was because not all trees were actually burned. This assumption may lead to an under-estimation of fire frequency using the dendrochronological data set.

Apart from the Bimberi site, the modelled and observed data show a similar pattern (the rank-order of the fire frequency of sites in the modelled data approximately matches that of the observed data) except that the overall frequency of fire in the observed data set was more frequent (Figure 3). This is not unexpected given that the observed data include the effects of anthropogenic ignitions while the simulated data do not.

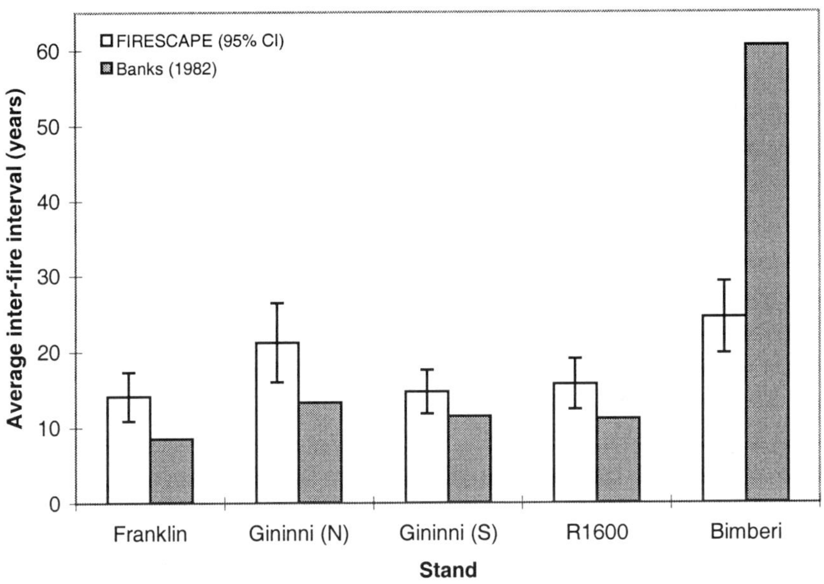

Figure 3. Average inter-fire interval (years) for all stands of *Eucalyptus pauciflora* in the Brindabella Range study (Data from Banks, 1982) and simulated average inter-fire intervals generated using the FIRESCAPE model.

The anomaly of the Bimberi site might be explained by it being occupied by mature woodland. Mature trees have thicker bark and therefore a lower sensitivity to individual fires. Also, mature woodland often has an understorey with a lower biomass than that of a developing woodland, resulting in a lower average fire intensity. We hypothesise that these factors are likely to result in the frequency of fire being underestimated at the Bimberi site. However, in the absence of actual long-term fire records, there is at present no way of testing this hypothesis.

We conclude that the model is capable of generating fire regimes that are reasonable when compared to the scarce observational data and is therefore suitable for undertaking an analysis of the effects of global warming on fire regimes in this part of the world.

5.   ANALYSIS OF THE EFFECT OF A CHANGED CLIMATE ON FIRE BEHAVIOUR AND PATTERNS OF FIRE FREQUENCY

A considerable number of climate change scenarios have been produced which include information for the ACT region. In this section we analyse the effect of a relatively simple scenario for climate change (DEST, 1994) on fire behaviour and subsequent effects on patterns of fire frequency. The climate change predictions given by this scenario for the year 2030 are: an increase in temperature of + 2°C; an increase of summer rainfall of 0 to + 20 %; and a decrease in winter rainfall of -10 to +10 %.

5.1   Methods

Three FIRESCAPE simulations were performed. These involved simulating spatial patterns in fire frequency using: i) current climate (Current climate); current climate with daily temperatures increased by 2°C (+ Temperature); and current climate with daily temperatures increased by 2°C and summer daily rainfall increased by 20% (+ Temperature + Rainfall).

Temperature has a number of important effects on fire spread rates. Firstly, the empirical models of fire spread are somewhat sensitive to changes in temperature directly. Further, fire spread is reasonably sensitive to variations in relative humidity (Beer, 1988), a function of atmospheric vapour pressure and air temperature. Also, the evapotranspiration rate used in the calculation of the daily soil dryness index (an input for the fire danger model) depends on air temperature and the moisture deficit of the soil (Mount, 1972). For the current analysis, it is assumed that relative humidity will remain unchanged with a changed climate.

For each simulation, the intensity of each fire spread event was recorded and the distributions of these were compared. Further, a map of cell-by-cell tests of significance of differences (students t-test) in fire frequencies generated using the current and changed (+Temperature +Rainfall) climates was produced.

5.2   Results

Increasing the temperature alone (+Temperature) results in a decline in the frequency of lower intensity fires (< 400 kW m^{-1}) and an increase in the frequency of more intense fires (Figure 4). Fires with a fireline intensity of between 2000 and 5000 kW m^{-1} are approximately 4 times more frequent for the modified climate than for the current climate. This effect is moderated somewhat by an increase in summer rainfall. In particular, the frequency of fires with a fireline intensity of less than 500 kW m^{-1} is similar to that for the current climate. On the other hand, the frequency of fires in higher fireline intensity classes is considerably higher for the scenario which includes increased rainfall but not as high as that which involves an increase in temperature only. With respect to the significance of the differences between mean fire frequencies under current and changed climate, fire frequency is predicted to be significantly higher for around half of the cells in the study region while no significant difference is expected for the remainder of the cells (Figure 5).

These findings agree with other studies of the effect of climate change on fire frequency. For example, Gardner *et al.* (1996) conducted a rather similar study for the Yellowstone National Park, Wyoming, USA. They elected to modify thousand hour time lag fuel moisture (THFM) to mimic the effect of climate change. THFM is the percentage moisture content of large ($\geq$ 7.6 cm) fuel particles that require approximately 1000 hours to respond to changes in precipitation, relative humidity and temperature (Deeming *et al.*, 1978). Gardner *et al.* (1996) drew THFM values from distributions with a minimum and maximum percentile value for THFM of 0.20 and 0.75 for a nominal climate. They defined a drier scenario as having minimum and maximum values of 0.15 and 0.70 and a wetter scenario as having minimum and maximum values of 0.25 and 0.80. They found that the drier scenario had 30% more fires than the nominal scenario while the wetter scenario had 20% less. Further, the wetter scenario had a longer average return interval but had more frequent extreme events, a function of the generation of large patches of highly flammable forest.

In addition, Clark (1990) used the FOREST model to study the effects of a drier climate on fire frequency for separate stands of mixed conifer forest in north-western Minnesota, USA. The model, based on monthly water balances, live and dead fuel dynamics, and lightning incidence, predicted that in the absence of fire suppression there would have a been a 20 to 40% increase in fire frequency in the warmer–drier 20th century, compared to the 19th century, for those forest stands.

These findings indicate that the overall increase in fire frequency resulting from climate change observed in this study is probably of general relevance to fire-prone plant communities. However, there is little information about changes to the fine-scale patterns of fire frequency with which to compare the fine-scale patterns produced in this study. Therefore we can make no statement about the generality of this aspect of our findings, except that because of the process-based nature of the FIRESCAPE model, there may be reasonable grounds to suggest that the results have some implications for fire-prone forest environments world-wide. This hypothesis requires a considerable amount of further testing.

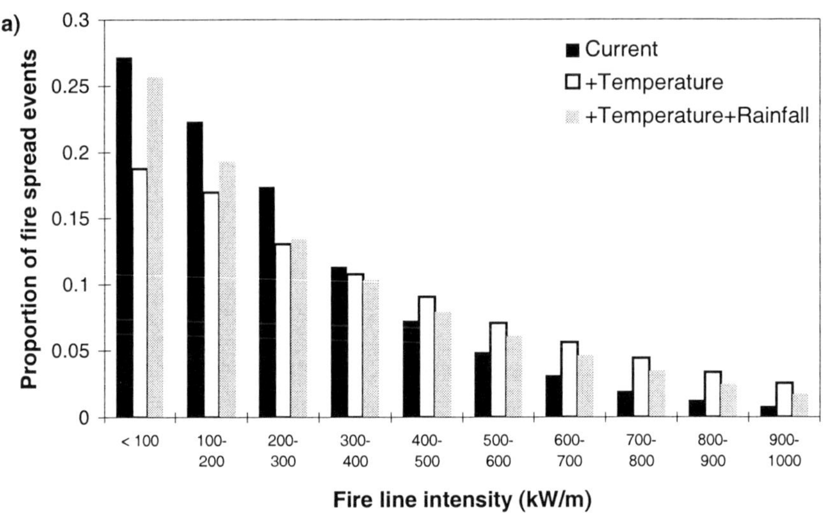

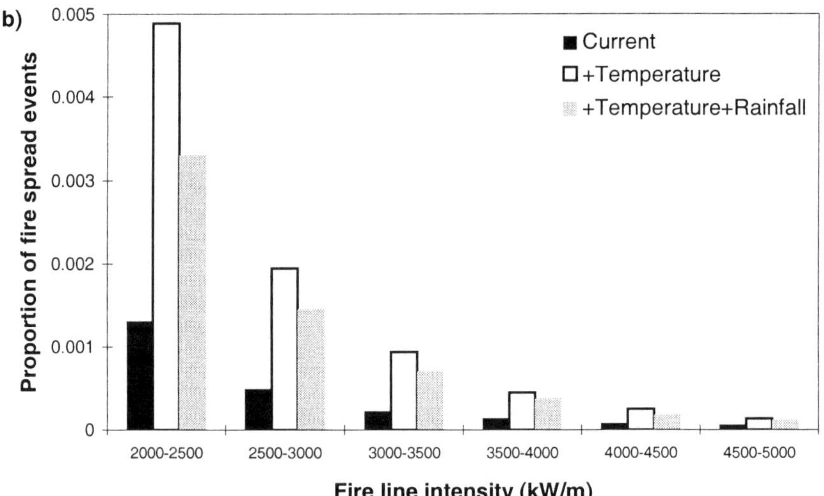

Figure 4. Frequency of fireline intensities of potential (including non-successful) fire spread events in FIRESCAPE simulations for different climate scenarios of the Australian Capital Territory region, Australia for a) fireline intensities between 0 and 2000 kW/m and b) fireline intensities between 2000 and 5000 kW/m. 'Current climate' refers to current climate; '+ Temperature' refers to current climate plus 2 °C added to daily temperatures; and '+ Temperature + Rainfall' refers to current climate plus 2 °C added to daily temperatures and 20 % added to daily rainfall quantities during summer months.

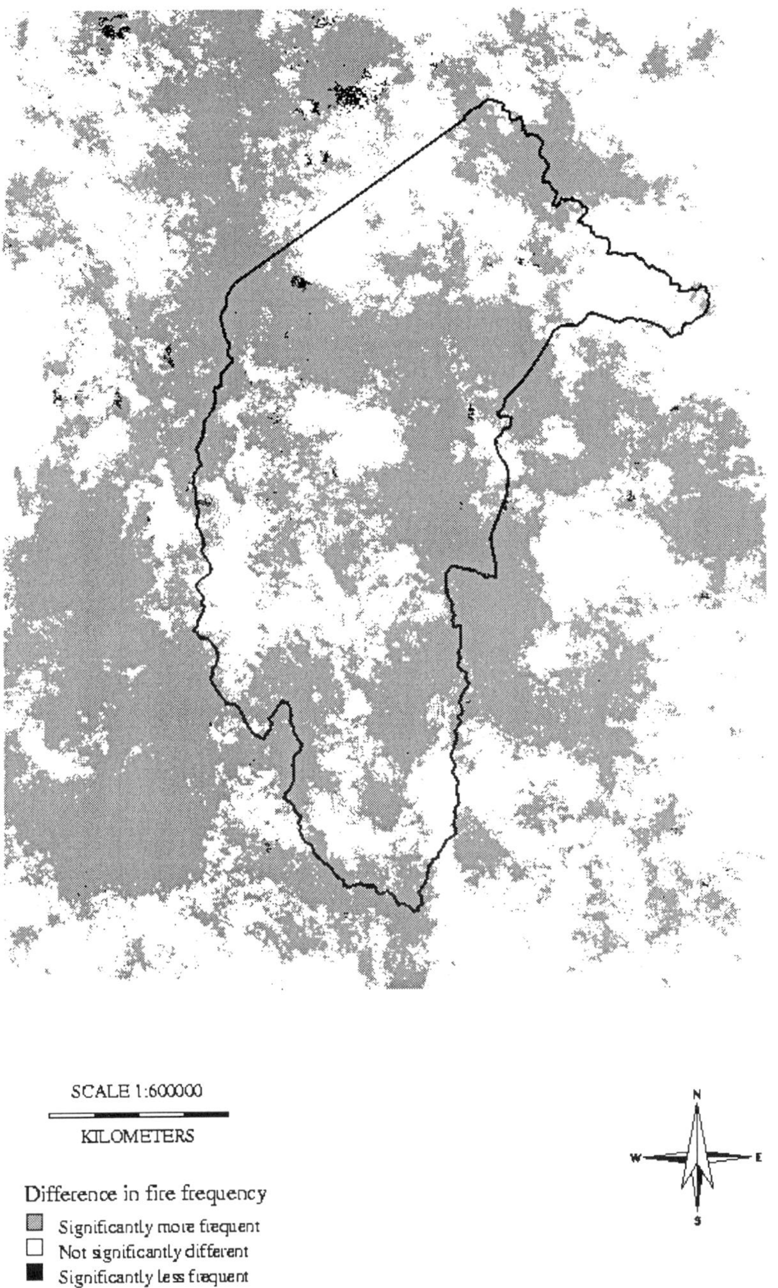

Figure 5. Spatial pattern in cell-by-cell significance of difference between average inter-fire interval simulated for current and changed (+Temperature+Rainfall) climates.

6.   IMPLICATIONS OF CHANGED FIRE REGIMES

FIRESCAPE predicts what might be called 'natural fire regimes'. They arise from the occurrence of lightning-ignited fires that spread and extinguish in the absence of human intervention, largely because the model was originally designed to investigate patterns of fire regimes that may arise from topographic complexity.

According to the dendrochronological data of Banks (1982), the fire frequency in the sub-alpine forests of the study region have changed considerably as a result of anthropogenic ignitions and changed management practices since 1750 (Figure 6). During the pre-European period (before 1830), fire frequency was low but increased after the arrival of Europeans in the 1840s until the 1970s when fire frequency has declined to the pre-European levels.

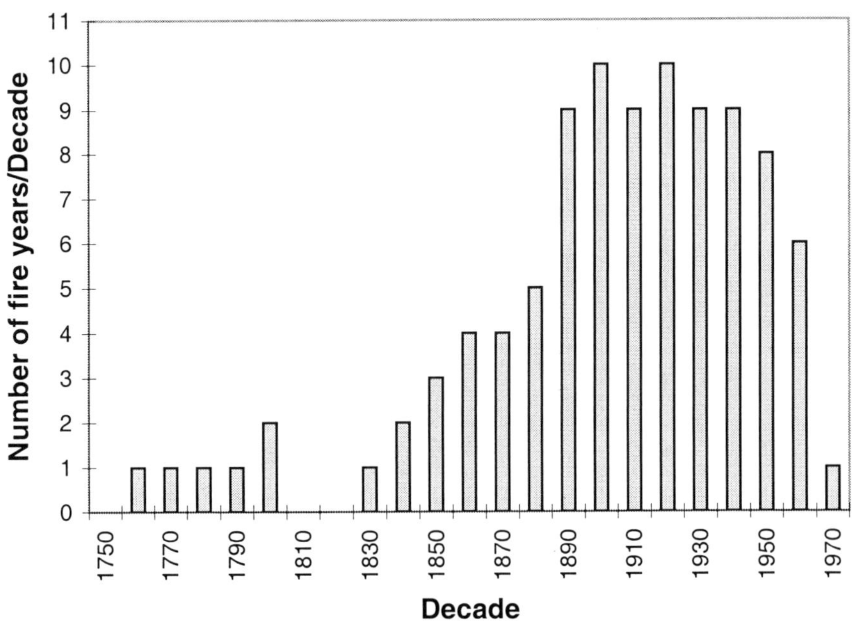

Figure 6. Temporal trends in the total number of fires determined from all stands in the dendrochronological study of fire frequency of the Brindabella Range (Banks, 1982).

It would appear, therefore, that the influence of humans on the fire frequency in south-eastern Australia may be at least as important, and probably more so, than changes in climate resulting from global warming. Further discussion involving experts in modern fire management and suppression is required before the effect of a changed climate can be fully understood.

7. ACKNOWLEDGEMENTS

The digital elevation model used for this research was supplied by the New South Wales National Parks and Wildlife Service. The Australian Bureau of Meteorology and the Commonwealth Scientific and Industrial Research Organisation supplied daily meteorological data. The development of the FIRESCAPE model was partly supported by an Australian Postgraduate Research Award. Professor Ian Noble, Dr Malcolm Gill and John Gallant are acknowledged for their discussion in the development of the FIRESCAPE model.

8. REFERENCES

Anderson, D.H., Catchpole, E.A., De Mestre, N.J. and Parkes, T. (1982) Modelling the spread of grass fires. *Journal of the Australian Mathematical Society (Series B)*, **23**, 451–66.

Banks, J.C.G. (1982) *The Use of Dendrochronology in the Interpretation of the Dynamics of the Snow Gum Forest.* Ph.D. Thesis, Department of Forestry, Australian National University.

Banks, J.C.G. (1997) Trees the silent fire historians. *Bogong*, **18**, 9–12.

Beer, T., Gill, A.M., and Moore, P.H.R. (1988) Australian bushfire danger under changing climatic regimes. In: *Greenhouse Planning for Climate Change* (G.I. Pearman, Ed.). CSIRO Division of Atmospheric Research Publication, 421–7.

Brown, P.M. and Swetnam, T.W. (1994) A cross-dated fire history from coast redwood near redwood National Park, *Californian Journal of Forest Research*, **24**, 21–31.

Byram, G.M. (1959) Combustion of forest fuels. In: *Forest Fire: Control and Use* (K.P. Davis Ed.). McGraw-Hill New York.

Cary, G.J. (1998) *Predicting fire regimes and their ecological effects in spatially complex landscapes.* Doctoral Thesis. Australian National University, Canberra, Australia.

Cary, G.J. (1997 a) FIRESCAPE – a model for simulation theoretical long-term fire regimes in topographically complex landscapes. *Proceedings of the Australian Bushfire Conference, Bushfire '97*, 8 – 10th July, Darwin, Northern Territory.

Cary, G.J. (1997 b) Analysis of the effective spatial scale of neighbourhoods with respect to fire regimes in topographically complex landscapes. Modsim 97, International Congress on Modelling and Simulation. 8 – 11th December, Hobart, Tasmania.

Cary, G.J. and Morrison, D.A. (1995) Effects of fire frequency on plant species composition of sandstone communities in the Sydney Region: Combinations of inter-fire intervals. *Aust. J. Ecol.* 20, 418 – 426.

Cary, G.J. and Gallant, J.C. (1997) Application of a stochastic weather generator for fire danger modelling. *Proceedings of the Australian Bushfire Conference, Bushfire '97*, 8 – 10th July, Darwin, Northern Territory.

Clark, J.S. (1990) Twentieth-century climate change, fire suppression, and forest production and decomposition in northwestern Minnesota. *Canadian Journal of Forest Research*, **20**, 219–232.

Deeming, J.E., Burgan, R.E. and Cohen, J.D. (1978) *The National fire Danger Rating System – 1978.* USDA Forest Service General Technical Report INT – 39.

DEST (1994) *Climate Change: Australia's National Report Under the United Nations Framework Convention on climate Change.* Department of Environment Sport and Territories, Canberra.

Fox, M.D. and Fox, B.J. (1986) The effect of fire frequency on the structure and floristic composition of a woodland understorey. *Aust. J. Ecol.* 11, 77–85.

Gardner, R.H, Hargrove, W.W., Turner, M.G. and Romme, R.H. (1996)Climate change, disturbances and landscape dynamics. In: *Global Change and Terrestrial Ecosystems* (B. Walker Ed.) pp. 149–72. Cambridge University Press.

Gill, A.M. (1975) Fire and the Australian flora: a review. *Aust. For.* 38, 4–25.

Hutchinson, M.F. (1995) Stochastic space–time weather-models from ground-based data. *Agricultural and Forest Meteorology*, **73**, 237–64.

Keane, R.E., Morgan, P. and Running, S.W (1996) FIRE–BGC – a mechanistic ecological process model for simulating fire succession on coniferous forest landscapes of the northern Rocky Mountains. *USDA Forest Service Intermountain Research Station, Res. Pap. INT–RP–484.*

Keeley, J.E. and Zedler, P.H. (1978) Reproduction of chaparral shrubs after fire: a comparison of sprouting and seeding strategies. *The American Midland Naturalist* **99**, 142–161.

McArthur, A.G. (1967) Fire behaviour in eucalypt forests. *Common. Aust. For. Timb. Bur. Leafl.* No.107.

McBride, J.R. and Lewis, H.T. (1984) Occurrence of fire scars in relation to the season and frequency of surface fires in Eucalyptus forests of the Northern Territory, Australia. Forest Science, 30, 970–6.

McRae, R.H.D. (1992) Prediction of areas prone to lightning ignition. *Int. J. Wildland Fire* 2, 123–30.

Morrison, P.H. and Swanson, F.J. (1990) Fire history and pattern in a Cascade Range landscape. *Gen. Tech. Rep. PNW_GTR–254. Portland, OR:U.S Department of Agriculture, Forest Service, Pacific Northwest Research Station.*

Mount, A.B., (1972) The derivation and testing of a soil dryness index using run-off data. Tasmanian Forestry Commission Bulletin.

Olson, J.S., 1963. Energy balance and the balance of producers and decomposers in ecological systems. *Ecology*, 44, 322–31.

Richardson, C.W.(1981) Stochastic simulation of daily precipitation, temperature, and solar radiation. *Water. Resources Research*, **17**, 182–90.

Rothermel, R.C. (1972) A mathematical model for predicting fire spread in wildland fuels. *U.S. Department of Agriculture, Forest Service. Res. Pap. INT–11.*

Taylor, A.H. (1993) Fire history and structure of red fir (*Abies magnifica*) forests, Swain Mountain Experimental Forest, Cascade Range, northeastern California. *Canadian Journal of Forest Research*, **23**, 1672–78.

Vankat, J.L. (1983) General patterns of lightning ignitions in Sequoia National Park, California. In: Proceedings – symposium and workshop on wilderness fires, Missoula, MT, USA. USDA Gen. Tech. Rep. INT–182. pp. 408–11.

Van Wagner, C.E. (1969) A simple fire-growth model. *For. Chron.*, 45, 103–4.

Wallace, G. (1993) A numerical fire simulation model. *Int. J. Wildland Fire*, **3**, 111–6.

The Interaction Between Forest Fires and Human Activity in Southern Switzerland

MARCO CONEDERA[1] and WILLY TINNER[2]

[1]*FNP Sottostazione Sud delle Alpi, Bellinzona, Switzerland*
[2]*Geobotanisches Institut der Universität Bern, Bern, Switzerland*

Abstract: The impact of human activities on the fire regime in southern Switzerland was studied using (pre)historical charcoal and pollen data from lake sediments and statistical data from the 20th century. The cultural impact on forest fire was established by correlating charcoal-influx data with pollen percentages of anthropogenic indicators such as *Plantago lanceolata*, the Cerealia (sum of *Avena* t., *Triticum* t. and *Hordeum* t.) and *Secale*. During the 20th century, fire frequency was correlated with precipitation, dry and very dry periods and landscape management indicators. The effects of human activity on the fire regime are clearly recognisable since at least the Neolithic period. Using palaeoecological or statistical data, the variations in fire regime originating from anthropogenic actions may be differentiated from those due to climatic changes if they are sufficiently conspicuous.

1.   INTRODUCTION

Physical environmental factors, such as climate, affect fire behaviour. Yet, since humans first tamed fire, the story of fire has also been a cultural phenomenon (Pyne *et al.* 1996). Humans have used fire as a powerful tool for making the world suitable for their needs. Although natural fire regimes persisted after Man's first mastery of fire, humans forced the biota to adjust to new anthropogenic fire regimes which varied in time and in different regions (Pyne and Goldammer 1997). The effects that past human societies had on fire regimes primarily depended on the type of culture and the population density. Therefore, the transition from natural to anthropogenic fire regimes differed over time and space. At the beginning of the mid-Pleistocene, the earth's fire regimes were neither wholly natural nor wholly anthropogenic (Pyne *et al.* 1996). In southern Switzerland, pre-historical data obtained by paleo-botanical methods and statistical data from the fire data base of the FNP Sottostazione Sud delle Alpi have allowed us to analyse fire history data on a different time scale. Using this source of information as a basis, the following questions have been examined: What kind of impact did human activities have on the fire regime in southern Switzerland? How did anthropogenic influence on the fire regime change over time?

2.   MATERIAL AND METHODS

2.1   Study region

Southern Switzerland is a region encompassing 4000 km^2 and is located south of the Alps (Figure 1). 48% of the area is covered by forest. In the lower hilly region (< 900 m a.s.l.) the dominant tree is *Castanea sativa*. This species was introduced into the southern Alps by the Romans nearly 2000 years ago (Zoller 1960; Tinner and Conedera 1995). Other important woodland trees found at these elevations are *Quercus petraea, Q. pubescens, Alnus glutinosa, Fraxinus excelsior, Betula pendula,* and *Tilia cordata* (Antonietti 1968). *Fagus sylvatica* is the dominant tree species in the montane zone (up to 1300 m a.s.l.), but is also mixed with *Abies alba* on the north-facing slopes. At higher elevations, the forest vegetation is represented by conifers (mostly *Picea abies, Larix decidua, Pinus sylvestris,* and *Pinus cembra*).

Generally the broadleaf forests are located in the Insubric area where the climate is mitigated by the influence of the lakes. It is a warm–temperate climate, with a mean annual precipitation of 1600–1800 mm and a mean annual temperature of about 12° C (transition between Cfb and Cfa climate after Köppen, mean temperature of January *ca.* 2–3° C, of July *ca.* 21–22 °C; Maggini & Spinedi 1996). The incidence of summer rain (June–September *ca.* 800 mm of precipitation) contrasts with the Mediterranean climate. Because of these climatic factors this area is subject to surface fires with a high rate of spread during the vegetation dormant period (mainly December to April). Fires in coniferous forests at higher elevations mostly occur in the summer season from July to August. These soil fires have a very low rate of spread and are usually ignited by lightning during episodic drought periods. On average, 90 fire events occur per year, burning nearly 1000 ha of land.

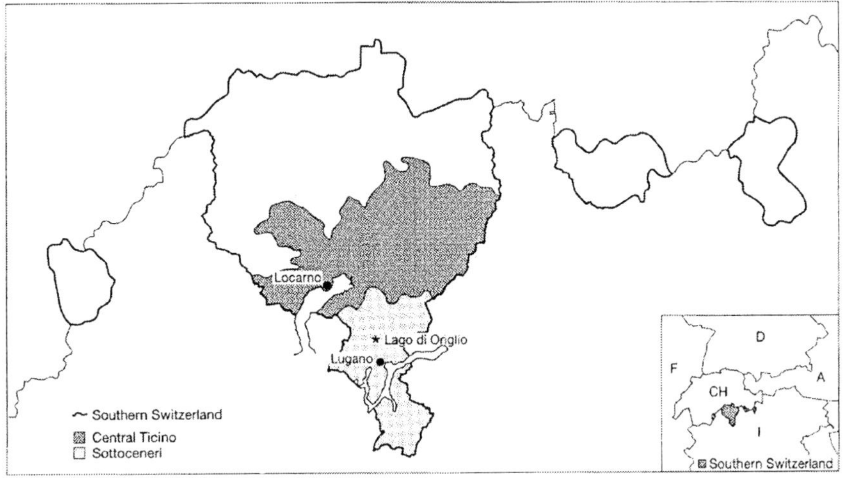

Figure 1. Map of southern Switzerland, showing the location of Lago di Origlio and the two sub-regions: Sottoceneri (reference meteorological station = Lugano) and Central Ticino (reference meteorological station = Locarno).

2.2　Lake sediment cores

Parallel cores were taken 1 m apart with a Streif modification of the Livingstone piston corer (Merkt & Streif 1970) from the deepest point of Lago di Origlio (416 m a.s.l.). This lake is located near Lugano, and has a surface area of about 8 ha and a hydrological catchment area of about 1.5 km^2 (Fig. 1). The water depth was 5.35 m during coring. The sediments older than 14500 BP (afforestation) consist of silt; afterwards silty gyttja (minero-organic sediments of lacustrine origin, mainly deriving from small plant and animal fragments) was deposited in the lake basin.

2.3　Palynology

Lycopodium tablets (Stockmarr 1971) were added to the 1 cm^3 sediment samples to estimate pollen concentration (pollen grains cm^{-3}) and pollen influx (pollen grains cm^{-2} yr^{-1}). The samples were treated chemically with HCl, KOH, HF and acetolysis). Later the same samples were passed through a sieve (0,5 cm) and decanted. We mainly used the key of Moore *et al.* (1991) for pollen identification.

2.4   Charcoal analysis

Charcoal was identified as black, completely opaque, angular fragments (Swain 1973; Clark 1988). The number of charcoal particles larger than 10 μm (or > 75μm^2) on pollen slides was counted and the regression equation proposed by Tinner *et al.* (1998) was used to estimate the charcoal area concentration (mm^2 cm^{-3}) from the particle number concentration (charcoal particles cm^{-3}). According to Tinner *et al.* (1998) this approach can be used to reconstruct regional fires within a radius of 20–50 km around Lago di Origlio. The diagrammatic zonation shown in Figure 2 is based on purely visual observation of changes in charcoal influx.

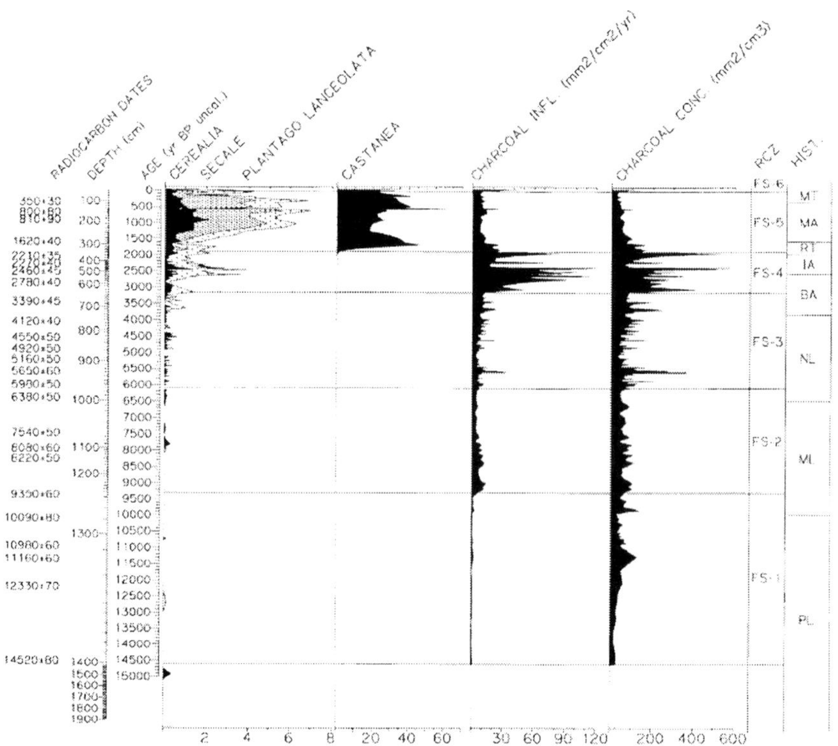

Figure 2. Pollen percentages, charcoal concentration and charcoal influx diagram for Lago di Origlio showing selected pollen types indicating human activities in relation to charcoal influx: RZ = regional charcoal zone, HIST. = historical and pre-historic periods: ML = Mesolithic, NL = Neolithic, BA = Bronze Age, IA = Iron Age, RP = Roman Period, MA = Middle Ages, MT = Modern Time. The charcoal particles were counted from 14600 BP onwards.

2.5   Chronology

Twenty-five terrestrial macro-fossils from Lago di Origlio were dated using AMS-techniques (Tinner *et al.* 1999). The age–depth curves of both study sites were smoothed by locally-weighted regression (lowess) and were calibrated as AD/BC cal. using the Calib Version 3.03c program (Stuiver and Reimer 1993). The results were combined with 7 ^{210}Pb dating results (Tinner *et al.* 1998) to calculate charcoal area (mm^2 cm^{-2} yr^{-1}) and pollen influx (pollen grains cm^{-2} yr^{-1}). Because of the calibration difficulties of the Late Glacial (Wohlfarth 1996) all Late Glacial data are presented with a uncalibrated BP age scale, although the presumed calibrated ages are indicated in the text with a query (?).

2.6   Cross correlations

The lag effect one variable has on another variable (e.g. pairs of pollen types, charcoal particles vs. pollen) can be studied by cross correlation analysis (Green 1981, 1982; Clark *et al.* 1989; Dodson 1990). This statistical analysis compares the separate values of two variables by shifting the value chains for a specified number of time lags and by calculating correlation coefficients at each time lag (Bahrenberg *et al.* 1992). The sequence chosen for cross-correlation analysis (1.73 m of sediment, 5100–3100 BP cal., 6200–4500 BP uncal.) has a sample interval of 11.6 years. For the computations of 20 time lags this results in a total time span of 230 years. Further details on cross-correlation methodology are given in Tinner *et al.* (1999).

2.7   Historical data

Fire data information collected by the Forest Service since 1900 (in standardised form after 1930) helped re-construct a fire history for southern Switzerland in this century. Fire data, incorporating date, time, duration, ignition, area burned, fire type, forest habitat, and other variables, were organised in a relational database (Conedera *et al.* 1996). The significance of these factors was then verified by comparing the results with charcoal concentrations in recent sediments from Lago di Origlio (Tinner *et al.* 1998).

3.   RESULTS

3.1   Fire history

Six major charcoal zones (regional charcoal zones, RCZ) delimit the regional fire history (FS) of southern Switzerland (Fig. 2). The first zone FS–1, ranging from at least 13000 BC cal.? to 8300 BC cal. (12700 to 9300 BP, FS–1), shows very low charcoal influx values. The charcoal values (expressed as both concentration and influx) reached a minimum during the Younger Dryas cooling. Afterwards, charcoal values increased suddenly to about 9200? BC cal. (10000 BP). At around 8300 BC cal. (9300 BP), the charcoal influx increased even more than over the Younger Dryas/Pre-Boreal transition, reaching low to medium values (FS–2). However, this increase is not represented in the charcoal concentration, so we are unsure whether it is the result of a sedimentation increase, leading to stable concentration values, or a function of chronological uncertainties. During FS–3 (6100–3200 BP, 5000–1400 BC cal.) the charcoal influx was medium. The charcoal influx and concentration values of zone FS–4 (3200–1900 BP, 1400 BC–100 AD cal.) are high and reflect frequent, severe forest fires. The regional charcoal zone FS–5 (100 AD cal. to 1960 AD cal.) displays a reduction to moderate charcoal influx and charcoal concentration. The period *ca.* 1400 AD cal. (550 BP) and the period from the 1960 to the present exhibit high charcoal sedimentation. The charcoal peak after 1960 delimits the start of charcoal zone FS–6 and can be also clearly recognised in the annual fire frequency, according to the historical forest fire information (Figure 3).

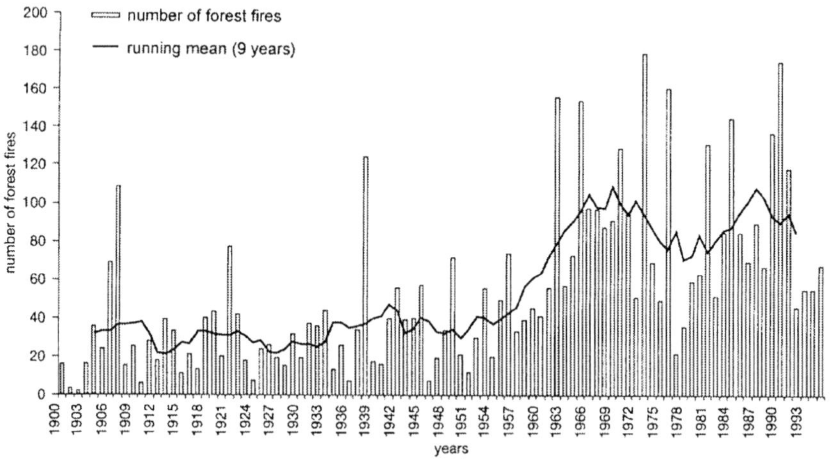

Figure 3. Course of development of number of fires in southern Switzerland since 1900 (source: forest fires data base of southern Switzerland).

3.2   Indicators of human activity

The long-term cultural impact was determined on the basis of the most frequent and decisive anthropogenic indicators *Plantago lanceolata*, the Cerealia (sum of *Avena* t., *Triticum* t. and *Hordeum* t., according to Faegri & Iversen 1989), and *Secale* and may be roughly summarised as follows (Fig. 2). The pollen indicating anthropogenic activities during the Palaeolithic (*Avena* t., *Secale*) is probably due to native wild grasses or to long-distance transport from the Mediterranean region. During the Mesolithic only a few anthropogenic pollen indicators suggest a human presence near the lake (for anthropogenic indicators during the Mesolithic, see Haas (1996) and Erny-Roadmann *et al.* (1997)). The presence of anthropogenic indicators increases during the Neolithic. The first *Triticum* t. pollen was found at around 4900 BC cal. (6000 BP), probably indicating the beginning of farming in the vicinity of the lake (all older Cerealia findings belong to the *Avena* type and might also derive from wild grasses). A distinct shift to more anthropogenic indicators occurred at the beginning of the Bronze Age. During the Iron Age and the Roman Time the presence of anthropogenic indicators is regular and uninterrupted and twice as high as during the Bronze Age. After the Migration of the Peoples at the end of the Roman period, the portion of anthropogenic indicators increases again and reaches maximal values between 900 and 1750 AD (1150–200 BP).

3.3   Fire and human activity

Correlograms of pollen percentage and charcoal influx show four different patterns (Fig. 4): fire-prone species (i.e. *Abies alba*), fire-adapted species (i.e. *Alnus glutinosa*), fire-opportunists (i.e. Rosaceae) and fire-precursors (i.e. Poaceae). It is interesting to note that all the major taxa (Poaceae, Caryophyllaceae, *Pteridium aquilinum*, *Plantago lanceolata*, *Artemisia* sp., and *Quercus* sp.) belonging to the fire-precursors group share a common characteristic: they are considered to be favoured by anthropogenic activities, though for different reasons. *Plantago lanceolata*, the strongest cultural indicator among them, is an adventive plant occurring on anthropogenically disturbed sites and in meadows (Behre 1981, Lang 1994). The Poaceae, the Caryophyllaceae, *Pteridium aquilinum*, *Artemisia* sp., and *Quercus* sp. indicate an increase in light conditions on the site. Their occurrence may be interpreted as an indication of grazing in slightly open forests (forest pasture).

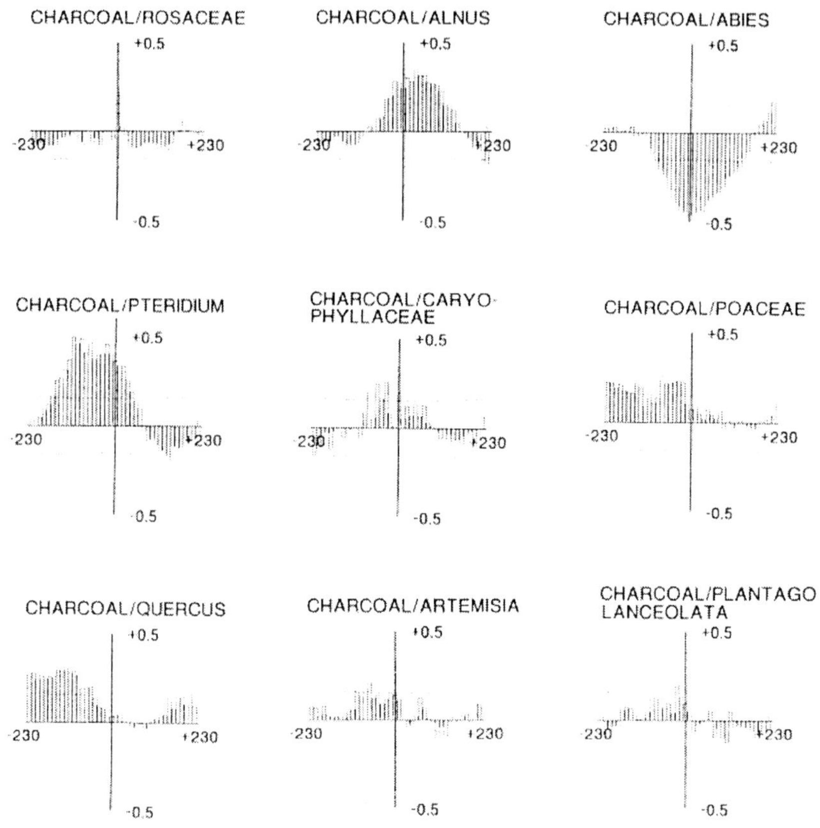

Figure 4. Charcoal influx vs. pollen percentage correlograms of selected taxa from Lago di Origlio (5100–3100 BC cal.). One lag (bar) = *c.* 12 yr. Positive and negative correlation coefficients of 0.5 indicated by +0.5 and –0.5. 230 = 230 years corresponding to 20 lags of 11.6 years. Significance level α=5%, two-sided, source: Tinner *et al.* 1999).

Finally, Figure 5 shows the relationship during this century between forest fire frequency and the evolution of indicators for agricultural activities or forestry activities. As shown in Figure 6, dry and very dry periods during the main fire season (December to April) do not show a systematic increase in number since the sixties. There is only a slight increase in the number of short very dry periods (30 to 45 days without rain) and long dry periods (more than 75 days without significant rain). In addition, no clear trends towards an increase in fire frequency during dry or very dry periods after the sixties can be detected (Figs. 7 and 8). Thus, we deduce that the sudden increase in fire frequency in the late 1950s is not mainly the result of changes in climatic conditions (general flux of precipitation and dry and very dry periods in particular). The constantly high level of fire frequency since the sixties seems much more related to the abrupt changes in landscape management initiated during this period (Fig. 5).

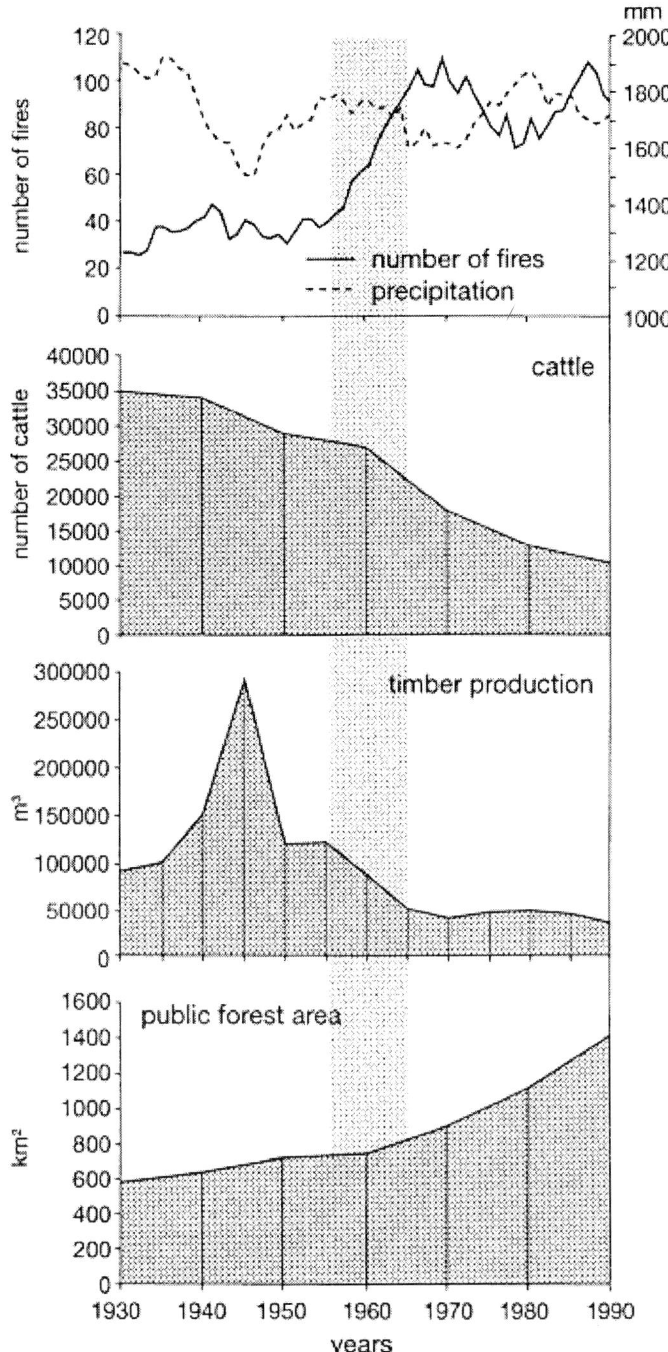

Figure 5. Landscape management and fire history of the past 70 years in the Canton of Ticino, southern Switzerland (Source: Conedera *et al.* 1960).

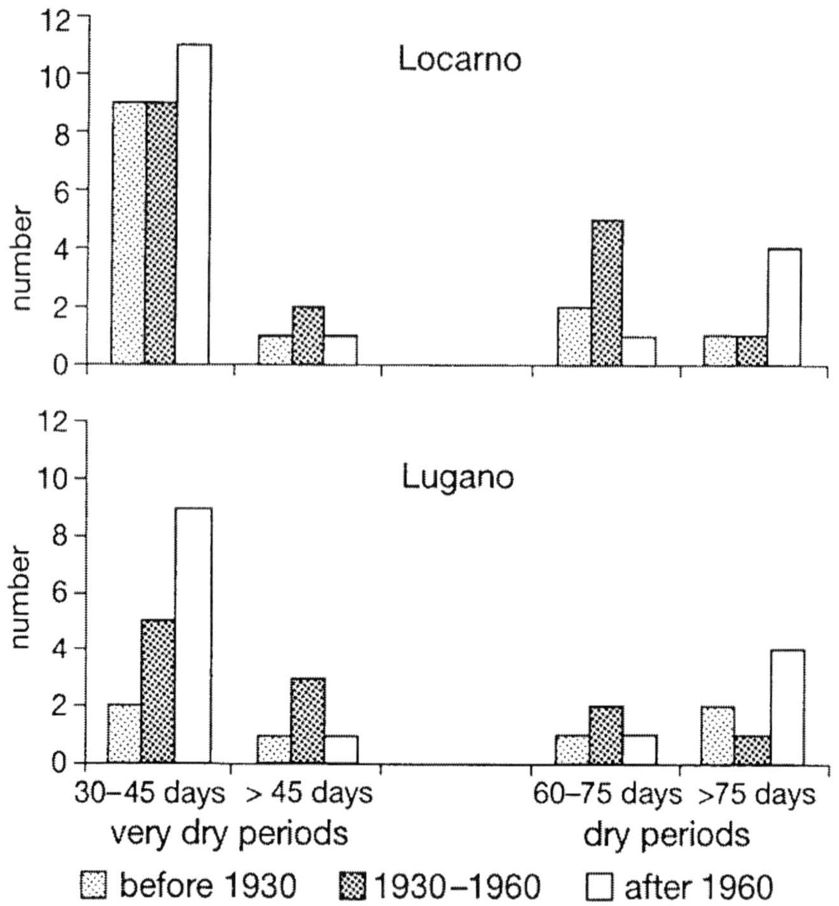

Figure 6. Frequency of drought and dry periods in southern Switzerland since 1900.
Source: SMI MeteoSwiss; Reference stations: Locarno and Lugano.
Very dry period = period of at least 30 consecutive days without rain,
dry period = period of at least 60 consecutive days without significant rain (less then $10 \, \text{l m}^{-2}$).

4.   DISCUSSION

The data referring to the charcoal influx (Fig. 2) show how the history of
the fires south of the Alps has been characterised by abrupt variations in fire
frequency. Tinner *et al.* (1998), taking the Mesolithic (FS–2) as reference,
quantify the increases for the zone FS–3 (corresponding more or less to the
Neolithic and the early Bronze Age) in the mean orders of a factor of 2 and a
factor of 8 (with peaks corresponding to a factor of 40) for the zone FS–4 (in

practice corresponding to the late Bronze Age and the Iron Age). The good correspondence between the trend of the curve of the charcoal influx and the anthropogenic indicators indicates the possible anthropogenic origin of these abrupt and significant variations in the fire regime. As pointed out by Tinner *et al.* (1999), this hypothesis is substantiated by the presence of a group of fire-precursors closely connected with the presence of humans. We suggest that, possibly, the charcoal peaks represent periods of additional forest clearance in order to increase the agricultural area. Nevertheless, between 5100 and 3100 BC cal. (6200–4500 BP uncal.), the simultaneous presence of anthropogenic indicators and undisturbed recovery of fire-sensitive plants can be registered over 100–500 years of time intervals (Tinner *et al.* 1999). This could be an indication that, over a short period, Neolithic farmers burned areas much larger than needed for agriculture. Whether this was intentional remains an open question. In fact, they were probably unable to keep the extended areas open by regular landuse. For this time period the charcoal minima partly coincided with periods of cold–humid climate, so that climate probably had a predisposing effect, diminishing the fire susceptibility of the woodlands during cool and wet periods (Tinner *et al.* 1999).

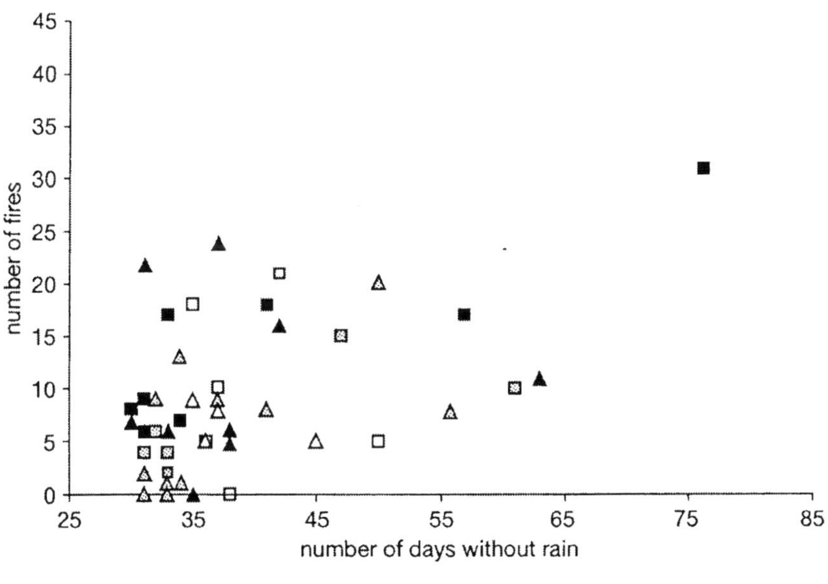

Figure 7. Number of fires as a function of the duration of very dry periods during winter time (December to April) in southern Switzerland. Very dry period = period of at least 30 consecutive days without rain.

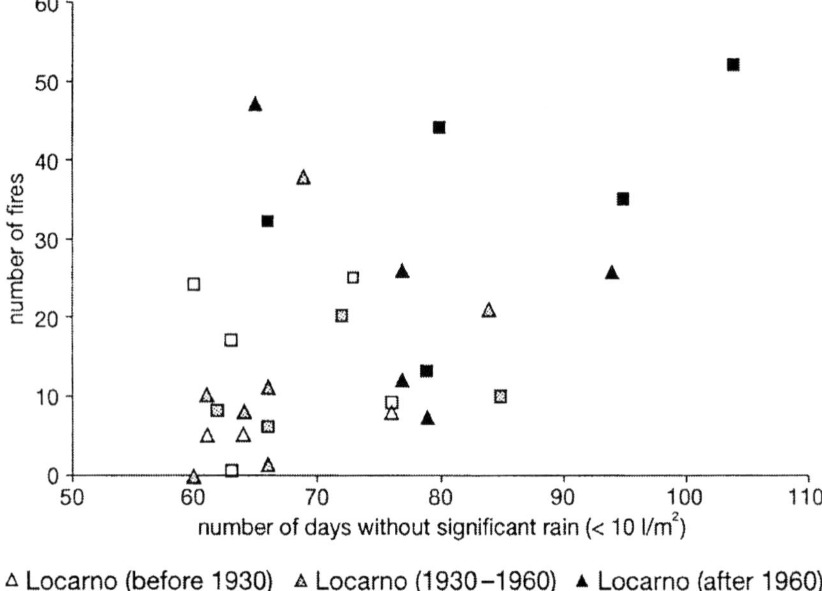

Figure 8. Number of fires as a function of the duration of dry periods during winter time (December to April) in southern Switzerland. Dry period = period of at least 60 consecutive days without significant rain (less then 10 l/m^2).

The strong charcoal peak at the end of the FS–4 zone precedes the introduction of *Castanea sativa* (Fig. 2). Fire was probably used to open the forests stands, in order to provide space for the introduction of chestnut. Due to the high value of chestnuts as an important subsistence crop, chestnuts groves rapidly became a very common form of woodland cultivation (Zoller 1961). This led to a fundamental change in landuse, with fire being abandoned as a method for forest clearance. This change started a new era that endured until the middle of the 20th century. For the first time, the high values of the indicators for anthropogenic activity no longer coincided with the high fire frequencies. Throughout the territory, all land was used to produce timber or wood, leaving no fuel in the forest, the fuel being reduced to a minimum because of the complete utilisation of the biomass and the fire regime being reduced proportionally (zone FS–5).

The progressive abandonment of land management, which started at the end of the 1950s, raises the issue of fire as a regulator of the biomass in forestry ecosystems. This is a clear anthropogenic influence, which can, for a medium-term period, obscure the influence of climate. Conedera *et al.* (1996) have shown how, starting from this period, the anthropogenic factor can in reality influence multiple aspects of the fire regime, such as the causes of ignition, the reduction of the burned surface as a function of better fire fighting strategies and the concentration of events of anthropogenic origin during weekends. This last aspect had to be integrated into fire risk forecasts south of the Alps, having been shown to be significant (Mandallaz and Ye, 1997). It is also surprising to note that the greater susceptibility to fire since the 1960s is not manifest during dry or very dry periods. As fires during the winter periods are mainly caused by humans, people are probably more careful during periods of evident fire danger.

5.   CONCLUSIONS

Our results clearly show that in southern Switzerland, fire regimes have been frequently and considerably modified by anthropogenic actions, from as early as the Neolithic, and maybe even from the Mesolithic.

The combined study of fire frequency and anthropogenic indicators has revealed itself as a very useful tool for this type of analysis. The approach also enables the distinction, with relative certainty, of the variations in fire regime originating from climatic changes from those due to anthropogenic actions, provided that the latter are sufficiently conspicuous or abrupt. Long-term historical information gained through a paleo-botanical approach may be particularly important. Knowledge about the degree of anthropogenicity of the fire regime and of the reaction patterns of the vegetation can constitute a fundamental premise for the study and ecological interpretation of terrestrial ecosystems and their degree of naturalness.

6.   ACKNOWLEDGEMENTS

We are grateful to Fosco Spinedi, SMI MeteoSwiss Locarno–Monti, for providing meteorological data and the reviewers for their useful recommendations.

7.   REFERENCES

Antonietti, A. (1968) Le associazioni forestali dell'orizzonte submontano del Cantone Ticino su substrati ricchi di carbonati. Mitteilungen Schweizerische Anstalt für das forstliche Versuchswesen, 44, 85–226.

Bahrenberg, G., Giese, E. & Nipper, J. (1992) Statistische Methoden in der Geographie. Band 2, Multivariate Statistik. Teubner, Stuttgart.

Behre, K. E. (1981) The interpretation of anthropogenic indicators in pollen diagrams. *Pollen et Spores,* **23**, 225–245.

Clark, J. S. (1988) Stratigraphic charcoal analysis on petrographic thin sections: Application to fire history in northwestern Minnesota. Quaternary Research, 30, 81–91.

Clark, J. S., Merkt, J. & Müller, H. (1989) Post-glacial fire, vegetation, and human history on the northern alpine forelands, south-western Germany. Journal of Ecology, 77, 897–925.

Conedera, M., Marcozzi, M., Jud, B., Mandallaz D., Chatelain, F., Frank, C., Kienast, F., Ambrosetti, P. & Corti, G. (1996) Incendi boschivi al Sud delle Alpi: passato, presente e possibili sviluppi futuri. Rapporto di lavoro PNR 31. Hochschulverlag an der ETH, Zürich.

Dodson J. R. (1990) Fine resolution pollen analysis of vegetation history in the Lough Adoon Valley, Co. Kerry, western Ireland. Review of Paleobotany and Palynology, 64, 235–245.

Erny-Rodmann, Ch., Gross-Klee, E., Haas J.N., Jacomet, S. & Zoller, H. (1997) Früher "human impact" und Ackerbau im Übergangsbereich Spätmesolithikum-Frühneolithikum im schweizerischen Mittelland. Jahrbuch der Schweizerischen Gesellschaft für Ur- und Frühgeschichte, 80, 27–56.

Faegri, K. & Iversen J. (1989) Textbook of Pollen Analysis. (4th edition by K. Faegri, P.E.

Green, D. G. (1981) Time series and postglacial forest ecology. Quaternary Research, 15, 265–277.

Green, D. G. (1982) Fire and stability in the postglacial forests of southwest Nova Scotia. Journal of Biogeography, 9, 29–40.

Haas, J.N. (1996) Pollen and plant macrofossil evidence of vegetation change at Wallisellen-Langachermoos (Switzerland) during the Mesolithic–Neolithic transition 8500 to 6500 years ago. *Dissertationes Botanicae, 267*, 1–67.

Lang, G. (1994) Quartäre Vegetationsgeschichte Europas. Gustav Fischer, Jena.

Maggini, L. & Spinedi, F. (1996) Misurazioni meteorologiche al Parco botanico delle Isole di Brissago, 1962–1995. Bollettino della Società Ticinese di Scienze Naturali, 84, 65–71.

Mandallaz, D., Ye, S.(1997) Prediction of forest fires with Poisson model. Can J. For. Res. 27, 10:1685–1694.

Merkt, J. & Streif, H.J. (1970) Stechrohr-Bohrgerät für limnische und marine Lockersedimente. Geolog. Jahrb., 88, 137–148.

Moore, P. D., Webb J. A., & Collinson, M. E. (1991) Pollen analysis. Blackwell Scientific Publications, Oxford.

Pyne, S.J., Andrews, P.L., Laven, R.D. (1996) Introduction to Wildland Fire. John Wiley and Sons Inc. New York, 769 p.

Pyne, S.J., Goldammer, J.G. (1997) The culture of fire: an introduction to anthropogenic fire history. Sediment Records of Biomass Burning and Climate Change. (eds. J.S. Clark, H. Cachier, J.G. Goldammer, & B. Stocks. NATO ASI Series I: Global Environmental Change, Vol. 51, 71–114. Springer, Berlin.

Stockmarr, J. (1971) Tablets with spores used in absolute pollen analysis. Pollen et Spores, 13, 615–621.

Stuiver, M. & Reimer, P.J. (1993) Extended 14C data base and revised Calib 3.0 14C age calibration program. Radiocarbon, 35, 215–230.

Swain, A. M. (1973) A history of fire and vegetation in northeastern Minnesota as recorded in lake sediments. Quaternary Research, 3, 383–396.

Tinner, W. & Conedera, M. (1995) Indagini paleobotaniche sulla storia della vegetazione e degli incendi forestali durante l'Olocene al Lago di Origlio (Ticino Meridionale). Bollettino della Società Ticinese di Scienze Naturali, 83: 91–106.

Tinner, W., Conedera, M., Ammann, B., Gäggeler, H.W., Gedye, S., Jones, R. & Sägesser, B. (1998) Pollen and charcoal in lake sediments compared with historically documented forest fires in southern Switzerland since AD 1920. The Holocene, 8, 31–42.

Tinner, W., Hubschmid, P., Wehrli, M., Ammann, B. & Conedera M. (1999) Long-term forest-fire ecology and dynamics in southern Switzerland. Journal of Ecology, in press.

Wohlfarth, B. (1996) The chronology of the last termination : a review of radiocarbon-dated, high-resolution terrestrial stratigraphies. Quaternary Science Reviews, 15, 267–284.

Zoller, H. (1960) Pollenanalytische Untersuchungen zur Vegetationsgeschichte der insubrischen Schweiz. Denkschrift der Schweizerischen Naturforschenden Gesellschaft, 83, 45–156.

Zoller, H. (1961) Die Kulturbedingte Entwicklung der insubrischen Kastanienregion seit den Anfängen des Ackerbaus im Neolithikum. Ber. Geobot. Inst. Eidgenöss. tech. Hochsch., Stift. Rübel 32:263–279.

Indirect and Long-Term Effects of Fire on the Boreal Forest Carbon Budget

ERIC S. KASISCHKE[1], KATHY O'NEILL[2], LAURA L. BOURGEAU-CHAVEZ[1] and NANCY H.F. FRENCH[1]
[1]ERIM International, P.O. Box 134008, Ann Arbor, USA
[2]Nicholas School of the Environment, Duke University, P.O. Box 90328, Durham, USA

Abstract: The landmark paper of Seiler and Crutzen (1980) clearly laid out the scientific rationale as to why the study of biomass burning was essential in terms of completely understanding greenhouse gases emissions from the land surface to the atmosphere. While this analysis fueled much of the fire and biomass burning-related research in the tropical biomes (forests and savannas), it turned scientific attention away from the boreal region, as one conclusion drawn from the Seiler and Crutzen analysis was that on a global scale, the amount of carbon released into the atmosphere was very small relative to other biomes. Recent research has shown that the Seiler and Crutzen estimates of carbon released into the atmosphere from fires in the boreal forests may be out by at least an order of magnitude. In addition, because fires consume organic soils which take thousands of years to form, biomass burning in the boreal forest may actually represent a direct net release of carbon to the atmosphere over intermediate time scales (hundreds of years). Finally, fires in boreal forests have significant impacts on the physical climate of the soil substrate in boreal forests, which in turn, affect soil respiration and plant growth. Thus, fires have important indirect effects on sequestration of atmospheric carbon in the boreal region.

In this paper, we review the overall short- and long-term effects that fire has on exchanges of carbon between the atmosphere and boreal forest ecosystems. Using examples from recent research in the boreal forests of Alaska and elsewhere, this paper addresses the following questions: (1) how much fire occurs in the boreal forest? (2) how much biomass is consumed during fires in the boreal forest and what types and what levels of greenhouse gases are released during these fires? (3) how much carbon was released during fires in the North American boreal forests from 1970 to present? (4) how do fires influence patterns of soil respiration? and (5) how do fires influence the patterns of forest succession?

Reference: Seiler, W. and Crutzen, P.J. (1980) Estimates of gross and net fluxes of carbon between the biosphere and atmosphere. *Climate Change* **2**, 207–247.

1.   INTRODUCTION

In a companion paper in this volume, Kasischke *et al.* (1999a) presented estimates of direct carbon release during fires in the boreal forest. As fires dramatically change the biological and physical characteristics of a forested landscape, they also have a strong indirect influence on carbon cycling. In particular, because of high levels of tree mortality, fires represent a dominant disturbance force, initiating secondary succession in most boreal forest ecosystems. Fires also change the physical characteristics of a site in the following ways:

1. consumption of canopy biomass and the loss of live foliage increases the direct solar insolation of the ground layer;
2. elimination of live vegetation, including mosses, significantly changes both the albedo and thermal emissivity and conductivity characteristics of the ground layer; and
3. the killing of the vegetation results in a loss of plant transpiration, which not only removes a significant latent heat loss source from the landscape, but also results in less water being removed from the soil.

A common characteristic of many boreal forest ecosystems is the presence of a permanently frozen soil below a certain depth, known as permafrost. Permafrost significantly impedes soil drainage, resulting in water-saturated conditions in the uppers layers of mineral soil and the lower layers of the organic soil profile during much of the growing season. As a result of changes caused by fire, the ground-layer of boreal forests experiences a significant warming during the first several years after a fire. This results in a deepening of the active layer, e.g., the depth to which permafrost thaws during the growing season. The increase in active layer depth, in turn, affects soil moisture conditions. In general, the organic soil layers become drier because of improved drainage and direct warming of the ground surface. Mineral soils may actually increase in moisture during the initial years after fire because of the elimination of the plant transpiration that removes a significant amount of water from the soil. The changes in soil temperature and moisture are significant factors in determining the patterns of post-fire soil respiration and plant succession which, in turn, will affect carbon storage in these ecosystems.

In this chapter, we discuss how fires indirectly influence carbon cycling in Alaskan boreal forests. Using recent field observations in the boreal forests of Alaska, we first present evidence of how fire affects soil temperature and moisture conditions. We next discuss how patterns of total soil carbon dioxide emissions vary between burned and unburned forest stands in our study sites. We then show how fire may be influencing longer-term patterns of forest succession. The last section of this chapter discusses the implications of the combined effects of climate change and fire on the boreal forest carbon budget. This discussion accounts not only for changes in direct emissions, but also the indirect influences of fire on carbon cycling.

2. EFFECTS OF FIRE ON SOIL TEMPERATURE AND MOISTURE

In mid-August 1998, a series of burned and adjacent unburned black spruce (*Picea mariana* [Mill.] B.S.P.) forest stands were visited in interior Alaska. These stands were located near Fairbanks, Delta Junction, and Tok, Alaska, within or near fires that occurred in 1972, 1987, 1990, and 1994. At the 1990 fire site, the unburned and burned stand were separated by a distance of approximately one kilometre, whereas at all the other sites this distance was less than 100 meters. The site examined in this study had fairly similar average tree height, diameter and density in both the burned and unburned areas. Examination of pre-burn aerial photographs for the sites were also used to confirm that the sites were similar.

Each one of the burned sites had between 10 and 20 cm of organic soil remaining after the fire, compared to 20 to 30 cm of organic soils in the unburned sites. At each pair of sites (burned and unburned), a soil pit was dug to a depth of one meter into the mineral soil or until the permafrost layer or a layer of gravel was reached. Previous studies have shown that the maximum soil temperature in this region is reached during the mid- to late-August time period (Adams and Viereck 1997). The maximum soil temperature lags the maximum air temperature (which occurs in mid-July) by four to six weeks. All the stands were located on flat topography ($< 5°$) to minimise the effects of slope and aspect on soil temperature and slope on soil moisture. Temperature and volumetric moisture measurements were obtained at 10 cm intervals throughout each soil profile. The accuracy of the temperature measurements was $\pm 0.5°$, whereas the accuracy of the soil moisture measurements was $\pm 5\%$.

Figure 1a presents the soil temperature profiles from the 1994 burn and illustrates the distinct warming that occurs throughout the soil profile. These measurements are based on single data points. The depth to the permafrost layer has been greatly deepened by the fire. Figure 1b presents a plot of temperature differences (burned temperatures minus unburned temperatures) for the four different study areas. A paired, two-sample t-test for means indicated that the temperature differences were significantly different ($P <$ .00001). These data illustrate that the effects of fire on temperature last for at least 16 years after the fire.

Soil moisture profiles were not collected in the 1972 burn because of rain at the time of sampling. In the other sites, a trend was present in the data, with the burned sites having slightly higher soil moistures in the upper 30 cm of mineral soil than the unburned sites (Figure 2). The average soil moisture was 2.1% higher in the burned stands than in the unburned stands, as expected because of elimination of the forest canopy, which removes soil water via transpiration. This difference was significantly different ($P <$ 0.007) based on a paired, two-sample t-test for means. As seasonal variations in soil moisture are highly dependent on precipitation events, much more extensive measurements would be required to completely assess the effects of fire on soil moisture.

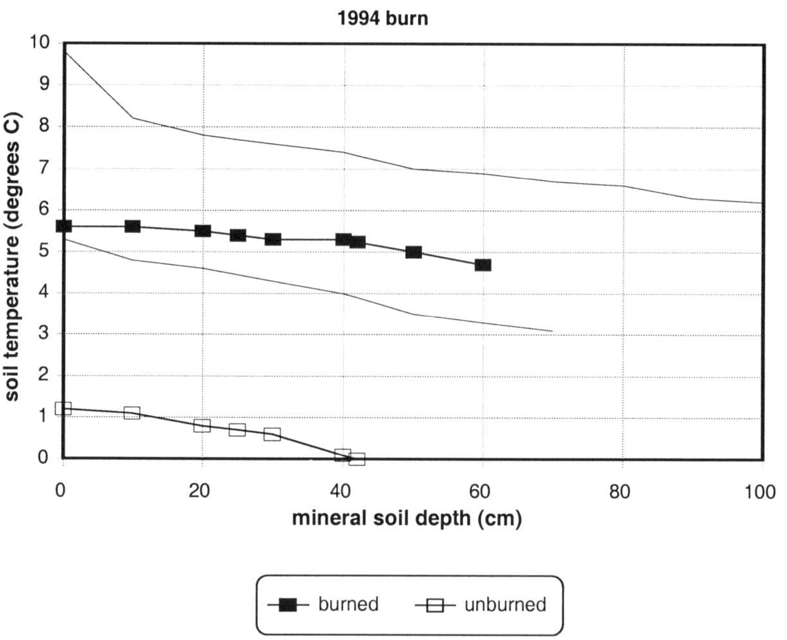

Figure 1a.

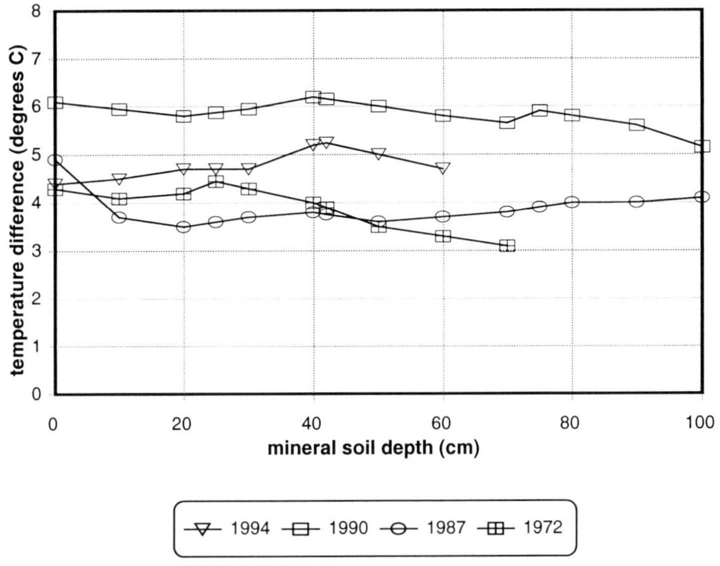

Figure 1b.

Figure 1a-1b. Effects of burning of organic soils on mineral soil temperature in an Alaskan boreal forest. (a) mineral soil temperature profiles collected in mid-August 1998 in a black spruce stand burned in the summer of 1994 and adjacent unburned stand. In the unburned stand, permafrost was reached at a depth of 42 cm. (b) temperature differences (temperature in burned stand minus temperature in unburned stand) in a series of black spruce forest test sites that burned between 1972 and 1994. The data in these points are from single measurements whose accuracy is ± 0.5° C.

3. EFFECTS OF FIRE ON TOTAL SOIL RESPIRATION

Microbial respiration in the soils of boreal forests is very sensitive to variations in temperature and moisture (Schlentner and Van Cleve, 1985). It has been hypothesised that increases in ground temperature would result in significant increases in fluxes of CO_2 as a result of this increased respiration (Bonan and Van Cleve, 1992; Kasischke *et al.*, 1995). To evaluate this hypothesis, total soil respiration was measured in different aged black spruce stands (both burned and unburned) during the summers of 1996 and 1997 (O'Neill *et al.*, 1997; Richter *et al.*, 1999). The focus of this study was to measure how fire affects the rates of soil respiration and to determine how these patterns are influenced by variations in soil moisture and temperature. Soil respiration rates were measured using a dynamic closed gas exchange system (EGM-1 environmental gas monitor and SRC-1 soil respiration chamber from PP Systems). A set of measurements was obtained every two weeks during May through August of 1997. A total of 10 samples were obtained in five black spruce stands that were burned in 1987, 1990, 1994 (2 sites), and 1996. Ten samples were also collected in adjacent unburned stands.

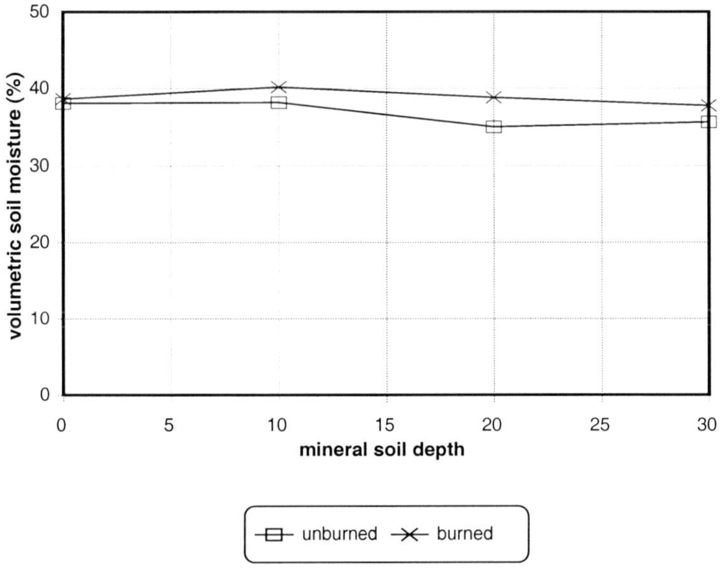

Figure 2. Volumetric moisture profiles collected in the mineral soil of a black spruce forest burned in 1994 and an adjacent unburned stand. The data in these points are from single measurements whose accuracy is ± 5%.

The surface CO_2 fluxes measured in this study originated from three sources: (1) respiration from the roots and trees occupying the site; (2) respiration from the mosses covering the ground surface; and (3) respiration from microbes and fungi living in the soils. Fires in Alaskan black spruce forests are extremely destructive to all vegetation at the site. Fires kill all overstory trees and consume all needles and small branches. They kill and consume all the living and dead plant material in the canopy understory, including the moss layer. Depending on fire severity, they typically consume between 20 and 90% of the organic soil layer, including any living roots present in this layer. Despite the intensity of the above-ground and surface fires, the rate of vertical heat transfer down into the soil layer is fairly slow because of the high moisture content of the soils. Therefore, microbes and fungi living a few centimetres below the burned surface are not affected by a fire. Because of the death of the living plants, all respiration immediately after a fire is attributable to respiration of soil microbes and fungi.

Figure 3 presents surface CO_2 emissions from one of the 1994 burn sites, illustrating the seasonal pattern of carbon flux (a more thorough discussion of these data are presented in Richter *et al.*, 1999). Previous studies have shown that total soil respiration is directly proportional to both soil temperature and soil moisture in Alaskan forests (Schlentner and Van Cleve, 1985), and the data in Figure 3 support these previous observations. Because of the dependence on soil temperature, rates of soil respiration should increase throughout the summer and reach a maximum during mid- to late August. This seasonal pattern would be modified by variations in soil moisture, which are largely controlled by precipitation patterns. The dip in soil respiration rates in late June is associated with low soil moisture levels (which reduce the rates of total plant and root respiration), while the sharp rise during early July is associated with increased soil moisture from heavy rain during this period.

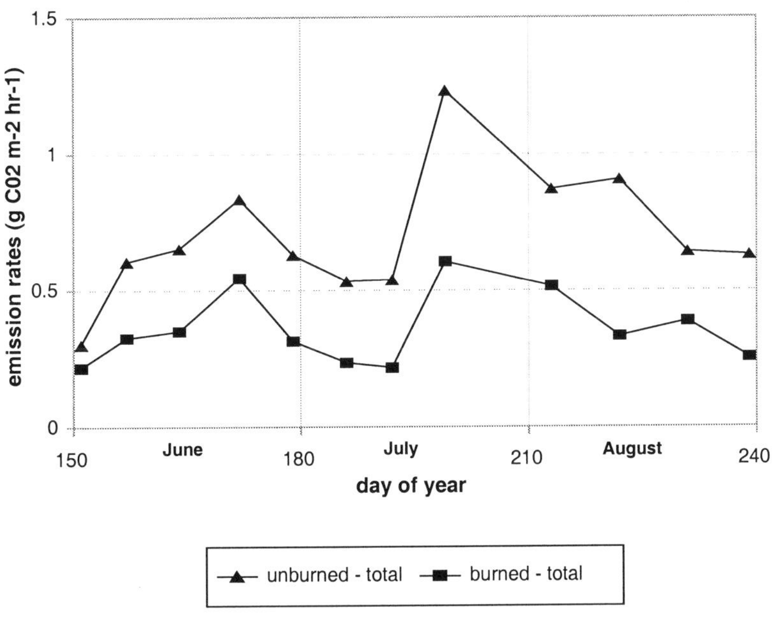

Figure 3. Measured soil surface CO_2 emission rates collected in a black spruce forest that burned in 1994 and an adjacent unburned stand during the summer of 1997.

Figure 4a presents average total CO_2 fluxes for the five different study sites. In all instances, the higher total surface respiration rates are in the unburned black spruce stands. However, total surface respiration has two components: microbial soil respiration and root respiration. Schlentner and Van Cleve (1985) estimate that in mature black spruce stands, 80% of the total surface CO_2 emissions originate from root respiration and 20% from microbial soil respiration. Figure 4b presents an estimate of the rates of microbial soil respiration. In the 1996 burn site (BS-1), no living vegetation existed, so all the surface CO_2 flux in this case was from microbial soil respiration. In the other stands, vegetation regrowth had occurred, but the amount of above-ground biomass present in this stands was no more than 10 – 20% of that found in the control stands (whose ages ranged between 70 and 110 years). To estimate the rates of microbial soil respiration, we estimated that root respiration contributed 5% of the total in the 1994 burn (BS-3a and 3b), 10% in the 1990 burn (BS-7) and 20% in the 1987 burn (BS-10). In all cases, the microbial soil respiration rate was higher in the burned stands (Figure 4b).

The respiration measurements in the 1994 burn were obtained from one site that had a remnant organic soil layer (after the fire), while the other site had very little organic soil remaining. In the latter case, the increase in CO_2 emissions originated from microbial respiration in the mineral soil.

Richter *et al.* (1999) estimate that increased microbial soil respiration after fires in Alaskan black spruce forests result in a net loss of 20 t C ha^{-1} from the soil layer during the first 15 to 20 years after a fire. This loss is in addition to direct carbon emissions through burning of organic soil during the fire, which has been measured at 36 t C ha^{-1} for black spruce forests (Kasischke *et al.*, 1999b).

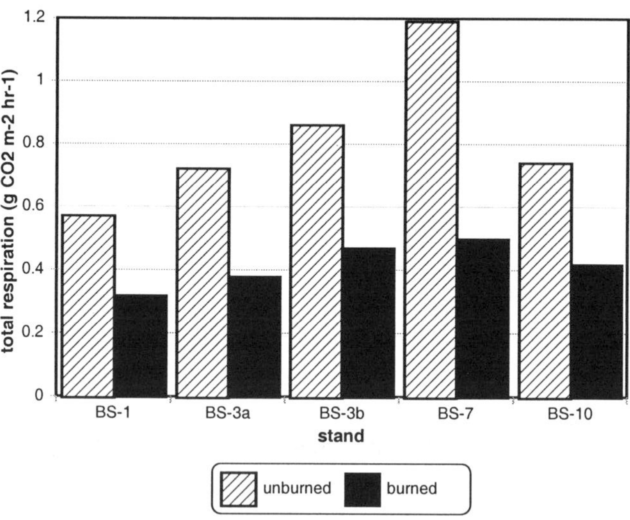

Figure 4a.

These results illustrate that fires have a strong direct and indirect influence on carbon storage in the soils of boreal forests. We estimate that during the fire itself and due to increased microbial respiration after a fire, the black spruce forests in interior Alaska are losing 56 t C ha^{-1} on average. The average ground layer biomass present in the black spruce stands was 228.7 t C ha^{-1} (the mineral soil had 150 t C ha^{-1} and the moss–litter–organic soil layers had 78.7 t C ha^{-1}). This means that fires result in a loss of nearly 25% of the total carbon stored in black spruce forests.

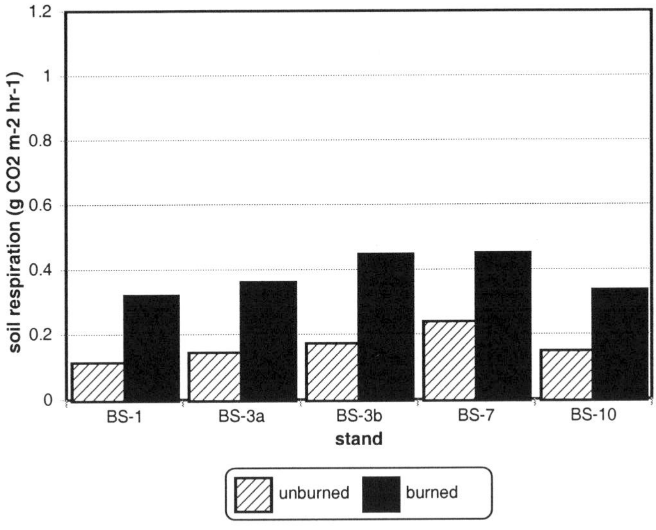

Figure 4b.

Figure 4a-4b Average CO$_2$ emissions collected between June and August 1997 in a series of burned and adjacent unburned black spruce stands in interior Alaska.
The stands burned in 1996 (BS-1), 1994 (BS-3a,b), 1990 (BS-7) and 1987 (BS-10).
(a) total surface emission rates. (b) estimated emissions from soil respiration.

4. EFFECTS OF FIRE SEVERITY ON FOREST SUCCESSION

In their classic studies of Alaskan boreal forest ecology, Keith Van Cleve and Les Viereck showed that the distribution of black spruce forests is strongly correlated with patterns of soil moisture and temperature. Figure 5 is based on data collected in white and black spruce forests found in interior Alaska. (Van Cleve *et al.*, 1983a,b; Van Cleve and Viereck, 1981). Soil moisture and temperature were measured in a series of spruce forest stands over a growing season. The seasonal patterns in these data were expressed in the average percent water present in the upper layers (20 cm) of the mineral soil and the degree days above 0° C at a 10 cm depth in the mineral soil. The oblong shapes in Figure 5 illustrate those regions where the different black spruce and white spruce forest stands were located in terms of the seasonal temperature–moisture space. White spruce forests are found on warmer, drier sites whereas black spruce forests inhabit cooler, wetter sites.

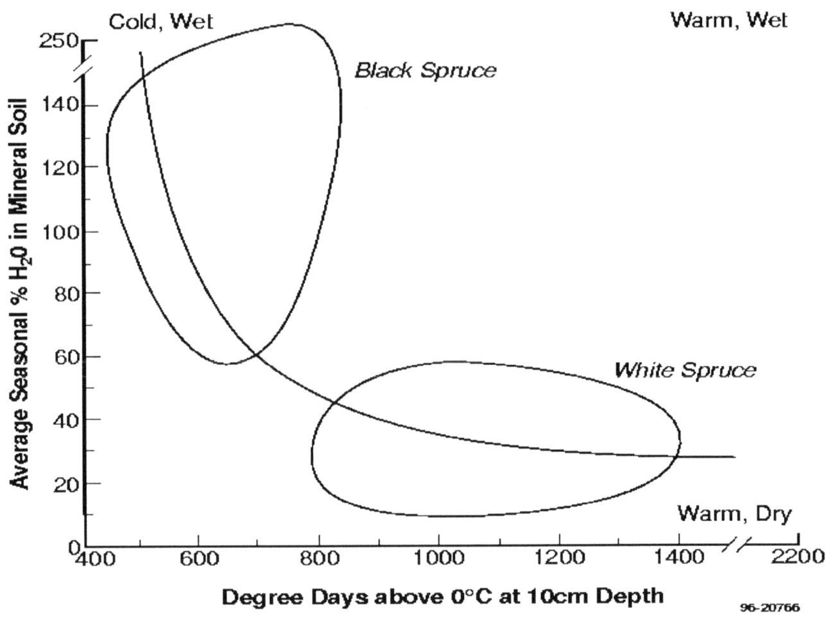

Figure 5. Influence of seasonal patterns of soil temperature and moisture
on forest ecosystem distribution in interior Alaskan boreal forests
(after Van Cleve *et al.*, 1983a,b; Van Cleve and Viereck, 1981).

Researchers believe that fire strongly influences patterns of soil temperature and moisture, with organic soils becoming warmer and drier immediately after a fire. While soil moisture and temperature data were not collected in burned forest sites, Van Cleve *et al.* (1983a) and Viereck 1983) hypothesised that soil temperature increases and soil moisture decreases immediately after a fire (Figure 6a). The relationships displayed in Figures 5 and 6a have been used to understand forest succession on upland white and black spruce forests in interior Alaska. On the warmer drier sites where white spruce forests are found, deciduous species (aspen – *Populus* spp. and birch – *Betula* spp.) dominate early in the successional chronosequence and are gradually replaced by white spruce after 60 to 100 years. In black spruce forests, shrubs dominate early in the chronosequence, but black spruce becomes the dominant canopy species in 20 – 30 years. In upland sites, deciduous species are only occasionally found in black spruce forests.

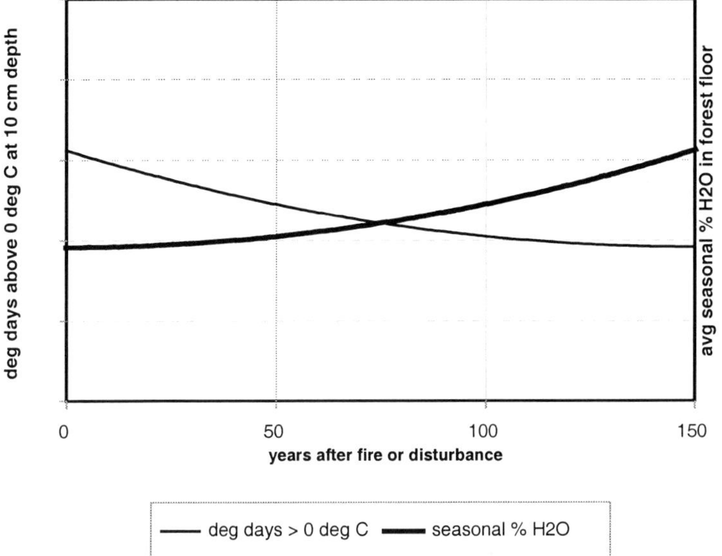

Figure 6a. [Enter a caption for this figure]

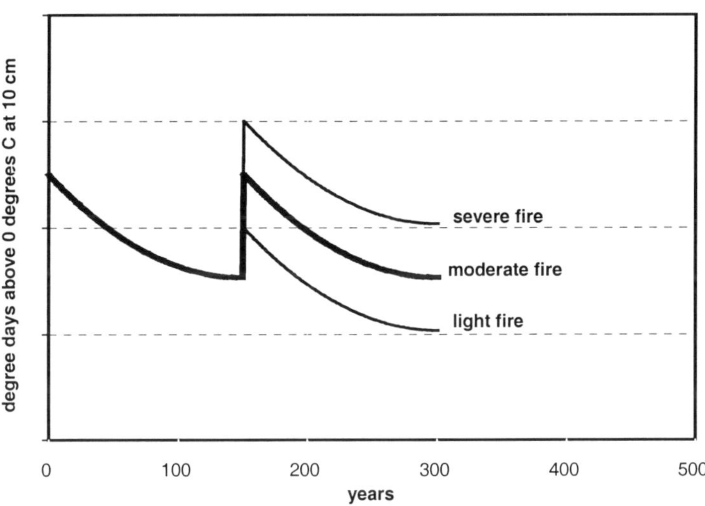

Figure 6b. [Enter a caption for this figure]

Figure 6a-6b. Influence of fire on post-fire temperature and moisture profiles. The y-axes on this plot are relative changes in soil temperature and moisture (a) Hypothesised changes based on a baseline fire severity (after Van Cleve *et al.*, 1983a; Viereck, 1983). Soil temperature immediately increases and soil moisture decreases after a fire. As vegetation re-grows during succession and the depth of the organic soil increases as well, the soil temperature decreases and soil moisture increases; (b) Hypothesised patterns of ground temperature based on different levels of fire severity. In this example, the fire occurs at year 150. The level of fire severity controls the magnitude of warming after a fire.

274   *Eric S. Kasischke et al.*

In recent Alaskan studies in forest sites with a flat topography, we have found a transitional forest type to occur, one where aspen or birch invades early in the successional process. Figure 7a presents a 40 year-old stand with a black spruce understory and an aspen overstory. Immediately adjacent to this mixed stand is a 40 year-old stand of pure black spruce (Figure 7b). The difference between the two stands is the depth of the organic soil, with the mixed stand having an organic soil depth of less than 5 cm and the black spruce stand having a depth of nearly 13 cm.

Figure 7. Examples of effects of fire severity on post-fire succession. (a) 40 year-old mixed aspen – black spruce forest in area of severe burn; (b) 40 year-old black spruce stand in area of less severe burn; (c) 3 year old burn with aspen/black spruce regeneration in area of severe burn; and (d) 3 year-old burn with black spruce regeneration in area of less severe burn.

In a more recent fire, we came across direct evidence of the establishment of aspen at a site that had previously been occupied entirely by black spruce (Figure 7c). At this site, a recent fire (in 1994) had consumed the entire organic soil mat down to the mineral soil. In these areas, aspen seedlings were establishing themselves and thriving, reaching a height of 50 to 80 cm by the summer of 1998. Black spruce seedlings were also present in this site. In another region of the burn, where the fire was less severe and 10 to 20 cm of organic soil remained, no aspen seedlings were present (Figure 7d). Field measurements in the summer of 1998 showed that the burned areas with organic soils remaining were 3° to 4° C cooler than sites with no organic soils.

These observations along with others have resulted in a hypothesis of how fire severity may be controlling post-fire succession in some regions of the Alaskan boreal forest (Kasischke *et al.*, 1999b). This hypothesis is based on the increase in ground temperature being dependent on fire severity, e.g., the amount of organic soil consumed during the fire. More severe fires result in warmer post-fire ground temperatures, while less severe fires result in cooler temperatures (Figure 6b). We hypothesise that a similar pattern occurs with soil moisture: more severe fires can result in drier soils and less severe fires can result in wetter soils. Taken together, our hypothesis is that a series of less severe fires will eventually lead white spruce forests to be replaced by black spruce forests while a series of more severe fires will result in black spruce forests being replaced by aspen – black spruce forests and eventually, white spruce forests (Figure 8).

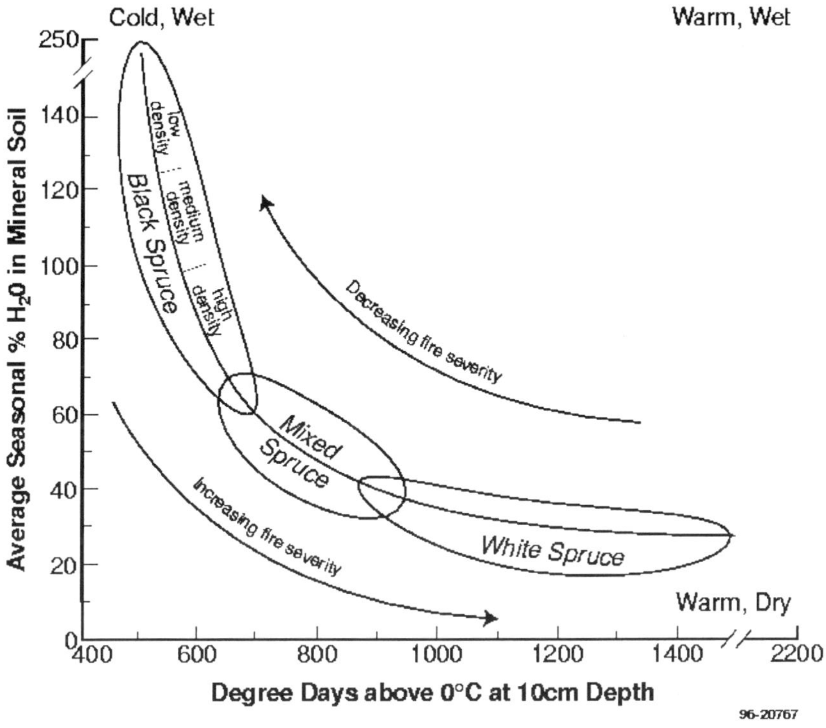

Figure 8. Hypothesised effects of fire severity on forest ecosystem distribution in the Alaskan boreal forest.

The hypothesis that black spruce forests replace white spruce forests is identical to the one developed by Van Cleve and Viereck (1981) to explain white spruce – black spruce succession in the floodplains of interior Alaska. There is a critical difference between these hypotheses, however. In our hypothesis, depending on the long-term fire regime, a site can first be occupied by black spruce, revert back to white spruce, and again be occupied by black spruce. The floodplain chronosequence only acts in one direction, e.g., white spruce eventually being replaced by black spruce.

This hypothesis has important implications for carbon dynamics in the boreal forest, especially in a warming climate. Most carbon-dynamic – climate change studies conducted for the boreal forest suggest the eventual replacement of many carbon-rich forest ecosystems (e.g., black spruce forests) with less carbon-rich ecosystems (e.g., white spruce) (Smith and Shugart, 1993). The hypothesis discussed above states that the rate of change in many ecosystems will be driven by patterns of fire severity. Recent modelling studies show that patterns of fire severity in the boreal forest are likely to increase proportionally to increases in temperature in this region forecasted by general circulation models (Stocks *et al.*, 1998). If this modelling study is accurate, then the rates of carbon loss from the boreal forests are likely to be much quicker than predicted by the current generation of forest succession models.

5.   COMBINED EFFECTS OF FIRE AND CLIMATE CHANGE ON THE BOREAL FOREST CARBON BUDGET

While it is believed that overall the boreal forests presently serves as a net sink of atmospheric carbon, as shown in this chapter and its companion (Kasischke *et al.*, 1999a), individual forest stands act alternately as atmospheric carbon sources and sinks because of the influence of fire. Stand-replacement fires are the norm in most boreal regions, resulting in the death and partial burning (oxidation) of most overstory trees, complete burning of understory vegetation, and partial burning of the mosses, lichens, litter and organic soil present in the forest floor. In fires with a long smouldering phase, the organic soil can be burned all the way to the top of the mineral soil. Thus, a fire can result in a virtually instantaneous release of large amounts of carbon in the form of greenhouse gases (between 10 and 200 t C ha $^{-1}$) to the atmosphere. The standing dead tree boles and branches remaining after the fire eventually decompose or become part of the undecomposed dead organic matter present in the forest floor.

A generalised pattern of carbon accumulation in a boreal forest after a fire (after Kasischke *et al.*, 1995) is presented in Figure 9. The plots in this figure assume the fire occurred at year 0. These plots also assume the tree biomass remaining after a fire becomes part of the forest floor or ground layer biomass. Typically, the amount of carbon present in living vegetation after a fire continuously accumulates until a mature forest stand is present. At this time, there may be a slight decrease in living-vegetation carbon because the biomass lost through mortality of older trees is greater than net annual production of living biomass. In the ground layer (forest floor) there is a net loss of carbon during the first several years after a fire because of increased soil respiration, as discussed in Section 3 of this chapter. Modelling studies by Bonan (1992) support the pattern of ground-layer carbon presented in Figure 9.

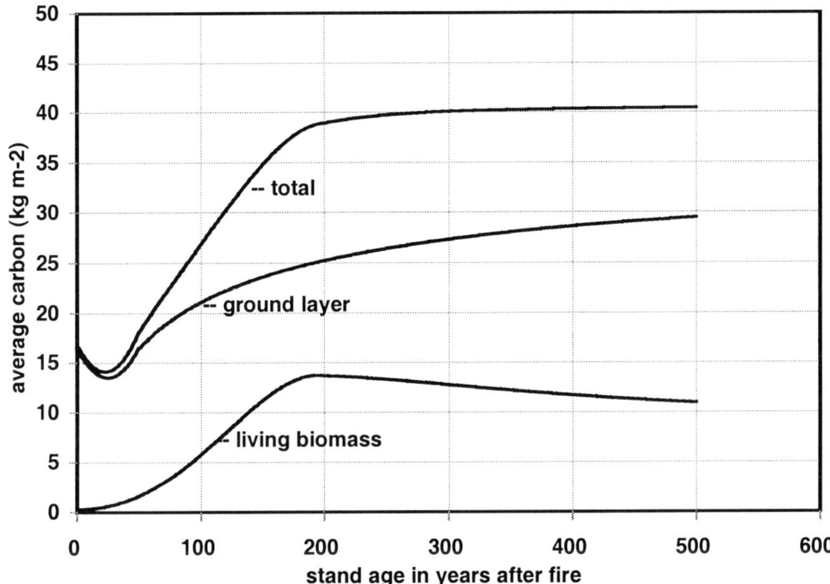

Figure 9. Patterns of average biomass – carbon accumulation as a function of stand age for the boreal forest region.

The other manner in which fire strongly influences the carbon budget in boreal forests is through its control on stand-age distribution. Van Wagner (1978) showed that using fire frequency along with relative stand flammability in a Weibull distribution adequately describes stand age distribution in regions such as boreal forests where fire is the dominant disturbance. The principle behind this relationship is simple and illustrated in Figure 10: areas with a high fire frequency will have a larger number of young stands, while areas with a low fire frequency have a larger number of older stands. When the stand age distribution curves are integrated with the curves describing the patterns of carbon accumulation, then an average carbon level as a function of fire frequency (or fire return interval) can be calculated. Figure 11 illustrates the relationship between fire return interval and the average carbon present in a boreal forest. From this figure it can be seen that as fire frequency increases (fire return interval decreases), the amount of carbon stored in boreal forest decreases.

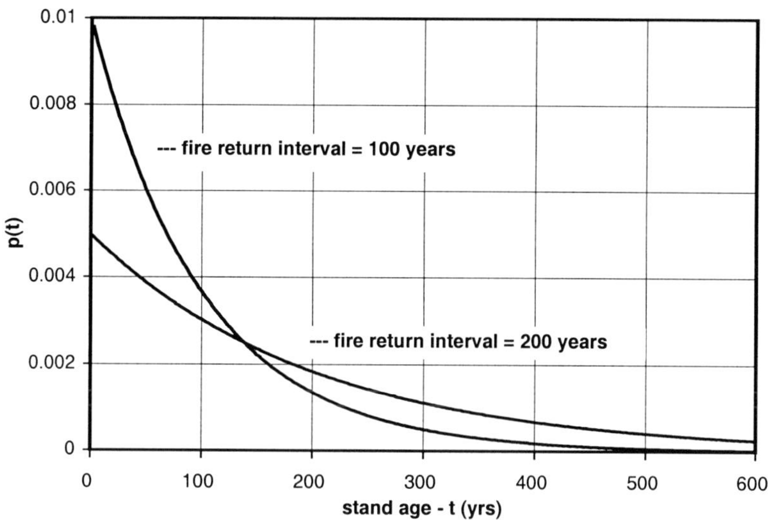

Figure 10. Stand age distribution for boreal forests based on fire return interval generated using a Weibull distribution.

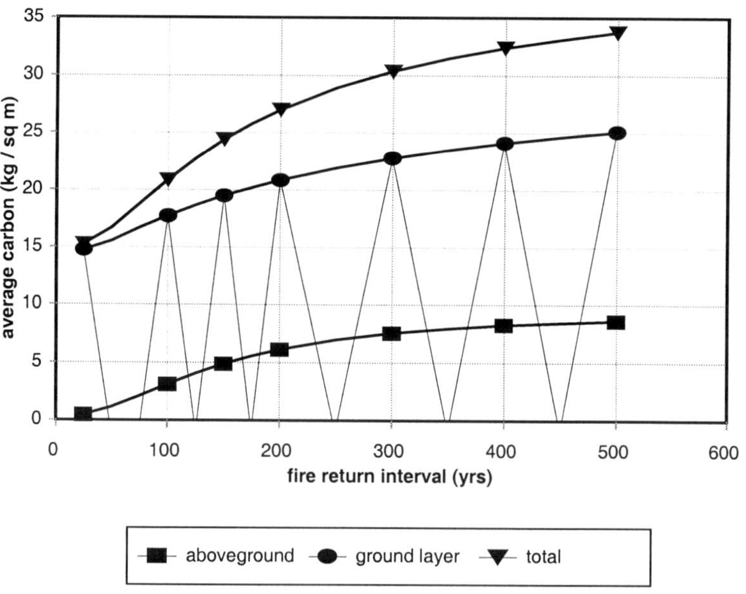

Figure 11. Model estimates of average carbon present in boreal forests as a function of fire return interval.

Kasischke *et al.* (1995) used this basic approach to investigate the influence of patterns of climate change on the fire regime and patterns of carbon accumulation in the boreal forest. In this study, they assumed that climate warming would result in a 25% increase in above-ground biomass present in the boreal forest due to changes in forest types and increases in net primary productivity. They estimated there would be an approximately equal decrease in the below-ground biomass because of increased burning of organic soil and higher rates of soil respiration. In total, they estimated the net effect would be a loss of 25 to 50 gigatons of carbon overall from the boreal forest under the climate scenarios currently projected by general circulation models.

6.   ACKNOWLEDGEMENTS

This research summarised in this chapter was supported by a series of grants from the National Aeronautics and Space Administration and the U.S. Environmental Protection Agency (EPA). It should be noted that the research discussed in this chapter has not been subjected to review by these agencies and therefore does not necessarily reflect the views of these agencies and no official endorsement should be inferred.

7.   REFERENCES

Adams, P.C. and L.A. Viereck, 1997. Soil temperature and seasonal thaw - controls and interactions in floodplain stands along the Tanana River, interior Alaska. *In Proceedings of the International Symposium on Physics, Chemistry, and Ecology of Seasonally Frozen Soils, Fairbanks, Alaska, June 10–12, 1997*, 105–111.

Bonan, G.B. 1992. Soil temperature as an ecological factor in boreal forests. pp. 126–143 in H.H. Shugart, R. Leemans, and G.B. Bonan, editors. *A Systems Analysis of the Global Boreal Forest*, University Press, Cambridge, England.

Bonan, G.B. and K. Van Cleve. 1992. Soil temperature, nitrogen mineralization, and carbon source–sink relationships in boreal forests. *Canadian Journal of Forest Research* **22:**629–639.

Kasischke, E.S., N.L. Christensen, Jr., and B.J. Stocks, 1995. Fire, Global warming and the mass balance of carbon in boreal forests. *Ecological Applications 5:* 437–451.

Kasischke, E.S., B.J. Stocks, K.P. O'Neill, N.H.F. French and L.L. Bourgeau-Chavez 1999a. Direct effects of fire on the boreal forest carbon budget, this volume.

Kasischke, E.S., K.P. O'Neill, N.N.H.F. French, L.L. Bourgeau-Chavez, and D. Richter, 1999b. The influence of fire on long-term patterns of forest succession in Alaskan boreal forests. in *Fire, Climate Change and Carbon Cycling in the North American Boreal Forests*, Kasischke, E.S., and B.J. Stocks (editors), Ecological Studies Series, Springer-Verlag, New York, (in press).

O'Neill, K.P.; Kasischke, E.S.; Richter, D.D.; Krasovic, V. 1997. Effects of fire on temperature, moisture and CO_2 emissions from soils near Tok, Alaska. *In Proceedings of the International Symposium on Physics, Chemistry, and Ecology of Seasonally Frozen Soils, Fairbanks, Alaska, June 10–12, 1997*, 295–303.

Richter, D., E.S. Kasischke, and K.P. O'Neill, 1999. Stimulation of soil respiration in burned black spruce *(Picea mariana L.)* forest ecosystems. in *Fire, Climate Change and Carbon Cycling in the North American Boreal Forests*, Kasischke, E.S., and B.J. Stocks (editors), Ecological Studies Series, Springer-Verlag, New York, (in press).

Schlentner, R.E. and K. Van Cleve. 1985. Relationships between CO_2 evolution from soil, substrate temperature, and substrate moisture in four mature forest types in interior Alaska. *Canadian Journal of Forest Research 15:*97–106.

Smith, T.M. and H.H. Shugart 1993. The transient response of carbon storage to a perturbed climate. *Nature* 361: 563–566.

Stocks, M.A. Fosberg, T.J. Lynham, L. Mearns, B.M. Wotton, Q. Yang, J-Z. Jin, K. Lawrence, G.R. Hartley, J.A. Mason, and D.W. McKenney 1998. Climate change and forest fire potential in Russian and Canadian boreal forests. *Climatic Change* 38: 1–13.

Van Cleve, K. and L. Viereck 1981. Forest succession in relation to nutrient cycling in the boreal forest of Alaska. Pages 184–211 *in* D.C. West, H.H. Shugart and D.B. Botkin, editors. *Forest Succession, Concepts and Application.* Springer-Verlag, New York.

Van Cleve, K., C.T. Dyrness, L.A. Viereck, J. Fox, F.S. Chapin III, and W. Oeche 1983a. Taiga ecosystems in interior Alaska. *Bioscience 33:* 39–44.

Van Cleve, K., L. Oliver, R. Schlentner, L.A. Viereck, and C.T. Dyrness, 1983b. Productivity and nutrient cycling in taiga forest ecosystems. *Canadian Journal of Forest Research 13:* 747–766.

Van Wagner, C.E. 1978. Age-class distribution and the forest fire cycle. *Canadian Journal of Forest Research 8:* 220–227.

Viereck, L.A., 1983. The effects of fire in black spruce ecosystems in Alaska and northern Canada. in *The Role of Fire in Northern Circumpolar Ecosystems*, Wein, R.W., and D.A. MacLean, editors, John Wiley and Sons, New York, 322 pp. 201–220.

Sustainable Forestry as a Source of Bio-energy for Fossil Fuel Substitution

MURARI LAL and ROMA SINGH

Centre for Atmospheric Sciences, Indian Institute of Technology, New Delhi, India

Abstract: In tropical countries, anthropogenic pressures have led to deforestation and degradation of forests and pasture lands. Realising the large potential and also the importance of producing biomass for energy as a substitute for fossil fuel, using degraded land for plantation forestry has been emphasised in recent years and could become one of the most important counter-agents to deforestation. In India, the area under forests has been reported to be stable at 65 Mha since 1982, although the area under dense forests (> 40% tree crown cover) has been increasing, which suggests an increase in carbon stocks sequestered by Indian forests. The current rate of afforestation in India is one of the largest in the world (about 2 Mha per annum). However, rural households in India depend largely on forests for their basic biomass needs such as medicines, fuelwood, livestock feed and raw materials for various products. Looking to the future needs of biomass in the country and the extent of land available for biomass production, the rate of afforestation needs to be further increased to meet the future demands for biomass. Bio-energy strategies offer the prospect of reduced CO_2 emissions to the atmosphere through storage of carbon in the biosphere and use of biofuels to replace fossil fuel use. A package of practices for high yields from productive tree species and short rotation tree crops suited for different agro-climatic regions of India is therefore crucial.

According to recent landuse – land cover statistics for India generated by remote sensing techniques, the area under non-forested degraded lands is 93.68 Mha and 35.89 Mha under forested degraded lands. The available land area which could be effectively utilised for biomass production in India amounts to 65.45 Mha. If a conservative productivity of 4 tonnes per hectare per year could be attained on only about half of the available surplus degraded land in India, it would be possible to obtain a carbon emission reduction of about 8 Gt in 100 years, compared to 4.4 Gt through carbon sequestration and storage options. Substitution of biofuels for coal reduces CO_2 and also the emissions of NO_x and SO_x In addition to obtaining higher carbon abatement benefits, the development of sustainable forestry for bioelectricity is also likely to lead to significant rural employment. This calls for a viable financial and institutional mechanism to promote sustainable forestry for bio-energy in developing countries.

1.   INTRODUCTION

Major environmental issues such as global warming, deforestation, decline in biodiversity, pollution of air and water and degradation of soil and water resources are facing the mankind today. Carbon dioxide (CO_2) has been identified as the single most important greenhouse gas, currently trapping about half of the total heat contributing to global warming. The atmospheric concentration of CO_2 had remained nearly constant at about 280 ppm from the end of the last Ice Age to the beginning of industrialisation, i.e., the 19th century. However, it has been increasing since then, from 315 ppm in 1958 to 358 ppm in 1996, at a rate of about 1.5 ppm each year. It is estimated that currently about 2 Gt of carbon each year is released to the atmosphere due to tropical deforestation. Until the mid-twentieth century, temperate deforestation and the loss of organic matter from soil were more important factors contributing to atmospheric CO_2 than the combustion of fossil fuels (Houghton and Skole, 1990; Keeling *et al.*, 1989). Since then, the burning of fossil fuels is the main contributor (almost 60% of the total during the 1990s) for the increase in CO_2 concentrations. Fossil fuels re-introduce CO_2, carbon monoxide (CO), hydrocarbons and some other (nitrogen and sulphur) oxides from geological sources to the biogeochemical cycles of the earth.

A further serious environmental problem that mankind faces today is biomass burning in forests. Biomass burning includes the burning of forests and savanna grasslands for land clearance and conversion, burning of agricultural residues, and burning of biomass fuel. The burning of forests is occurring the world over as a natural phenomenon, irrespective of forest type and region. If fires occur occasionally, they may help to maintain healthy ecosystems, wildlife populations and the biodiversity of forests. Controlled burning is carried out sometimes for natural regeneration and to prevent large scale damage caused by accidental fires. For example, burning of Chir pine (*Pinus roxburghii*) forests in spring provides an excellent seedbed for the regeneration of Chir. The practice is to burn the forest in spring before seeds fall in the area planned for regeneration. Burning in Sal (*Shorea robusta*) forests is considered essential to thin out the underwood and replace it with light grass, and to remove the covering of dead dry leaves (Joshi, 1991). Teak (*Tectona grandis*), one of the most important artificial forest trees in India, sheds huge amounts of leaves in winter, resulting in a thick layer of leaves forming on the ground. No other vegetation can grow on such ground because of the inhibitors in the leaves (Mandal & Brahmachary, 1998). Natural regeneration of teak is therefore more abundant in burned forests. In some cases, however, biomass burning is undesirable and highly detrimental to species diversity.

The biomass burning activities in the world have also led to significant emissions of anthropogenic greenhouse gases. The total biomass burned annually is estimated to be about 8.7×10^{15} gm of dry matter. Annual emissions of about 3.5×10^{15} gm carbon as CO_2 and 4.0×10^{13} gm carbon as CH_4 results from this burning (Andreae, 1990). Hence, biomass burning contributes about 40% of the world's annual production of CO_2 (Kambis and Levine, 1996).

Biomass has been used as a source of energy since historical times. The non-sustainable use of forest biomass has led to the loss of a range of forest products and plant diversity (Whitmore & Sayer, 1992). The demand for fossil fuels and forest products all over the globe is likely to rise in the future as world's population increases. More than one third of the world's population depends on wood for cooking and heating, and fuelwood is the fourth largest contributor to the world's energy supply after petroleum, coal and natural gas. Biomass continues to be the largest source of energy in the developing countries, accounting for about 38% of the primary energy use. Fuelwood, the main source of energy for cooking among the rural masses in most developing countries, is reported to be in under-supply for 1.4 billion people and this figure may rise to 2.5 billion by 2010 (FAO, 1994). To counter the situation of increased biomass demand, the need to increase the rate of afforestation, reforestation or tree planting outside forest areas is crucial. Intense forest conservation measures have recently been taken up in several tropical countries.

The current and future energy scenarios for developing countries indicate the urgent need for exploring alternative energy sources. Sustainable harvesting of existing forests, plantation forestry and the efficient use of wood products have the potential to substitute fossil fuel energy and could be a CO_2 mitigation option in the future. Plantations often have a much higher productivity for specific products through selection of fast-growing or high-yielding species. The substitution of sustainably managed wood fuels for non-renewable fossil fuels could reduce the net emissions of CO_2 to the atmosphere indefinitely and reverse the century long accumulation of the greenhouse gases in the atmosphere (Houghton, 1991). In this paper, we examine the prospects of sustainable biomass plantations for providing renewable energy sources in India. These plantations could be adopted effectively as a potential substitute for fossil fuels and could contribute towards mitigating strategies for the reduction of greenhouse gas emissions.

2. INDIA'S FOREST COVER AND BIOMASS CONSUMPTION

Growth and development of any country is sustainable only when the environment is protected and an ecological balance is maintained. Increases in the World's population and the corresponding increases n energy needs are threatening our existing biological resources. The forest resources of the World are limited and the demand for forest products such as fuelwood, fodder and industrial wood is increasing every year (Lefevre *et al.*, 1997). Forests constitute the base for agriculture, animal husbandry, industrial development and other developmental activities. In several developing countries, energy demand has increased abruptly in recent years because of population increases and rising per capita energy consumption, with subsequent increases in CO_2 emissions (Table 1). In India, biomass fuel has been one of the largest sources of energy in the domestic sector for decades. However, the increasing cost and non-availability of fuelwood in recent years has forced the rural population to shift towards the use of fossil fuels. The quality of life in rural households depends on the availability and access to biomass, particularly woody biomass that is used as fuelwood, industrial wood and structural wood, thereby increasing pressure on forests.

Table 1. Average Energy Consumption and Carbon Dioxide Emissions in Selected Regions*

Country	GDP per unit of energy use		Traditional fuel use	CO_2 emission			
	(1987 $ per kg oil equiv.)		(% of energy use)	Total (Million metric tonnes)		Per Capita (Metric tonnes)	
	1980	1995	1995	1980	1995	1980	1995
Low Income Countries	0.9	1.1	**19.0**	2037.8	4503.7	0.9	1.4
Middle Income Countries	1.2	1.1	7.7	2775.7	7073.8	**2.9**	**4.5**
High Income Countries	2.9	3.4	2.5	8772.1	11,122.7	12.0	12.5
India	1.9	1.7	**23.3**	347.3	908.7	**0.5**	**1.0**
South Asia	2.0	1.7	**25.6**	392.4	1024.1	**0.4**	**0.8**
World	2.2	2.4	**6.8**	13,585.7	22,700.0	**3.4**	**4.1**

* Source : World Development Indicators 1998 (World Bank)

Indian forests occupy about 2% of the world's forest area but have to meet the fuelwood needs of almost 15% of the world's population. It has still been possible to maintain approximately 65 million ha (Mha) of forest cover (about 22% of the country's geographic area) over the last decade (Table 2). The rate of afforestation in India was 1.67 Mha yr^{-1} during the 1980s and currently it is 2 Mha per annum. Table 3 lists the progress of afforestation in India since 1951. During the Ninth Five Year Plan period (1997–2002), about 10 Mha of land area is targeted for afforestation in India. This rate of afforestation is one of the highest among tropical countries. The annual productivity of Indian forests has increased from 0.7 m^3 ha^{-1} in 1985 to 1.37 m^3 ha^{-1} in 1995 (FSI, 1988; 1996). In India, fuelwood clearing from forests has been banned since the enactment of the Forest (Conservation) Act in the 1980s; this Act has also prevented the indiscriminate conversion of forest lands. These changes, together with better management practices, have helped to increase the productivity of Indian forests in recent years. Increases in annual productivity indicate an increase in the total standing biomass of Indian forests. However, the demand for forest products in India is substantial, as more than 70% of the people there live in rural areas and they mostly rely on forests for their basic biomass needs, such as cattle feed, fuelwood, food and shelter etc.

Table 2. Average Land and Forests in Selected Regions*

Country	Land Area (10^3 km^2) 1995	Land Use (% Land Area)			Forests (10^3 km^2) 1995	Annual Deforestation	
		Crop Land 1995	Perma. Pasture 1994	Other lands 1994		(km^2) 1990-95	(Avg. % change) 1990-95
Low Income Countries	39,294	13	32	54	6,227	38,690	0.6
Middle Income Countries	59,884	10	23	67	19,985	74,598	0.4
High Income Countries	30,951	-	24	-	6,501	-11,564	-0.2
India	2,973	57	4	39	**650**	-72	**0.0**
South Asia	4,781	45	10	45	744	1,316	0.2
World	130,129	11	26	62	32,712	101,724	0.3

* Source : World Development Indicators 1998 (World Bank)

Table 3. Progress of Afforestation in India

Plan Period	Area Afforested in Plan Period (Mha)	Cumulative Area (Mha)
First Plan (1951-56)	0.05	0.05
Second Plan (1956-61)	0.31	0.36
Third Plan (1961-66)	0.58	0.94
Post Third Plan (1966-69)	0.45	1.39
Fourth Plan (1969-74)	0.71	2.10
Fifth Plan (1974-79)	1.22	3.32
Annual Plan (1979-80)	0.22	3.54
Sixth Plan (1980-85)	4.65	8.19
Seventh Plan (1985-90)	8.88	17.07
Eighth Plan (1992-97)	7.84	24.91
Ninth Plan (1997-2002)	10.00 (target)	34.91 (target)

Indian forests are currently able to produce only 40 Mt (Million tonnes) of fuelwood annually on a sustainable basis. An additional supply of 60 Mt of wood annually comes from village farm lands. The fuelwood demand in the country is currently more than 300 Mt per year. This puts Indian forests under immense pressure and there may be a net loss of forest wood in future. Apart from the sustainable harvesting of existing forests, plantation forestry seems to be an option for meeting biomass needs. As our existing forests are inadequate to meet the future demands of biomass and land availability for plantation forestry is not a constraint in the immediate future, the establishmnet of forests outside the current forest area is inevitable.

Commercial energy demands in India are also increasing daily and have been projected to reach a value of 600 Mtoe (Million tonnes of oil equivalent) by 2010, from about 200 Mtoe in 1990 (TERI, 1992). Although the per capita commercial energy consumption in India has increased gradually since the early 1970s, it still remains less than 10% of the levels achieved in the industrialised world (Fig. 1). The challenge for most developing countries today is how to fulfil their energy demands without damaging the economic and ecological environment. Can biomass be used as a eco-friendly substitute for fossil fuel to meet future energy requirements in developing countries?

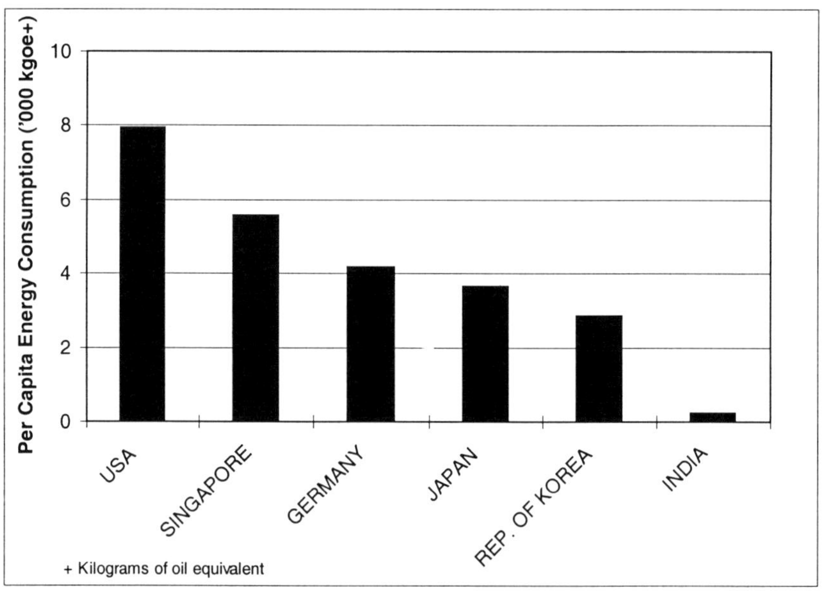

Figure 1. Per capita commercial energy consumption in a few selected countries.

3.   ENERGY NEEDS OF INDIA

The power sector in India is facing serious problems, as the gap between demand and supply has been increasing annually by about 8%. The shortage of power during the Ninth Five Year Plan period (1997–2002) in India is estimated to be about 55000 MW (MegaWatt). Coal is one of the primary energy resources, accounting for about 67% of the total commercial energy consumption in the country. Thermal power plants at present account for 73% of the total power generation in India; hydroelectric power plants contribute 25% and the remaining 2% comes from nuclear plants. Of the total 394 BkWh (Billion kiloWatt hour) of power generated in the country during the year 1996–1997 (Fig. 2), 265 BkWh was generated by coal burning (about 163 Mt of coal was burned to produce this power). Consumption of fossil fuels in India has also increased abruptly during the last few years (Fig. 3). In 1996–1997, the total fossil fuel consumption in India was 408 Mt, which contributed to about 300 Mt of carbon emissions (Table 4). Between 1981 and 1991, increases in the consumption of commercial energy in India was at a rate of 4.5% per annum and projections for energy use suggest that the current growth rates of 5.5% are likely to increase further in future.

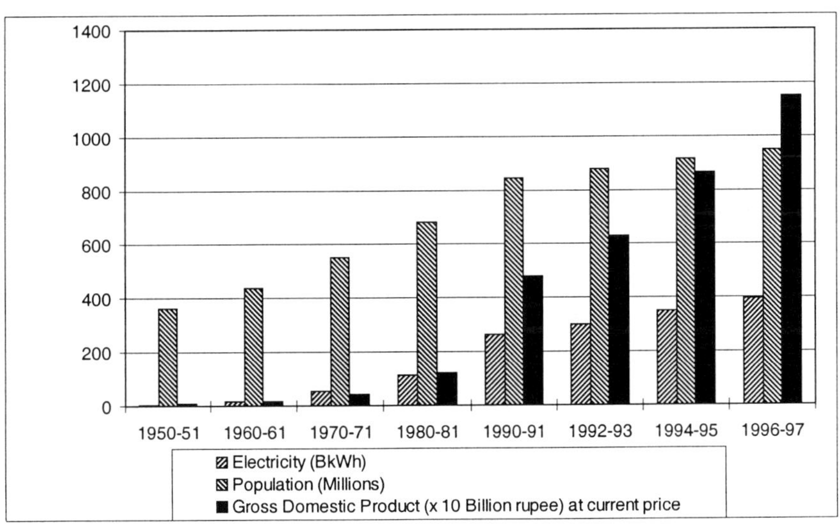

Figure 2. Observed increases in national electricity production, population and gross domestic product (at current prices) in India during the period from 1950-51 to 1996-97.

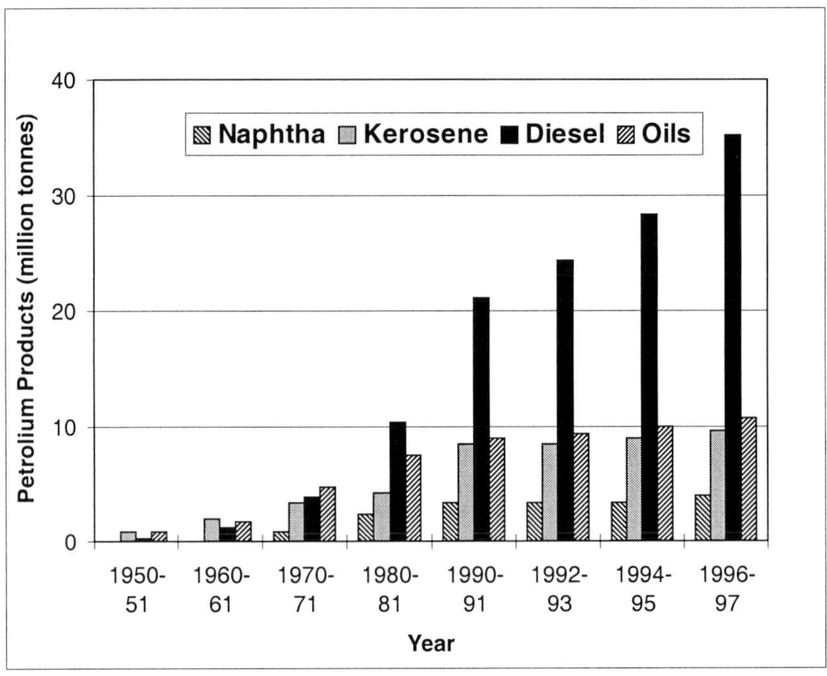

Figure 3. Observed increases in consumption of some of the petroleum products in India during the period from 1950-51 to 1996-97.

Table 4. Carbon Emission due to Fossil Fuel Consumption in India (1996-97)

Fossil Fuel	Consumption		Carbon Emission
	(Million tonnes)	(Mtoe)*	(Million tonnes)
Petroleum Products	**77.2**	**79.45**	**66.39**
Naphtha	4.0	4.29	3.61
L.P.Gas	4.2	4.74	3.42
Petrol	5.0	5.35	4.23
Kerosene	9.6	10.25	8.42
Diesel Oil	35.2	36.41	30.81
Fuel Oil	10.7	10.26	9.07
Others	8.5	8.15	6.83
Natural Gas	**22.8****	**19.05**	**12.20**
Coal and Lignite	**308.2**	**205.98**	**221.90**
TOTAL	**408.2**	**304.48**	**300.49**

*   Million tonnes of Oil Equivalent

** Billion cubic meters

Biomass energy represents 36% of the primary energy use in India. It is mostly used in rural areas where electricity and fossil fuels are not in abundant supply. Rural industries and services rely heavily on biofuels because of the limited availability of commercial fuels. About 120 million rural households use traditional wood stoves for cooking in India with an efficiency of only about 10–15%. Improved stoves, however, could provide a thermal efficiency of up to 40%. In areas where fuelwood is not available, dung cakes and crop residues are being used as alternate sources of energy. Use of dung cakes as fuel is, however, a direct loss of potential fertilizer. Moreover, its use as fuel produces smoke which creates serious health hazards. Instead, if dung could be used in biogas production, it would provide a more efficient fuel for cooking and lighting purposes. Slurry, a by-product of biogas plants, also serves the purpose of fertilizer for crop fields. Fuelwood use in rural households could be replaced with biogas to reduce the emission of smoke in the households.

The Indian government has made serious efforts during the last decade to fulfil rural energy needs. Out of 0.58 million villages, 0.51 million have already been electrified primarily to meet their domestic and agricultural needs such as the operation of tube-wells and other equipment. However, these villages are connected to an electricity supply for only 2–3 hours each day. The provision of regular energy services to the rural population needs the careful attention of energy planners as about 70% of the population lives in rural areas and the large majority of them are poor. Distribution of commercial energy is also irregular in India. Only 18.9% of the generated electric power goes to domestic use (Fig. 4).

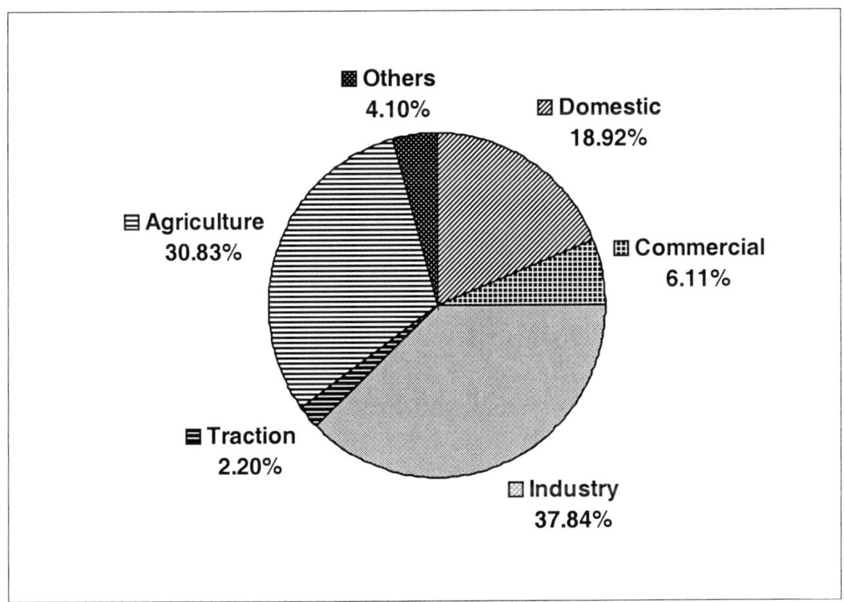

Figure 4. Major sectors of electricity consumption in India (based on data for 1995-96).

Electrical power shortages are more pronounced in rural India as a result of technical problems with the connectivity of the central grid. In rural India, biomass could become a major clean and convenient source of energy if it could be efficiently converted into modern energy carriers such as fluid fuels and electricity. Bio-energy plantations on rural degraded land could make local populations self-sufficient in energy (i.e., they could meet their energy requirements for domestic purposes, agriculture and industries) through a decentralised power-generating system.

4.   ECO-FRIENDLY POWER GENERATION

Ecologically important distinctions exist between the non-fossil fuel derived energy (solar power, hydroelectric power and wind power etc.) and energy from fossil fuels and nuclear power. Each source of commercial energy (hydroelectric, thermal and nuclear) has major environmental consequences. The utilization of these resources with the least environmental damage remains a challenging task as it involves determining the optimal technology, scale and location of such projects. Coal mining is responsible for the degradation of land area. Open-cast mining leads to degradation of land, deforestation, soil erosion, land slips and the disruption of aquifers. Many social and economic issues also need to be taken into account, such as the displacement of people, their rehabilitation and the health of the workers. Indian coal has high ash contents and the sites surrounding most power plants are degraded due to fly ash. Even modern plants with electrostatic precipitators cannot adequately deal with the high ash content because of financial constraints and other limitations (NR to UNCED, 1992). Coal-based thermal plants pollute the atmosphere through gaseous emissions, causing acid rain, which damages the soil, vegetation and aquatic life of the region. Hydroelectric power plants replace natural biological habitat and hence are not always encouraged.

The potential of non-conventional eco-friendly energy is quite high in India but appropriate planning and technology is required for its full exploitation (Table 5). In general, no incidental chemical by-products are released into the environment when renewable energy sources are used. The volume of waste products depends on combustion technology but differs from nearly 1 kg MWh^{-1} for wood to almost 24 kg MWh^{-1} for coal. Biomass, if used as a fuel on a sustainable basis, would simply redistribute the nutrients absorbed during growth. The sustainable harvesting of bio-energy plantations would protect the environment against degradation and might help in conserving biodiversity. Plantations can be established on degraded land, marginal crop land, degraded forests or pastures. The absence of sulphur in most biomass fuels also renders them relatively clean when compared to coal. Fewer nitrogen oxides are emitted from biomass combustion than from coal combustion. Wood ash from power and heating plants contains all the nutrients withdrawn at harvesting except for nitrogen (Brinck *et al.*, 1992). Fertilising plantations with this ash in areas with high nitrogen deposition protects crops from leaching and also helps to maintain soil productivity. The strategy should be to grow biomass on village fallow land and use a wood gasifier to generate motive power. These gasifiers use biomass five times as efficiently as conventional methods. Growing biomass is a labour-intensive activity which could create jobs in rural areas whilst providing a convenient energy source that would promote other rural industries, thereby helping to reduce urban migration.

Table 5. Renewable Energy Potential and Achievements in India

Sources/Technologies	Unit	Approximate Potential	Achievements (upto 31st March, 1997)
Power Generation			
Wind Power	MW	20,000	900
Small Hydro Power (upto 3 MW)	MW	10,000	141
Biomass Power	MW	17,000	83
Solar Photovoltaic Power	MW/km^2	20	28 (MW)
Urban and Municipal Waste	MW	1,700	
Installed			3.75
Being taken up			90.00
Thermal Applications :			
Solar Thermal Systems	MW/km^2	35	-
Water Heating Systems	Area (m^2)		4,00,000
Solar Cookers	Nos.		4,30,000
Biogas Plants	Nos (Millions)	12	2.5
Improved Biomass Stoves	Nos (Millions)	120	23.7

5. BIOMASS ENERGY TO REPLACE COAL ENERGY

Biomass could make major contributions to the global commercial economy in ways that would help promote rural development, reduce local environmental problems and reduce greenhouse gas emissions through fossil fuel substitution. It is a traditional source of energy in many developing countries and can be converted into modern energy forms such as liquid and gaseous fuels, electricity and process heat. Modern biomass energy systems could be set up in virtually any location where plantations can be grown.

Biomass power has the potential to be the most important renewable energy option for the next quarter of a century and it has been projected that an installed capacity of 25 GW (Gega Watt) of biomass power could be generated in the USA alone by 2010 (USDOE, 1993).

Various studies have demonstrated that substituting biomass for fossil fuels in electricity and heat production is, in general, less costly and provides larger CO_2 reductions per unit biomass than substituting biomass for power needs in the transport sector (e.g., Gustavsson *et al.*, 1995). In India, almost 70% of all electric power is generated through coal. Indian coal is of relatively poor quality and the burning of almost 0.62 tonnes coal is required to generate 1 MWh (Million Watt hour) of electricity. As much as 163 Mt coal has been burned in the year 1996–1997 to generate 265 BkWh (Billion kilo Watt hour) of electricity. Future projections indicate an increase in coal consumption for electricity generation of at least 5% per year. Coal is responsible for almost 75% of the total carbon emissions from fossil fuel burning in India. If the use of coal is continued at the projected level of increase, it is estimated that the available reserves will be exhausted within the next 245 years (TERI, 1992). From the various options available, power generation through the conversion of biomass to gas seems to be the most economical and suitable from the point of view of environmental protection. Of the various forms of biomass available, two major categories, agricultural residues and energy plantations, have scope for energy applications in India.

The amount of agricultural crop residues has increased from 88.6 Mt in 1950–1951 to 341.9 Mt in 1996–1997, which is more than the total amount of fuelwood consumed in India during the year 1996–1997. The current predominant use for agricultural crop residues in India is for cattle feed. In some cases, crop residues are left in the fields as they help to restore soil productivity and retain rain water. For better crop management, crop residues must be fed to live stocks which in turn provide dung to be used as manure. Hence, agricultural crop residues are not the best potential source of bio-energy for commercial purposes.

During the seventh Five Year Plan, the Government of India introduced, for the first time, a massive energy plantation programme based on the wide spectrum of benefits likely to occur from such plantations. The natural forest cover in India will continue to sustain its present productivity and area. India has the potential to increase its forest plantations by at least 25 Mha in the next 50 years. In commercial plantations, much higher productivity could be attained than in natural forests. The productivity of plantation forest is about 3.2 t ha^{-1} yr^{-1} (tonnes per hectare per year) (Sudha, 1997). Extraction of wood from forests is expected to double in the next 25 years and treble in 50 years. Land, money, technology and plant species with high biomass yields are some of the basic requirements for energy plantations. The land categories that can be conveniently exploited for plantation are degraded forests, marginal crop land, wastelands and the area under shifting cultivation. There continues to be a large potential for expanding energy plantations in India. According to one estimate (PC, 1992), the area of degraded non-forest land is 93.68 Mha and there are 35.89 Mha of degraded forests.

A comparison of the economics of small capacity gasifier-based power generation systems versus diesel systems (5 KW and 96 KW) of identical capacity shows that producer gas systems are cheaper to operate (Rs/KWh) than diesel systems (Ravindranath & Mukunda, 1993). Reddy *et al.* (1990) conducted a detailed financial analysis of decentralised power generation, conventional centralised power generation and conservation options for India. The installed costs of power for other bio-energy options, namely biogas and producer gas, are almost 10% lower than the dominant coal and large hydroelectric options. A summary of the results obtained from various case studies is useful for an effective review of the current economic viability of technologies for power generation from biomass. Table 6 clearly demonstrates the economic viability of biomass based power generation in India, even at a technology level that is far behind the advanced technologies of thermal or hydroelectric plants and diesel engine pumpsets and gensets. Biomass can be converted into a combustible gas called producer gas through the gas conversion process. For this, the two main technologies used in India are steam turbine cycle technology and gas turbine cycle technology.

Table 6. Economics of Power Generation from Biomass

Applications	Conversion Technology	System Rating (MW)	Assumed Biomass price ($/tonnes)	Cost of generation ($/MWh)	Conventional options and typical cost ($/MWh)
Village Electrification	Biomass Gasification	0.024	7.14 & 14.29	39.76-47.62	DG set : 55.95-60.71 Grid Extension : 53.57-58.33
Pumpset Energisation	Biomass Gasification	0.016	7.14 & 14.29	80.95-93.33	Diesel Pumpsets : 57.14-61.90 Grid Extension : 71.43-78.57
Captive Industrial Power Generation	Biomass Gasification	0.300	7.14 & 14.29	30.95-39.05	DG set : 47.62-52.38
Power from Energy Plantations	Biomass Gasification	2.000	5.24 & 9.52	14.52-24.28	Thermal/Hydro/ Nuclear power 23.81-33.33
Power from Energy Plantations	Dendro-thermal power generation	3.000	5.24 & 9.52	26.19-37.14	Thermal/Hydro/ Nuclear power 23.81-33.33

The Ministry of Non-conventional Energy Sources has set up biomass research centres in eleven agro-climatic regions in India. When the mean productivity of high yielding species at different ages of plantation is taken, it appears to increase up to about 4 years on a national average basis and then declines (Sudha, 1997). Taking into consideration these experimental results, a scenario has been developed here to determine the possibilities of reducing carbon emissions by replacing coal energy with biomass energy.

6.   COAL REPLACEMENT CARBON EMISSION SCENARIOS

India has almost 129 Mha of degraded land, of which at least 50% is highly degraded and unsuitable for afforestation. Moreover, because of the high cost of raising plantations and then maintaining them, some available land cannot be considered for an energy plantation programme. The current availability of various categories of land for afforestation in India is therefore taken as about 65 Mha (Fig. 5). If reforestation and afforestation is undertaken on degraded forests and other available land totalling about 40 Mha, a standing biomass of about 188 t ha^{-1} could be achieved in 55 years (assuming a productivity of 4 t ha^{-1} yr^{-1}). The forests would only be able to sustain themselves without any further increase in biomass if a longer time period of 100 years is considered. This would lead to a carbon emission reduction of about 4.4 Gt in 100 years.

The major constraint in bio-energy plantations is the cost of raising plantations, which is estimated to be approximately US$ 240–360 per hectare at current exchange rates (NWDB, 1991). For degraded lands, land preparation costs are also high. The first preparation is the most expansive as it involves conversion work (clear cuts), which account for 25% of the total cost (FAO, 1985). With such costs, economic constraints make it difficult to afforest large areas of degraded land. In this study, we consider a land availability of 40 Mha for biomass plantations on a sustainable basis. It is assumed that a sum of US$ 290 per hectare is needed to establish plantations on degraded land at the currently prevailing rates. A total sum of US$ 12 billion would be required to cover a land area of 40 Mha with biomass plantations. If in each year 5 Mha of degraded land is planted, we would be able to cover the entire available land area with biomass plantations in eight successive years. Initially, at least US$ 1.5 billion per annum would be required to establish and subsequently maintain these plantations. There would be no harvesting for the first 8 years, after which 5 Mha could be harvested each year. After 8 years, at a conservative productivity of 4 t ha^{-1}

yr^{-1}, these plantations would be able to produce 160 Mt of biomass per year on a sustainable basis.

At any given time, the total biomass present on this land would be approximately 560 Mt and the carbon stored would be 252 Mt (a conversion value of 0.45 is taken. following IPCC (1996)). A number of studies suggest that one MWh of electricity can be generated from one tonne of wood (Jain, 1993). 160 Mt of wood from biomass plantations could thus produce 160 BkWh of electricity, for which we currently need to burn about 99 Mt of coal in India. The burning of 99 Mt of coal would release about 76.29 Mt of carbon as CO_2 and CH_4. It is thus evident that eight years after the start of the biomass plantations, we would be able to reduce emissions of 76.29 Mt of carbon each year by replacing coal with wood. In addition to this, some carbon would be taken up by vegetation around the trees and by the soil, and this is not accounted for here. If we assume a biomass partitioning of 0.4 t ha^{-1} yr^{-1} into the roots (below ground productivity) as compared to 4 t ha^{-1} yr^{-1} in the stem (above-ground productivity), total carbon emission reductions of about 1.75 Gt, 3.86 Gt and 8.07 Gt are thus possible in the next 25 years, 50 years and 100 years, respectively (Table 7). Substitution of biofuels for coal would thus help reduce CO_2 and also the emissions of NO_X and SO_X. It is also interesting to note here that timber plantations reduce atmospheric CO_2 levels by about 10% while biomass energy plantations can reduce atmospheric CO_2 levels by as much as 15% (Vitousek, 1992).

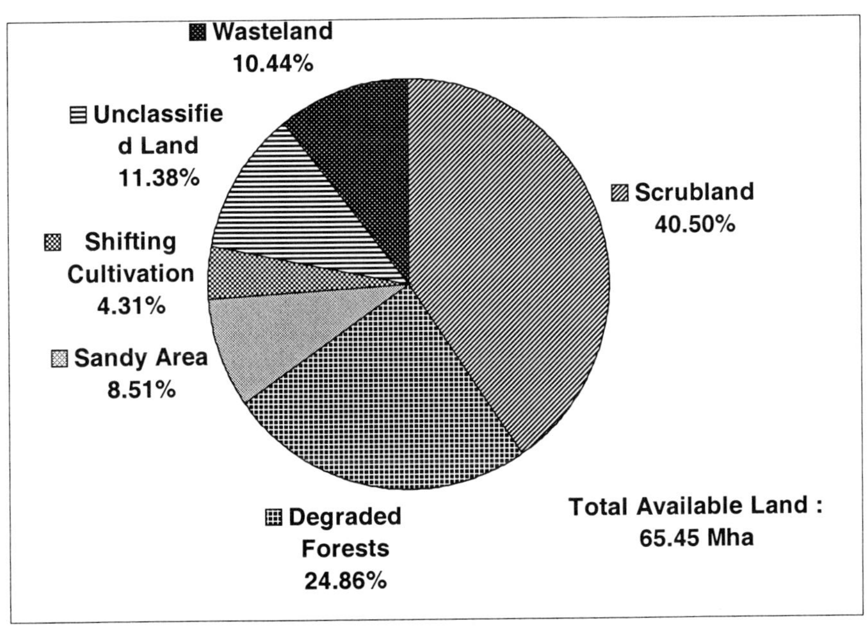

Figure 5. Land categories available in India for afforestation purposes.

Table 7. Carbon Emission Reduction by Replacing Coal with Energy Plantations in India

Years	Carbon Emission Reduction (Gega tonnes)	CO_2 Emission Reduction (Gega tonnes)
25	1.75	6.42
50	3.86	14.12
75	5.96	21.85
100	8.07	29.59

7.   CONCLUSIONS

Urgent policy decisions are needed at regional and national levels on greenhouse gas emission abatement strategies to arrest future global warming. However, in developing countries, environmental conservation is largely hampered due to economic and social constraints. In India, almost 36% of the total population is still below the poverty line and the government has the primary responsibility to provide them with at least their basic needs, including water, food, shelter and healthy environmental conditions. Bio-energy plantations could provide electricity in and around rural areas through decentralised power generation. The remaining tree tops, twigs and branches from the energy plantations that are not used in gasifiers could fulfil fuelwood needs in rural areas. Small shrubs and grass can be grown between trees and used for cattle feed. Such a strategy would help in reducing pressure on natural forests and conserving biodiversity. With a shift in government policy towards sustainable bio-energy plantations as part of the afforestation programme, the rural masses of India could also be provided with employment opportunities which would improve their social and economic conditions.

Sustainable bio-energy plantations on about 40 Mha of degraded land in India could produce 160 Mt of biomass per year after 8 successive years of plantation establishment. The biomass could replace fossil fuels and thus have the potential to reduce total carbon emission of about 1.75 Gt, 3.86 Gt and 8.07 Gt in the next 25 years, 50 years and 100 years, respectively. Growing bio-energy plantations on 40 Mha of India's degraded land would also have environmental benefits as, with the bio-energy plantations, the total forest area would effectively be 105 Mha (nearly 32% of the total land cover of India). The biomass plantation programme is expected to have a significant positive effect on rainfall pattern, retention and recharging of underground water resources, arresting soil erosion and nitrogen run-off and raising habitat diversity as well as being an effective step towards CO_2 emission abatement strategies.

8. ACKNOWLEDGEMENTS

This study was funded by the Indian Council of Forestry Research and Education, Dehradun under the World Bank FREE Project. The authors acknowledge the support of Forest Survey of India for providing the data used in this study.

9. REFERENCES

Andreae, M. (1990): Biomass burning in the tropics: Impact on environmental quality and global climate, Chapter 1 in Proc. of Conf. on Global Biomass Burning – Atmospheric, Climatic and Biospheric Implications, Williamsburg, Va, 19–23 March.

Brinck, L., Emborg, L., Jaul-Kristensen, B., Kristiansen, A., Pulliainen, Selvig, E., Vaittinen, A. and Benestad, O. (1992): Environment and energy in the Nordic Countries, Energy scenarios for 2010, Nordic Council or Ministers Report 1992, 548.

FAO (United Nations Food and Agriculture Organization, 1985): Tropical forestry action plan. Committee on Forest Development in the Tropics, Rome, Italy.

FAO (United Nations Food and Agriculture Organization, 1994): Forest development and policy dilemmas, Rome, Italy.

FSI (Forest Survey of India, 1988): The State of Forest Report 1987, Ministry of Environment and Forests, Government of India.

FSI (Forest Survey of India, 1996): The State of Forest Report 1995, Ministry of Environment and Forests, Government of India.

Gustavsson, L., Borjesson, P., Johansson, B. and Svenningsson, P. (1995): Reducing CO_2 emission by substituting biomass for fossil fuel energy, Oxford, 20(11), 1097–1113.

Houghton, R. A. and Skole, D. L. (1990): Change in the global carbon cycle between 1700 and 1985 in the earth transformed by human action. (ed: Turner, B. L.), Cambridge University Press, Cambridge.

Houghton, R. A. (1991): The role of forests in affecting the greenhouse gas composition of the atmosphere In Global Climate Change. (ed: Ehrlich, P.R.), Chapman and Hall, New York, 43–55.

IPCC (1996): Revised Guidelines for greenhouse gas inventory workbook, Volume 2, Module 5 – Land Use Change and Forestry, Report prepared by UNEP, OECD, IEA and IPCC, 5.1–5.45.

Jain, B. C. (1993): Decentralized energy options and technology, Centre for Energy Studies Technical Report, IIT Delhi.

Joshi, V. (1991): Biomass burning in India, In Global Biomass Burning: Atmospheric, Climatic and Biospheric Implications, (ed: Levine, J. S.), The MIT Press, Cambridge, 185–193.

Kambis, A.D. and Levine, J.S. 1996. Biomass burning and production of carbon dioxide: A numerical study. In: Biomass burning and global change, Vol. 2, edited by J.S. Levine. MIT Press, Cambridge, 170–177.

Keeling, C. D., Bacatow, R. B., Carter, A. F., Piper, S. C., Whorf, T. P., Heimann, W. M., Mook, W. B. and Roeloffzen, H. (1989): A three-dimensional model of atmospheric CO_2 transport based on observed winds, In Aspects of Climatic Variability in the Pacific and the Western Americas, (ed: Peterson, D. H.), Geophysical Monograph 55, American Geophysical Union, Washington, D.C., 165–236.

Lefevre, T., Todoc, J. L. and Timilsina, G. R. (1997) : The role of wood energy in Asia, Food and Agriculture Organization of the United Nations, Rome, Italy, Working Paper, FOPW/97/2,107.

Mandal, S. and Brahmachary, R. L. (1998): Growth stimulators in the shed leaves of Teak, The Indian Forester, 124 (3), 267–269.

NWDB (1991): Guidelines for microlevel planning. National Wasteland Development Board, Government of India, New Delhi.

NR (National report) to UNCED, (1992): Traditions, concerns and efforts in India. Ministry of Environment and Forests, Govt. of India, New Delhi.

PC (Planning Commission, 1992): Report of the Eighth Five Year Plan. Government of India, New Delhi.

Ravindranath, N. H. and Mukunda, H. S. (1993): Rural energy centre based on energy forest wood gasifier system at Hosahalli village. ASTRA Report, Indian Institute of Science, Bangalore.

Reddy, A. K. N., Sumithra, G. D., Balachandra, P. and D'sa, A. (1990): Comparative cost of electricity conservation – Centralized and decentralized electricity generation. Economic and Political Weekly, 3, 1201–1216.

Sudha, P. (1997): Plantation Forestry – Land availability and biomass production potential in Asia. ARRPEEC Technical Report, AIT, Bangkok.

TERI (1992): Foolish trends and wise choices – Options for the future. Report, Tata Energy Research Institute, New Delhi.

USDOE (1993): Electricity from biomass – National biomass power programme. U.S. Department of Energy, Washington, USA.

Vitousek, P. (1992): An analysis of forests as a means of counteracting the buildup of CO_2 in the atmosphere. Unpublished report, Stanford University, USA.

Whitmore, T. and Sayer, J. (1992): Tropical deforestation and species extinction. Chapman and Hall, London.

Managing Smoke in United States Wildlands and Forests: A Challenge for Science and Regulations

DOUGLAS G. FOX[1], ALLEN R. RIEBAU[2] and RICHARD W. FISHER[3]

[1]*Cooperative Institute for Research on the Atmosphere, Colorado State University, Fort Collins, Colorado, USA*
[2]*Wildlife, Fisheries, Watershed and Air Research, USDA, Forest Service, Washington, USA*
[3]*Air & Watershed Management, USDA, Forest Service, Fort Collins, USA*

Abstract: In past decades the Forest Service and other land management agencies in the United States developed the image in the national popular imagination of forest fires being unnatural occurrences fraught with danger and personal loss. These policies, which were designed to save forests from fire, especially commercially valued forests and associated private property, have now left many ecosystems with fuel loadings that can easily result in large conflagrations unless these loadings are carefully and methodically removed by controlled burns.

New air pollution regulations for fine particulates less than 2.5 μm in diameter and visibility protection potentially conflict with our growing understanding of the ecological benefits of fires in many forest ecosystems. Seeking a new balance between air quality regulatory programs and ecosystem stewardship, state air agencies and federal land managers are grasping for a new paradigm incorporating improved processes for permitting open burns, higher technology approaches to assess fuel loading more accurately, improved use of fire weather information, the development of improved fire emission factors, new methods for cooperative fire planning, and the implementation of improved dispersion models to understand the implications of biomass burning emissions to regulatory programs.

This paper reviews models that are currently being used in smoke management in the United States. These modes run the gamut from simple, straight-line Gaussian dispersion models to sophisticated regional meteorological forecasting tools (e.g., MM5 and CSU RAMS) coupled with sophisticated puff dispersion models (e.g., Calpuff). Based upon an assessment of appropriate applications for this spectrum of tools including their operational data requirements, we present a strategy for implementing a proposed operational smoke management system. This system will be designed for field personnel of the Forest Service and other land management agencies, for both planning and burn project management.

1.   INTRODUCTION

Since the early decades of this century, the Forest Service and other land management agencies in the USA developed the image in the national popular imagination of forest fires being unnatural occurrences fraught with danger and personal loss. These policies, which were designed to save forests from fire, especially commercially valued forests and associated private property, worked very well. In a sense, too well, because the result is that we are now left with fuel-loaded ecosystems that can easily result in large conflagrations unless this loading is carefully and methodically reduced. This situation is exacerbated in many areas of the urbanising western US as people develop homes intermingled with natural vegetation complexes and wildland fuels in a growing urban/rural interface.

Forest fire is not an unnatural occurrence; it is a natural component of many forest ecosystems. Indeed most forest types require the regular presence of fire in order to maintain themselves in healthy, diverse, and productive condition. The unnatural suppression of wildfire, historically a result of lightning ignition, has led to ecosystems with reduced biological diversity, with increased susceptibility to devastation by uncontrolled insect and disease activities, and with the potential for catastrophic fires. Moreover, efforts to simulate the ecological benefits of fire using alternative mechanical and chemical treatments have met with limited success. Often, there is simply no substitute for fire.

Wildfire, however, negatively impacts human economic values and, obviously, cannot be tolerated in developed communities. Because fire is a natural element of most landscapes, it is not surprising that human attempts to control fire, especially in forests in Mediterranean climates, have not been successful. In Southern California, for example, attempts to suppress wildfire have effectively eliminated small, frequent, low-intensity fires but have not seriously reduced large, high-intensity, devastating fires. This clearly illustrates that while humans might manage fire, we can never eliminate it.

Management of fire includes the following elements:

– suppression of human-caused wildfire;
– monitoring and, if necessary, suppression of prescribed natural fire; and
– ignition, monitoring, and, if necessary, suppression of prescribed fire.

Forest fire fighters have developed an extensive set of techniques that facilitate fire management. Increasingly important among them are tools to estimate the location, magnitude and intensity of smoke and to minimise it. Smoke, of course, is composed of gases and particles emitted from the burn. To date, these tools for smoke management have been developed and applied locally or regionally where smoke was considered a problem. However, in the USA, the need for smoke management tools is changing. The US Environmental Protection Agency (EPA), in 1998, promulgated new national ambient air quality standards (NAAQS) setting limits on the concentration of particles smaller than 2.5 micrometers in diameter ($PM_{2.5}$) to protect human health and welfare. In early 1999, EPA drafted new Regional Haze regulations to achieve another goal established by the Clean Air Act to improve in clean air areas of the USA. In addition, concerns about smoke obscuring visibility on highways and causing a nuisance to people sensitive to wood smoke combine to raise conflicts between ecological and practical benefits from using managed fire and public health and associated welfare costs. In the USA, responsibility to administer these new regulations and to deal with the concerned public falls to State and local air quality managers.

Thus, a conflict exists between growing acceptance, by forest and grassland managers as well as the public, of the ecological and practical benefits of and the need for prescribed fire and the potential negative health and visibility consequences of smoke from this fire. To be more specific, in the past decade, the USDA Forest Service has been conducting about 1 million acres of prescribed burning annually. The Forest Service plans to increase this acreage threefold, to as much as 3 million acres, in the next decade. As many as 10 or more million acres of public lands are now burned under prescription each year in the USA.

Seeking a new balance between clean air objectives and ecosystem stewardship requires air managers and land managers to work together to improve their collective ability to manage fires and to reduce the negative impacts of fire smoke. In the USA new paradigms incorporating improved processes for permitting open burning are emerging. New technology-based approaches are increasing the abilities of managers to assess fuel loading accurately, use fire weather information, develop improved fire emission factors, co-operatively plan fire uses, and implement improved dispersion models. In turn, these all lead to better understanding, predictions and facilitate the mitigation of impacts of biomass burning on air quality, resulting in improved capabilities to manage smoke. The objective of this paper is to illustrate a potential application of these emerging tools to improve smoke planning and forecasting.

Based on an assessment of applications and their operational data requirements, we will present an example strategy for implementing a proposed operational smoke management system. This system is designed for field personnel of the US Forest Service and other land management agencies, to be used for both planning and operational burn project management. Fortunately at this time several new technologies have become widely available making such a fire management system more feasible than ever before, namely:

1. reasonable cost workstation class computers;
2. mesoscale meteorological forecast models with suitable (and continually improving) spatial resolution being run operationally at weather centres around the world;
3. widespread availability of networks with progressively larger bandwidth (World Wide Web) facilitating the exchange of large data sets and communication; and
4. standardised, self-documenting formats (GRB, NETCDF, HDF) allowing easier transfer of large data sets between models and machines.

2. EXISTING SMOKE MANAGEMENT SITUATION IN THE USA

The formal management of smoke from wildfire and prescribed natural fire has a long history in the federal land management agencies. It remains a component of most fire planning activities. Substantial fractions of the particulate in wild land smoke are in the size ranges that are regulated by the US Environmental Protection Agency. Ambient atmospheric concentrations of particulate matter less than 2.5 micrometers in diameter ($PM_{2.5}$) and particulate matter less than 10 micrometers in diameter (PM_{10}) are regulated under the US National Ambient Air Quality Standards (NAAQS) for fine particles. These standards require that concentrations over different averaging times must remain below numerical levels set to protect people who are especially sensitive to this form of air pollution.

Attainment or non-attainment of the NAAQS is determined by ambient air quality monitoring but has both scientific and practical/political components. Obviously, monitoring is a local activity; as officially approved monitors all collect an air sample, selection of locations, what area the sample represents, decisions about record length, and the like are all subjects of public debate. The US Clean Air Act gives the responsibility to develop monitoring and determine which areas attain and which do not attain the standards to States, Indian nations and other local air quality control districts. This responsibility is implemented through State (or Tribal) Implementation Plans (S/TIPS).

In many areas of the USA, the $PM_{2.5}$ standard level is either being exceeded (non-attainment areas) or in danger of being exceeded. In such circumstances, the law (Clean Air Act) requires that S/TIPs include strategies to either reduce emissions to attain the NAAQS or to manage pollution increases such that attainment is maintained (Egan *et al.*, 1981). As a result, smoke from forest burning is specifically regulated in some States and informally regulated in others. Almost everywhere in the USA where there are publicly managed wildlands, forest fire smoke is recognized as a potentially volatile public policy issue. The issue is confounded by recognition that forest wildfire is, to a large extent, an uncontrollable natural occurrence. As such, wildfire contributions to ambient concentrations are often exempt from attainment/non-attainment decisions. This conflicts with the concept that smoke from managed fire use, often done to diminish the wildfire threat and always resulting in lower net smoke emissions, is included as a contributor to attainment/non-attainment decisions.

Since the 1930s, the US Forest Service, along with academic co-operators and experts in other wildland management agencies, have conducted research on fire. Historically, this research was driven by the desire to understand wildland fire and, from that understanding, to predict fire ignition and fire behaviour. Eventually, fire research developed into a sophisticated effort to improve understanding of the ecological and associated environmental consequences of fire, including smoke production, transport and dispersion (Sandberg *et al.*, 1979). These research activities led to an expanded knowledge base and an inventory of available smoke prediction technologies.

Prescribed fires usually lack on-site meteorological data. Although remote, deployable meteorological stations exist, much of the time they are not used for economic and logistical reasons. Most site meteorological information comes from the nearest stations, which may or may not represent the burn site, or from estimates made on-site by burn personnel. Current local or regional wind field patterns are usually only estimated intuitively, and not by either formal modelling or measurement.

Local scale weather forecasts are rarely available for the prescribed fire area, except as "spot" fire weather forecasts in the immediate area of a fire. In some regions, experimental mesoscale models generate useful local forecasts but these forecasts are not widely used by prescribed fire managers. At present prescribed fire managers do not have the resources to generate or use local scale meteorological information. Thus, as weather service offices begin to produce regional scale forecasts, prescribed fire management teams will need to develop the capacity to gather and interpret this data and generate useful information from them.

Predicting smoke behaviour from wildland fires is a difficult task. To be more tractable, the problem has been divided into discreet elements such as gas and particle chemistry, transport and dispersion, and emissions and heat production. A number of entrepreneurial contributions by various US research and operations groups have provided solutions to many of these elements (NWCG, 1985; Fox *et al.*, 1986; Riebau *et al.*, 1988; Southern

Forest Fire Laboratory, 1976; Peterson and Ottmar, 1991; Harrison, 1996; Hummel and Rafsneider, 1995; Lavdas, 1996). For each of these activities, products have been developed to suit particular needs and customers (e.g., Forest and Rangeland planners, operational fire fighters, or regulators). Validation demonstrations often have not been fully satisfactory for a number of reasons, among them, that valid observational data have been scarce. Consequently, "model validation" has often consisted of an assessment of user friendliness or applicability rather than a measure of precision and accuracy of model performance.

Predicting precisely where smoke from a forest burn is likely to go, and in what amount it will get to a selected receptor in the complex mountain topography of many forests cannot be confidently done at present in any deterministic way. This is because smoke prediction involves a wide variety of components that take on quite uncertain, effectively random, values. Strung together in a modelling system, these lead to assured failure of any prediction scheme in at least some significant circumstances. Smoke prediction cannot be improved without more and better observational data to try to capture these uncertainties and improve understanding of limits of predictability.

The fundamentals of natural vegetation combustion are poorly characterized because fire behaviour is physically complex. It depends on the location, distribution, and condition of the fuels, the topography and micrometeorology of the vegetation/fuel complex, and the nature of the combustion. Experimentation has illustrated that different fuels burn differently, generate different chemicals at different stages of the combustion process and physically disperse those chemicals differently depending on all of the above variables. For example, predicting plume rise is not as straightforward an engineering calculation as it is for a stack emission because of complex turbulent interactions between the ambient airflow and the micrometeorology of the vegetation canopy. As the flame front progresses through the fuel bed, varying amounts of heat and momentum are released.

In the US, air quality regulators and land managers recently agreed on a non-binding Interim Air Quality Policy on Wildland and Prescribed Fires (EPA, 1998). In it, they described and agreed upon the need for both wildland fires and clean air. They also defined the roles of each entity in smoke and air quality management. Actions to minimise emissions were proposed along with ancillary tracking measures to be used to maintain accountability. Being a policy agreement, it did not include details of specific models or technologies for estimating smoke emissions or behaviour. It did, however, serve to set the stage for implementing advanced tools to help manage smoke from managed and wildfires.

3.   SMOKE MANAGEMENT NEEDS

The United States Congress recently established a Joint Fire Science Program to address a growing concern about the ability of forest managers to control wild fires, especially at the growing interface between urban and rural land, and the use of fire for managing vegetation in these regions. Congress acknowledged the potential for conflict in this urban interface between the use of fire to accomplish ecologically beneficial land management goals and negative impacts of smoke on clean air, human health and welfare goals. It recognized that fire managers must strive to accomplish both of these goals. To do so, they need a set of smoke management tools that are both readily accessible and scientifically appropriate. The Joint Fire Science Project management, in its request for proposals, in June 1998 stated that "... the capabilities of current models to predict emissions and transport of smoke from fires are inadequate in coverage and are incomplete in scope." As a result, it is necessary to co-ordinate the work of fire managers, fuels specialists, air quality managers (regulators) and fire, air quality and meteorology researchers, to develop and implement technical advances in smoke evaluation tools, otherwise known as smoke management modelling.

Specifically, users of smoke management tools include:

- fire managers – responsible for burn activities;
- fire behaviour analysts – responsible for predicting fire evolution;
- fire meteorologists – responsible for predicting meteorological conditions relevant to the fire;
- public information officers – responsible for providing fire information to the public;
- fire effects analysts – responsible for studying fire, smoke, and their impacts;
- burn planners – responsible for choosing, permitting, and scheduling areas to burn; and
- State/local air quality agencies – responsible for ensuring the health and safety of their constituents by reviewing Burn Plans and deciding whether or not to authorise specific prescribed wildland fires.

The set of tools that they require for smoke management include:

1. technologies to predict particle size ranges, chemical and physical source characteristics of emissions from prescribed and wildfires in all appropriate fuel and fire types;

2. technologies to predict meteorological conditions at scales from the immediate vicinity of the fire up to the atmospheric mesoscale (up to 1000 km from the fire);
3. technologies to predict smoke dispersion, including smoke location, timing and amounts at sensitive points in the vicinity and downwind of the fire;
4. technologies to integrate all of these tools into an operational smoke management model and to communicate its results for fire manager's use in operational time frames;
5. methodologies for involving information service providers, e.g. public and not-for-profit private sector emergency information systems, in the distribution of resultant products.

4. SMOKE MANAGEMENT TECHNOLOGY INNOVATION PARTNERS

Traditionally, smoke management tools have been developed by the land management agencies in the USA. These agencies include the US Department of Interior (USDOI) Bureau of Land Management, the USDOI National Park Service, the USDOI Fish and Wildlife Service, and the US Department of Agriculture (USDA) Forest Service. Because of the importance of the contributions of fire smoke to local and regional particulate loading, and the consequent regulatory dilemma of allocation of emission amounts and rates, the issue now interests a wider variety of stakeholders.

A truly national approach to improve smoke management in the USA will require participation of a broad group of potential partners. New players in the development and supply of smoke management-related tools and data include, but are not limited to, the following organizations:

- The National Oceanic and Atmospheric Administration (NOAA) National Weather Service (NWS). It is clear that there are increasing capabilities in the realm of meteorological modelling becoming available from the main stream meteorological research establishments. NWS, through their National Center for Environmental Prediction (NCEP) is beginning to make fine scale high-resolution forecast model data available to users. Improved forecast tools continue to be available from the Forecast Systems Laboratory (FSL) (http://www.noaa.gov).
- US Air Force Weather Agency (AFWA). The Air Force Weather Information Network (AFWIN) routinely provides fine scale high-resolution operational weather data and forecasts for the USA and selected world-wide locations. Cooperation between the AFWA, NWS and the

US Navy Fleet Numerical Operations Center is providing shared and complementary weather products for use world-wide. This division of labour and cooperation is a model for the type of partnerships envisioned here.

– US Environmental Protection Agency. EPA's Meteorology Division in Research Triangle Park, NC is developing a universal environmental modelling framework called MODELS3. Intended to integrate all manner of environmental effect predictions by accommodating a comprehensive suite of atmospheric and other models, MODELS3 represents the regulatory modelling future for EPA. MODELS3 may be the ultimate target system for a resident national smoke modelling system. (http://www.epa.gov). Meanwhile, EPA's Office of Air Quality Planning and Standards (OAQPS) continues to list appropriate modelling technologies for application in formal regulatory actions.

– The University of Washington (UW), Atmospheric Sciences Department. UW is conducting special studies relevant to the US Pacific Northwest region using down scaling outputs from the MM5 mesoscale model runs developed and output by NCEP. This work is enhanced by the cooperative research with Forest Service scientists at the USDA/FS Pacific Northwest Research Station in Seattle, WA (http://www.sol.cfr.washington.edu).

– Colorado State University, Cooperative Institute for Research on the Atmosphere and Atmospheric Sciences Department (CIRA). CIRA is developing operational capabilities based on MM5, MAPS/RUC II (an FSL adapted version of MM5) and RAMS (a comprehensive model developed at CSU). This work is enhanced by cooperative research with the National Park Service (http://www.cira.colostate.edu).

– Florida Division of Forestry (FDF). In cooperation with Florida State University, the University of Florida and the Canadian Forest Service, FDF has developed an integrated GIS and mesoscale-forecasting-model-based smoke management system applied to the State of Florida (http://www.fl-dof.com).

– National Center for Atmospheric Research (NCAR) is archiving data for developing comprehensive model testing data sets (http://www.ncar.org).

– State Regulatory Agencies. As modelling needs in individual States become more complex, some of these States are giving consideration to establishing regional modelling centres. For example, the State of Washington co-operates with the University of Washington Atmospheric sciences group to obtain MM5 data on an operational basis (http://www.westar.org; http://www.stapalapco.org).

5. MANAGING SMOKE FROM PRESCRIBED FIRE: PLANNING, PERMITTING, EXECUTION AND MONITORING

The tasks of planning fires, obtaining burn permits from regulators, and safely and effectively executing a prescribed burning are complex and important. This is especially true as air quality regulations and standards are applied more broadly to prescribed fire smoke and as the urban–wildland interface grows.

5.1 Planning

Two activities are key to a successful prescribed fire planning process. The first is seasonal scheduling. Seasonal scheduling is the activity of identifying a window of time when a burn will occur during a given burn season. Scheduling will allow for better resource utilization, thus maximising the number of burns that can be completed during a year. Seasonal scheduling is mostly resource dependent. Schedulers need project management tools to effectively allocate resources. The project management tools need to be flexible enough to account for unplanned events such as wildfires. Seasonal range weather forecasts may also help determine when to schedule burns. Forecasts at the seasonal time scale are reported in terms of excess probability of exceeding normal values for precipitation and temperature. Knowledge of local "normal" values is therefore important.

The second key issue is short-term scheduling. Short-term scheduling is the activity of scheduling a burn with lead-times equal to or less than 10 days. It ensures that the fuel and weather conditions required for a successful burn will be present, as will the staffing needed for a safe burn. A successful burn accomplishes land management objectives of fuels reduction, ecological improvement and no unsafe conditions created, as well as air quality management objectives of no violation of ambient air quality standards, no impairment visibility in specially protected class I areas and few or no citizen complaints.

The short-term scheduler needs meteorological forecasts for winds, temperatures, humidity, planetary boundary layer (PBL) mixing, and precipitation. Typical forecasts would be for lead-times of ten days, five days, three days, two days, and one day. The scheduler also needs to understand the uncertainties associated with various types of forecasts. To fully understand these forecasts and their implications, the scheduler needs to be able to converse with those producing the forecasts.

5.2　　Permitting

Permitting is the activity of receiving or providing approval from state or local regulators to conduct a burn based on air quality impact assessments. Most states require dispersion modelling studies to show that impacts will not violate air quality standards, although the specific requirements vary. In many instances, the Simple Approach Smoke Estimation Model[1] (SASEM) is the tool of choice for permitting fires (Riebau *et al.*, 1988). Although permit requirements vary greatly between states, most states require a quantitative estimate of pollutant emissions and many States request evidence that the prescribed burn will not violate the National Ambient Air Quality Standards (NAAQS). To estimate accurately the amount of material emitted to the air, one needs to know:

1. what material will be burned;
2. the chemical composition of material that will be burned;
3. how much of each material will be burned;
4. the fire characteristics;
5. meteorological conditions during the burn; and
6. water content of the material to be burned.

Fuel amount and status, fire information as well as meteorological forecasts and ambient air quality data are all needed simultaneously to determine if the burn would violate NAAQS.

Each of the variables needed to make these predictions change in time and space over the burn area. The problem is complicated because the amount of material emitted to the air depends on the properties and evolution of the fire itself. Fire characteristics are needed as input to atmospheric dispersion models. These include: 1) fire temperature, 2) particle size distributions, 3) area being burned, and 4) buoyancy of emitted smoke. A geographic information system (GIS) can be useful to keep track of information about a fire and accumulate information about the entire fire season.

Estimates of material emitted and fire characteristics are inputs along with meteorological conditions to atmospheric dispersion models that estimate plume direction, dispersion and particulate concentrations. Atmospheric dispersion models range in complexity from simple straight-line Gaussian screening models able to run on any computer system (SASEM), to prognostic fully 3-dimensional models that require mainframe or advanced workstation class computer resources. In general, regulatory agencies will

---

[1]　Simple Approach Smoke Estimate Model, developed by the Bureau of Land Management as a tool for estimating whether or not smoke from wildland fire will exceed ambient air quality standards.

accept simple source estimates or results from simple "screening" models so long as these models are known to be conservative in the sense of overestimating concentrations. If the screening models project concentration levels that exceed standards, regulatory agencies are likely to require more complex modelling rather than simply prohibit the burn.

As an alternative to the use of screening and more complex models, air quality scientists within the regulatory agencies are considering using probabilistic assessments. To date this method has been applied to toxic air pollutant releases but this thinking and approach may also prove applicable for smoke management. Along with an estimate of the emission source strength, probabilistic assessments require:

1. simple statistical routines (to rank or describe modelling results);
2. large sets of meteorological data (a representative sample of weather conditions that may occur during a burn);
3. an atmospheric dispersion model (to estimate concentration values); and
4. moderate to large computer resources (depending on the complexity of the dispersion model and the number of meteorological data sets used).

When a probabilistic assessment has been used, regulatory agencies have provided a well-defined (standard) set of meteorology for all permit applicants' use. In this case, the meteorological database has to be of appropriate spatial extent, contain as much valid data as possible, and readily accessible to potential users.

Regulatory agencies in some regions of the USA are faced with the problem of issuing a large number of burn permits. Each burn by itself does not compromise air quality standards. However, the combination of many concurrent burns may exceed standards or otherwise cause unacceptable problems. Few agencies have the ability to simultaneously manage smoke from multiple burns. There is, therefore, a need to track the smoke from simultaneous events individually and collectively as well as to account for the emissions from other pollution sources and emissions transported from other regions.

5.3   Execution

Burn start/stop is the activity of determining if a planned burn should be initiated, allowed to proceed or be stopped. The start/stop decision is usually made with a lead-time of 24 hours from actual source ignition. This activity considers near-term weather conditions, fire-fighting resources, and local concerns over fire spread, safety, smoke impacts, and resource allocation. Meteorological forecasts are needed to identify both the potential for stable atmospheric conditions, good for fire behaviour but bad for smoke dispersion, and for strong winds or uncertain winds, good for smoke dispersion but bad for fire behaviour. The burn start/stop activity requires accurate meteorological and smoke forecasts somewhere within bounds for both atmospheric stability and wind to ensure safety, nuisance mitigation, and minimise air quality violations. Meteorological forecasts should be issued 24 hours, 12 hours, 6 hours, and 1 hour before a scheduled prescribed burn starts. While the burn continues, 48 hour, 24 hour and 12 hour forecasts should be given. Meteorological forecasts should include winds, temperatures, humidity, PBL mixing, and precipitation likelihood. Additionally, the forecaster should provide immediate notice of any anticipated or observed abrupt weather changes or any severe weather including approaching thunderstorms, lightning, hail, or strong wind gusts. As with short term scheduling, the burn start/stop activity requires the ability to immediately converse with forecasters.

The burn start/stop activity requires knowledge of where smoke is or will be going, what the pollutant concentrations are or will be, when the plumes will arrive and leave locations, and what the consequences will be. As smoke plumes are time dependent, smoke forecast maps need to depict the time progression of the plumes. "What if" scenarios may also be important as they give fire managers the ability to minimise impacts based on when the fire is set. Finally, those agents given the burn start/stop responsibility may wish to have plots of projected visibility and pollutant concentrations at selected locations as a function of time. The time plots are also useful for public information and air quality activities.

Smoke forecasts should be produced for the same time periods that are covered by available meteorological forecasts. The smoke forecasts may include all or some of the following pictorial maps: surface plume concentrations of PM_{10}, $PM_{2.5}$, and CO (in time), plume location and timing, surface visibility impacts, vertical cross section visibility impacts (important for scenic views, airports, etc.), projected consequences (areas where concentrations exceed specified limits), and local wind fields.

As estimates of smoke concentration depend on source term information, the burn start/stop activity requires knowledge of the fire characteristics and estimates of the source emission characteristics. This type of information is similar to that required for the permitting activity but must be provided in real time if they are to be of any value. Smoke forecast models and output maps must serve the user. Therefore, those using the products need access to the developers and analysts running the models for rapid modifications, updates, training, and discussions.

Nuisance minimisation is the collection of actions taken to reduce public annoyance due to prescribed fires. Generally smoke concentration levels that cause a nuisance are below accepted health thresholds. Nuisance concentration levels vary greatly depending on the receptor's health, age, attitude, and knowledge. The nuisance minimisation activity has the same needs as the permitting activity with the addition of knowing the location of sensitive populations. The locations of these populations may best be displayed on a GIS generated map that displays spatial relationships between sensitive population locations and projected smoke plumes. Knowledge of smoke concentrations that cause annoyance is also important for minimising nuisance complaints. With estimates of the source emission amounts and the local smoke dispersion patterns, fire managers can reduce complaints by informing the populations of expected smoke impacts or by scheduling the burn so that it does not impact sensitive populations.

5.4   Monitoring

After the fire, a continuing level of monitoring should be conducted for at least two purposes. In a sense, the fire generated an atmospheric burden of aerosol that can be tracked and identified as having been generated by this specific activity. This allows air regulators to consider the source of a measured increase in ambient particles. It also allows regulators to determine regional air quality impacts. A second consideration is for monitoring to evaluate how effectively the burn accomplished its planned objectives, both on short and long terms. Were anticipated amounts of fuel consumed? Did planned ecological benefits accrue? Did the tools used to forecast smoke work well enough? Can those tools be improved?

6.   USING COMPUTER MODELS IN SMOKE MANAGEMENT

Computer modelling is required by the United States Clean Air Act to determine the extent and duration of air quality impairment that is likely to result from the operation of new air pollution sources. As air agencies use these types of models for other source categories, it is logical to apply them to smoke from wildland fires. Several types of models have been used for smoke management including:

6.1   Wind field analysis

Wind field analysis is the activity of depicting local or regional airflow patterns. The wind field analysis can be from current meteorological conditions or from predicted conditions, if available. To perform a wind field analysis one needs meteorological wind data, an analysis program and a display mechanism. To obtain a current observation-based wind field analysis, one needs meteorological observations, a real-time data acquisition system, a computer code capable of incorporating these data into an analysis and generating a fully 3-dimensional wind field while accounting for the limited number of observations and the presence of complex topography and, finally, a way to display the complex resulting fields. For predicted wind field analyses, one needs predicted wind data and a way to display the fields. As predicted data usually only exists for coarse scale grids, it may be useful to process the coarse wind data with a program that generates full 3-dimensional wind fields from limited observations and topography. The graphical output should contain enough landmarks to identify all regions of interest.

6.2   Regional transport analysis

Regional transport analysis is the activity of evaluating the impacts from smoke transported into or out of the region. The impacts may include effects from more than one fire and more than one external region. Regional transport analysis requires the ability to track plumes over large, multi-state, regions. The ability to share information from other regions is important for combining regional pollutants with pollutants from local sources. Atmospheric models need to have the ability to combine background or regionally transported pollutants with locally generated pollution. To analyse regional transport one also needs to have regional scale meteorological data and dispersion models that will use the data. Wet deposition removal processes become very important at regional scales (otherwise concentrations may build up to unreasonable levels) so models need to use real-time precipitation estimates.

6.3   Fire behaviour analysis

Fire behaviour analysis is the activity that attempts to predict what the fire will do. This activity includes estimates of timing, consumption, intensity, location, and other characteristics of the fire. The fire behaviour analyst needs local and state scale meteorological observations and forecasts, terrain information, and detailed information concerning local vegetation. Weather forecasts with lead times of 5 days, 3 days, 2 days, 1 day, 12 hours, 6 hours, 3 hours, and 1 hour will be useful. Where available, detailed local wind field and precipitation patterns are also useful.

7.   EXISTING MODELS

Various types of models have been designed and implemented to support these analyses.

7.1   Source emissions models

Experimental efforts have been conducted to collect information about smoke emissions. Models have been developed with the capacity to estimate the nature and amount of specific emissions under a variety of burning conditions. While by no means completely solved, the quality of models that predict the physical and chemical nature and amounts of emissions from different types of fire in different ecosystems is reasonable and possibly good enough for introduction into dispersion and transport models.

Several computer models currently exist for estimating source characteristics. Models such as SASEM, EPM, and FOFEM[2] are readily available. However, source term estimate models applicable to all regions of the USA for all fuel types or all burn conditions are not available. Some source term estimate models have simple straight-line Gaussian dispersion models built into them (e.g., SASEM). A few have been integrated into more complex puff or particle models. Few models are integrated to use real-time or forecast meteorology for source estimate or for dispersion modelling.

7.2   Meteorological models

Mesoscale meteorological and regional dispersion models with ever greater resolution are being run operationally by meteorological data sources world-wide, for example, in the USA, NCEP and globally, AFWIN. In addition, understanding of and technology for linking mesoscale meteorological simulation models with local scale models is improving as meteorological applications routinely utilise these technologies. Research in the atmospheric science community has improved the ability to simulate atmospheric behaviour on all scales from global to local. Current progress is being made in moving simulation and prediction abilities down in scale along with development of the new tools discussed here. As a result, there is an ever increasing coalescing of approaches, technologies and capabilities, such that cooperation with the atmospheric sciences research community will be more fruitful in the future provided that it is co-ordinated.

---

[2]   Developed by the Intermountain Fire Sciences Laboratory to model the immediate effects of a fire including emission estimates and damage to trees.

7.3   Dispersion models

Currently there are a variety of models available and proposed for routine use in smoke management. In a Smoke Emission and Dispersion Modelling training program used by the Forest Service, models for fire fuel consumption, emissions, dispersion and visibility were presented. Specific models mentioned include: the Emissions Production Model (EPM); First Order Fire Effects Model (FOFEM), Simple Approach Smoke Estimation Model (SASEM), the Ventilated Box Model, VSMOKE, VSMOKE–GIS, TSARS Plus, CALPUFF and NFSPUFF. Specific training was conducted for using the emissions models, SASEM and NFSPUFF.

The Interagency Working Group on Air Quality Modelling (IWAQM) consisting of representatives from EPA, USDA/FS, USDOI/NPS, USDOI/FWS and State regulatory agencies, has developed preferred modelling schemes for dealing with the unique challenges presented to regulatory modelling by sources located in the mountainous terrain of the western USA, often causing impacts on class I areas farther than 50 km away. (Class I areas are a specific collection of national parks and wilderness areas in the United States which are afforded special air quality protection under the US Clean Air Act) The group acknowledges the need for both meteorological models as well as dispersion models. IWAQM is developing formal regulatory guidance on the use of CALMET and CALPUFF as the preferred models for this purpose. CALMET is a meteorological processor model driven by observations and simulated meteorological data from MM5. CALPUFF is an atmospheric dispersion model designed for use in complex terrain that accounts for aqueous phase chemical reactions and estimates visibility degradation. Soon to be accomplished will be the automatic integration of outputs from smoke emission models and the integration of forecast meteorology.

8.   AN EXAMPLE OPERATIONAL SMOKE MANAGEMENT SYSTEM

Certainly the most critical element in smoke management are the decisions that are made when and after the match is lit. Specialists must do their best to estimate the behaviour of the fire and the smoke that is and will be produced. Key information for these decisions is the forecasted weather. While weather has long been included in this process, never has it been available at a national level so that smoke behaviour can be forecasted rigorously and quantitatively. Recently, the National Forest Systems Air Resource Management Program of the USDA/FS supported a demonstration of a prototype GIS format web-based smoke forecasting centre developed by Alphatrac Inc. of Westminster, Colorado.

This demonstration smoke management centre included the following characteristics:

1. modellers and computer analysts associated with the centre continually update models, information, and technology and provide users with smoke management products;
2. computer resources generate and transmit output products to clients for multiple simultaneous and geographically separate fires; and
3. experts in emission source characterization, atmospheric dispersion modelling, numerical forecasting, computer science, air quality permitting, and meteorology are accessible on short notice.

If implemented, such a smoke management centre could maintain current information pertinent to smoke management. The information, quality assured, referenced, and as up to date as possible, could be easily accessible via a web site, phone and hardcopy publications. Specific information could, for example, include, but not be limited to:

1. supported smoke management models;
2. other (non-supported) smoke management models with references and contacts;
3. smoke management groups;
4. tables or listings for emission source estimates;
5. organizations that deal with wildfires;
6. permitting regulations and contacts;
7. weather forecasts from external groups;
8. current wildfires and prescribed fires;
9. current news in smoke management;
10. air quality and visibility issues;
11. model validation data sets; and
12. information about the smoke management centre.

The smoke management centre would also provide clients with meteorological information related to smoke management activities. Information limited to clients is proposed to include:

1. climatological information;
2. local current meteorological conditions;
3. regional current meteorological conditions;
4. local meteorological forecasts;

5. regional meteorological forecasts;
6. wind field analyses and forecasts; and
7. specialised meteorological forecasts (as arranged).

The smoke management centre would maintain Internet links to other meteorological data sources and will house quality assured, complete, meteorological data sets for model comparisons or for probabilistic hazard assessments.

9. JOINT FIRE SCIENCES PROGRAM

Twenty separate research and development initiatives collectively valued at about US$ 8 million and aimed at providing tools that will aid managers in US wildland fuels reduction, were awarded in 1998. Among the initiatives is the Technically Advanced Smoke Management Tool: Needs Assessment and Feasibility Investigation (Riebau and Fox, 1998). It is intended to develop a broad administrative and technical management solution for US smoke management.

10. REMAINING CHALLENGES

Competitive grass roots efforts in separate and, by-in-large independent, research and development groups around the USA and the world dilute the limited resources available to solve the smoke management problem. As has been suggested here, integration of the tools now available can easily yield product improvements that will advance smoke management skills. Further, an operational system or centre that integrates such disparate but complementary tools can serve as a nucleus around which continued research and development can flourish more effectively. Undoubtedly, there is a need to tailor products for regional and local consumption; central planning and forecasting is not always the best solution. Implicit in this evolution is discriminating between those activities, which are clearly research, and those, which are clearly operational. We cannot succeed in the long haul when research organizations are expected to conduct operational missions. Conversely, day-to-day operational units should not be expected to conduct research. Our challenge is to build on existing strengths while at the same time promoting necessary new technologies and delivering a superior product to smoke managers for protecting human health and public welfare.

11. ACKNOWLEDGEMENTS

The authors acknowledge the significant contribution made by John Ciolek, Jr., Michael R. McCarter, and C. Reed Hodgen from AlphaTrac, Inc. AlphaTrack, under contract to the US Forest Service, developed a comprehensive report regarding smoke management as well as demonstrating the concept of an operational smoke management centre. Their report was formative in developing this paper. The report is available from Richard Fisher at the above address. We also acknowledge many contributions made to this paper and to the field of smoke management by Dr. Mike Sestak, currently at the U.S. Navy Meteorology and Oceanography Center in Monterey California.

12. REFERENCES

Egan, B.; Fox, D.G.; Hanna, S.; Randerson, D.; White, F. 1981. Air quality modeling and the Clean AirAct: Recommendations to EPA on dispersion modeling for regulatory applications. 288p. American Meteorological Society.

Faber, B. Fox, D.G., Wallace, W.W. and Watts, R. 1993. The TERRA Laboratory: An Interagency decision Support Environment. Air and Waste management Association, 86th Annual Meeting, Nashville, Tennessee, June 1993. 9p.

Fox, D.G.; Dietrich, D.L.; Mussard, D.E.; Riebau, A.R.; Marlatt, W.E., 1986. The Topographic Air Pollution Analysis System. Envirosoft. Computational Mechanics Institute, UK, pp., 123–144.

Harrison H., 1996. A User's Guide to NFSPUFF. Unpublished final report. Available from USDA Forest Service, Northern Region, Air Resources Management, Missoula, MT.

Hummel, J & Rafsneider, J., 1995. TSARS plus smoke production and dispersion model user's guide. Unpublished. Available from A. Riebau, USDA Forest Service, Washington DC.

Lavdas, L., 1996. Program VSMOKE – Users Guide. GTR SRS–6. Asheville NC.

National Wildfire Coordinating Group, 1985. Prescribed fire smoke management guide. NEFES No. 1279 Boise Interagency Fire Center.

McCutcheon, M. H.; Fox, D. G, 1986. The effect of elevation on wind, temperature, and humidity. J. Applied Meteorology & Climate 25(12), pp. 1996–2013.

Peterson, J.L & Ottmar, R.D., 1991. Computer applications for prescribed fire and air quality management in the Pacific Northwest. Proceedings, 11th Fire & Forest Meteorology Conference, American Meteorological Society.

Riebau, A.R. and D.G. Fox, 1998. Technically Advanced Smoke Evaluation Tools (TASET): Needs Assessment and Feasibility Investigation. Accepted proposal to the Joint Fire Science Program. USDA Forest Service, Wildlife, Fisheries, Watershed, and Air Research, Washington, DC. 14p.

Riebau, A.R.; Fox, D.G.; Sestak, M. L.; Dailey, B.; Archer, S.F., 1988. Simple Approach Smoke Estimation Model. Atmospheric Environment.

Ross, D.G.; Smith, I.N.; Manins, P.C.; Fox, D.G., 1988. Diagnostic Wind Field Modeling for Complex Terrain: Model Development and Testing. J. Applied Meteorology & Climate 27(4).

Sandberg, D.; Pierovitch J.; Fox, D.G.; Ross, E., 1979. Effects of Fire on Air Quality. USDA Forest Service Research Paper GTR WO–9. Washington, D.C.

Sandberg, D. V and others, 1998. Wildland Fire and Air Quality: Modeling and Data Systems. Unpublished manuscript. Available from the lead author at http://www.fs.fed.us/r6/aq.

Southern Forest Fire Laboratory, 1976. Southern Forestry Smoke Management Guidebook. GTR SE–10 Asheville NC.

United States Environmental Protection Agency 1997. National Ambient Air Quality Standard for Particulate Matter: Final rule, July 18, 1997. (62 FR 38652). Available at http://www.epa.gov/ttn/oarpg/rules.html

United States Environmental Protection Agency 1998. Interim Air Quality Policy on Wildland and Prescribed Fires. Available at http://www.epa.gov/ttn/oarpg/rules.html

Area Burned Reconstruction and Measurement: A Comparison of Methods

CHRIS LARSEN
Department of Geography, State University of New York, University at Buffalo, Buffalo, USA

Abstract: Knowledge of temporal changes in the area burned by wildfires is required to understand their influence on global climate change. This paper reviews the primary methods of reconstructing and measuring area burned. The area burned by wildfires is typically reconstructed using historical records, satellite imagery, tree-ring records and sediment records. These methods are described and compared in terms of the spatial and temporal resolutions and extents of the observations made using them. The different ways of measuring area burned that they employ are also described and compared.

Tree-ring and sediment studies are uniquely useful over small areas and long time periods, while historical records and AVHRR images are more useful over large areas and shorter time periods. The methods do, however, overlap over several orders of magnitude of scales of observation. The different methods of measuring area burned, the fire rotation, mean fire interval and fire cycle, are also found to have similarities.

A comparison of the estimates of area burned obtained using different methods in the same area suggest that, although somewhat similar, calibration methods need to be developed to translate between them. It is thus suggested that although much current work is focussed on refining individual methods of estimation, some emphasis should be given to determining methods to compare the results obtained using different methods. This would allow the meaningful comparison of the estimates of area burned made over very different time periods and spatial areas.

1.   INTRODUCTION

Wild-fires, lightning- and human-caused, have long been recognized for their role in vegetation and ecosystem dynamics (e.g. reviews in Pyne 1995, Bond and van Wilgen 1996). More recently, fire-produced greenhouse gasses have been recognized as having a potential influence on climate change (Crutzen *et al.*, 1979). Wildfire in this role has been recast as biomass burning. Note that wildfire is only one component of biomass burning, a concept that includes many other forms of burning such as domestic fires and fossil fuel burning.

The production of greenhouse gasses through wildfire-caused biomass burning is measured as:

$$M = A * B * \alpha * \beta$$

where M is the mass of dry biomass (db) burned per year (g [db] yr^{-1}),
A is the area of vegetation burned each year (m^2 yr^{-1}),
B is the areal density of dry biomass (g [db] m^{-2}),
α is the proportion of dry biomass that is above-ground (dimensionless)
β is the proportion of the above-ground dry biomass that is combusted (dimensionless) (Seiler and Crutzen, 1980).

The emitted mass of different greenhouse gases is measured as the product of M and the emission ratio of each gas (e.g. Levine *et al.* 1991). The emission ratio for a gas varies with the types of fuel and fire (Cofer *et al.*, 1997; Cofer *et al.*, 1998).

An accurate assessment of the effect of biomass burning on the production of greenhouse gases requires that the four independent variables be known for all of Earth's ecosystems. Area burned is believed to be the variable that is least accurately known (Levine, this volume). As such, improved estimates of the influence of biomass burning on greenhouse gas production depend primarily on obtaining more accurate area-burned data.

Given the importance of accurately assessing area burned, this paper will review the different methods by which it is reconstructed. Note that this paper will not compare methods of measuring fire severity, a variable that might be considered equivalent to the product of B, α and β, although this would be a valuable project for future research. In this paper I first examine how the different methods of reconstructing area burned are employed. I also note the typical spatial and temporal resolution and extent of the observations made by the different methods. I then compare the scales of observation employed by different methods. This comparison should highlight the scales of observation over which some methods are uniquely suited, and the scales of observation over which methods overlap. An overlap in the scales of observation should provide a window in which the different methods could be calibrated, thus enabling a comparison of area burned over extended spatial and temporal scales. Different methods of reconstructing area burned often employ different methods of measuring area burned and or fire frequency. Thus, to further facilitate the translation of results obtained by different methods of reconstructing area burned, I describe and compare the common methods by which they measure area burned and fire frequency. Finally, I review several studies that compared estimates of area burned determined using different methods of reconstructing fire.

2. METHODS FOR RECONSTRUCTING BURNED AREAS

There are four general methods by which area burned is reconstructed: historical records, satellite imagery, tree-rings, and sedimentary records. Each method contains a variety of data types. Each method will be considered in terms of the manner by which it records or evidences fires, and the typical spatial- and temporal- resolutions and extents of its records.

I refer to the resolution of the data as the smallest unit of time or space over which fires are recorded. In the spatial context this is similar to the idea of the minimum mapping unit (Aronoff, 1993). Similarities or differences between the resolution of data and space will be discussed below. The extent of the spatial and temporal records is the total area or time over which observations were made. In most cases the minimum extent, temporal or spatial, over which a study could be conducted will be of the same size as the resolution. As such, unless stated otherwise, the minimum extent will be considered to be the same size as the resolution.

2.1 Historical records

There are a variety of types of historical records of wildfire activity. They include maps, databases of the characteristics of individual fires, databases of total area burned in different time periods, and narrative accounts.

a) Records may consist of maps of the actual area burned by individual wildfires (e.g. Chou *et al.*, 1993a; Chou *et al.*, 1993b). The burn-scarsespecially of large fires, are usually sketched onto a topographic map while over-flying the area in light aircraft (e.g. Kasischke and French, 1995). In areas where fires are usually small, burn-scars as small as 40 ha may be mapped (e.g. Chou *et al.*, 1993b). In areas where fires are large, these sketches are typically of fires larger than 200 ha (e.g. for Canada, B.J. Stocks pers. comm.) that occurred in the second half of this century. The burn-scars of smaller fires are typically not mapped in the boreal forest because there are so many of them. The spatial resolution of the data is thus usually 200 ha. The spatial extent of a study may be as small as the resolution and as large as Canada.

The data are usually grouped into one-month periods, so the temporal resolution is usually one month. Similarly, the minimum temporal extent is thus usually one month. The maximum temporal extent is typically 50 years since such historical records date back 50 years, although longer records are available for some European countries.

It is important to note that while the spatial resolution of the burn scars is typically 200 ha, the space in which these scars are mapped has an infinite resolution. For example, if the resolution of burn-scars was 200 ha, it is possible that two 200-ha burn scars might be mapped such that they slightly overlap each other. Thus, as space is not broken into unique 200 ha plots which are either burned or not, the spatial resolution is not 200 ha but is infinitely resolvable. As space is infinitely resolvable, it is possible for a study to employ a spatial extent that is smaller than the resolution of the data. However, as few areas have fires so frequent that the burn-scars overlap, it is highly unlikely that this will be done. As such, I consider the minimum spatial extent to be the same size as the minimum resolution.

b) Databases of fires may indicate the characteristics of individual wildfires such as where they started, their eventual size, and weather conditions. These databases are typically available for managed regions such as parks and timber holdings, although they are becoming available for large government districts (e.g. British Columbia Forest Service, 1996). The availability of these databases depends on the organization that created them. Some are published in CD format (e.g. British Columbia Forest Service, 1996), and others are available on request (e.g. from the Chief Park Warden of Wood Buffalo National Park, Canada).

These databases typically record fires as small as 0.01 ha and the date and time to the minute that a fire is first observed and extinguished. Although the location of the fire's ignition is geo-referenced, the complete area burned is not. These databases thus include small fires that would be too time-consuming to map as burn-scars. The databases frequently extend back to the year 1950. The maximum spatial extent may be as large as a Canadian province. The temporal resolution is usually the period over which an individual fire burned. The temporal resolution may be finer for large fires that burn over many months, or for experimental fires whose behaviour is closely monitored.

c) A database may indicate the number of fires and the total area they burned in different time periods (e.g. Canadian Council of Forest Ministers, 1995). These are typically compiled from databases from small organizational units, such as the fire district, in which all fires are listed. Note that these small units typically keep their records for only a matter of years. Larger organisational units usually aggregate the data from all of the smaller organisational units, but do not maintain the information at the scale of the smaller units. The spatial resolution will typically be as small as the smallest organisational unit from which a database can be found. For example, in Canada the smallest area for which a multi-decade database is available would be the smallest province (e.g. Harrington *et al.*, 1983). If country-scale records are employed, the coarsest resolution would be Russia. Note, however, that the Russian database contains many errors (B.J. Stocks, pers. comm.). As most countries report these data, the maximum spatial extent of the records is most of the Earth. The minimum temporal resolution is usually the year, but may be as fine as the month. The maximum temporal extent depends on the region, but may be up to 80 years (e.g. Van Wagner, 1988).

d) Narrative accounts may describe burned areas found by explorers, surveyors and other record keepers (Whitney, 1994). The spatial resolution of the observations is uncertain, but is likely to be of the order of 1 ha. The spatial extent of the studies is typically small, although 26 114 km^2 were surveyed in western New York in the last decade of the 18th century (Seischab and Orwig, 1991). The temporal resolution of this data is typically around 15 years as the evidence that the explorers used for fire occurrence is believed to have a temporal window of this length. Thus, if fire evidence is observed, it is unknown whether the fire occurred 2 or 15 years ago. The maximum temporal extent is related to how many times that area was explored. If the area was only explored once, then the temporal extent would be the same size as the temporal resolution; if it was explored in 1700 and 1750 then the temporal extent would be 50 years.

2.2 Satellite imagery

The sensing of burn-scars using satellite imagery, and the associated problems and limitations, is discussed in detail in this volume and in Levine (1991). The fire signals that can be detected include heat, light, smoke, vegetation change, and post-fire surface albedo (Robinson, 1991). The types of imagery used include Advanced Very High Resolution Radiometer (AVHRR), Geostationary Operational Environmental Satellite System (GOES) and Landsat. I limit my discussion to AVHRR since it is the type most commonly used for mapping fire occurrence, especially over large areas.

AVHRR imagery has been used to detect fires using temperature (e.g. Setzer and Pereira, 1991; Scholes *et al.*, 1996) or using multiple-seasonal changes in the Normalized Difference Vegetation Index (NDVI) (e.g. Kasischke and French, 1995). The AVHRR imagery is available at the Local Area Coverage (LAC) spatial resolution of 1.2 km^2. AVHRR imagery is archived, however, at the Global Area Coverage (GAC) spatial resolution of ca. 4.8 km^2 (the actual resolution depends on a variety of factors). Note that AVHRR can be used to detect active fires as a temperature signal, and burn-scars using NDVI.

The maximum spatial extent of AVHRR-based studies is all of Earth (e.g. Dwyer *et al.*, this volume). As satellite data is geographically referenced so that pixels can be temporally differenced, space is not infinitely resolved as it is with burn-scars mapped from aeroplanes. In this case, therefore, the minimum spatial extent really is the same size as the spatial resolution.

The temporal resolution of the AVHRR imagery is two to four times daily. The minimum temporal extent of the data is six hours; the maximum temporal extent is 21 years, as the first AVHRR sensor was put into orbit in 1978. Note that the actual spatial and temporal resolution and extent of the data will, as a result of problems with clouds, smoke and the uncertainty of burn-scar identification, be less than the ideal (e.g. papers in Levine, 1991).

2.3   Tree-rings

The annual growth-rings in trees may provide two types of evidence that can be used to reconstruct fire occurrence: fire-scars on the tree-trunk, and total tree-age.

a) Fire-scars may be created on the base of tree trunks by ground fires. The ground fire must be hot and persistent enough at a portion of the base of a tree to heat and kill the cambial layer that underlies the bark, but not of such an intensity as to destroy the whole tree (e.g. Gutsell and Johnson, 1996). The hotter portion of the fire typically occurs on the downwind side of the trees. The portion of the cambium that is killed does not produce growth-rings in subsequent years, while the rest of the tree continues to produce growth-rings. Counting and cross-dating of tree-rings allows the calendar year of the fire scar to be determined (e.g. Johnson and Gutsell, 1994). The season of the fire scar can be determined if the fire scar occurs within the annual growth-ring (e.g. Baisan and Swetnam, 1990). As a result, useful fire scars on trees are most common in regions where fires rarely reach the crown, since this would kill the trees, and on trees with bark that is thick enough to insulate most of the tree from the heating action of the fire. Fire scars may form on trees within tropical climates but, as these trees do not generally produce annual growth-rings, they would not provide a useful record of past fire activity.

The spatial resolution of fire-scar studies is the individual tree, the canopy of which may be 10 m^2, but will depend on the size of the tree. The spatial extent of fire-scar studies has been as large as the American Southwest (Swetnam and Betancourt, 1990). The temporal resolution of fire-scar studies may be as fine as the season. The maximum temporal extent of these studies has been 2000 years (e.g. Swetnam, 1993), but depends on the maximum age of a tree species.

b) Tree-ages may provide information on fire activity in areas such as the boreal forest where crown fires are common (Johnson and Gutsell, 1994). As crown fires kill essentially all trees in their paths, the age of post-fire tree establishment can be used to date the last crown fire that burned through an area. Multiple trees are aged to ensure that what is being dated is a cohort of establishment. This method assumes that tree establishment following fire occurrence can be differentiated from that following other disturbances such as wind-throw or insect outbreaks (see Johnson and Gutsell for these methods). Calibration of tree-ages against mapped burn-scars and fire scars on trees has shown that, in the boreal forest of Canada, tree-ages may underestimate the time since the last fire at a site by 9% (Larsen, 1996).

The time since the last fire can be determined for all forest patches in an area (e.g. Johnson and Larsen, 1991) or for a sample of points distributed through the study area (Larsen, 1997). The fire frequency, and temporal variations in it, can then be statistically estimated from the frequency distribution of time-since-last-fire ages (Reed, 1994; Reed *et al.*, 1998).

The spatial resolution of the tree-age studies is the same size as the spatial extent of the study area, unless analyses are conducted over a sub-section of the study area. The spatial extent has ranged in size from 35 km^2 (Agee *et al.*, 1990) to 45,000 km^2 (Larsen, 1997). The temporal resolution of the fire frequency estimates has varied from the decade (Larsen, 1996) to 500 years (Engelmark *et al.*, 1994). The temporal extent of these studies ranges from 100 to 700 years, decreasing in length in proportion to the fire cycle (Larsen, 1997; Engelmark *et al.*, 1994).

2.4   Sedimentary records

Evidence of past fires may accumulate in sedimentary basins such as lakes and soils. For example, charcoal records of fire have been detected in lakes (e.g. Clark, 1990), soils (e.g. Carcaillet, 1998), peat sediments (e.g. Kuhry, 1994), and glaciers (e.g. Taylor *et al.*, 1996).

a) Horizons of large charcoal particles in lake-sediments may be evidence of local fires, as suggested by correspondence with nearby fire-scar records (e.g. Clark, 1990; Millspaugh and Whitlock, 1995). The abundance of small charcoal particles can be related to the amount of fire in the region, as suggested by correlations with mapped records of large regional fires (e.g. MacDonald *et al.*, 1991). Evidence of fires can also be obtained from decadal-scale variations in pollen records that reflect local forest succession (e.g. Larsen and MacDonald, 1998). The creation of charcoal and pollen records from multiple depths in the lake sediments may thus provide a record of multiple fire events.

The spatial resolution of lake-sediment studies depends on factors such as the size of the lake basin and wind speed, as these influence the source-area of the pollen and charcoal (Clark and Royall, 1995; Sugita *et al.*, 1997). Smaller lakes have a smaller source area. The spatial resolution may be as small as 1 km^2 or as large as several 1000 km^2. The spatial extent will cover the same area as the resolution, unless a series of lakes is studied (e.g. Clark and Royall, 1996). The temporal resolution depends on the precision with which the study is undertaken, and the degree of mixing of the sediment. It may be as fine as the year, or as coarse as a millennium. The temporal extent may be as short as 100 years, and as long as the sedimentary record, some of which extend hundreds of thousands of years.

b) Radiocarbon (^{14}C) dating of large pieces of charcoal in soil sediments is used to provide a chronology of fire-events (e.g. Lavoie and Payette, 1996; Carcaillet, 1998). The spatial resolution of soil sediment studies is the size of the soil pit (ca. 1 m^2), assuming minimal transport of charcoal by overland flow. The spatial extent of the studies is limited to areas that have not been ploughed. The temporal resolution is limited by radiocarbon dating to several hundred years. The temporal extent may be tens of thousands of years.

c) Layers of large charcoal particles in stratigraphic sections of peat deposits may provide a record of fires that burned through the surface vegetation (e.g. Kuhry, 1994). The spatial and temporal resolutions and extents for these data are similar to those of soil sediments.

d) The abundance of charcoal particles and other combustion products of biomass burning in glacial ice may provide an index of the area burned within large regions over which the air- masses travel prior to reaching the glacier on which they are deposited (e.g. Taylor *et al.*, 1996). The spatial resolution and extent is likely to be sub-continental, the temporal resolution can be the year, and the temporal extent could be hundreds of thousands of years.

3.   A COMPARISON OF THE SCALES OF OBSERVATION

The foregoing review enables a comparison of the spatial and temporal resolutions and extents over which the different methods are applicable (Figs. 1 and 2). The spatial resolution and extent of area-burned reconstruction covers 14 orders of magnitude, and the temporal resolution and extent covers 6 orders of magnitude.

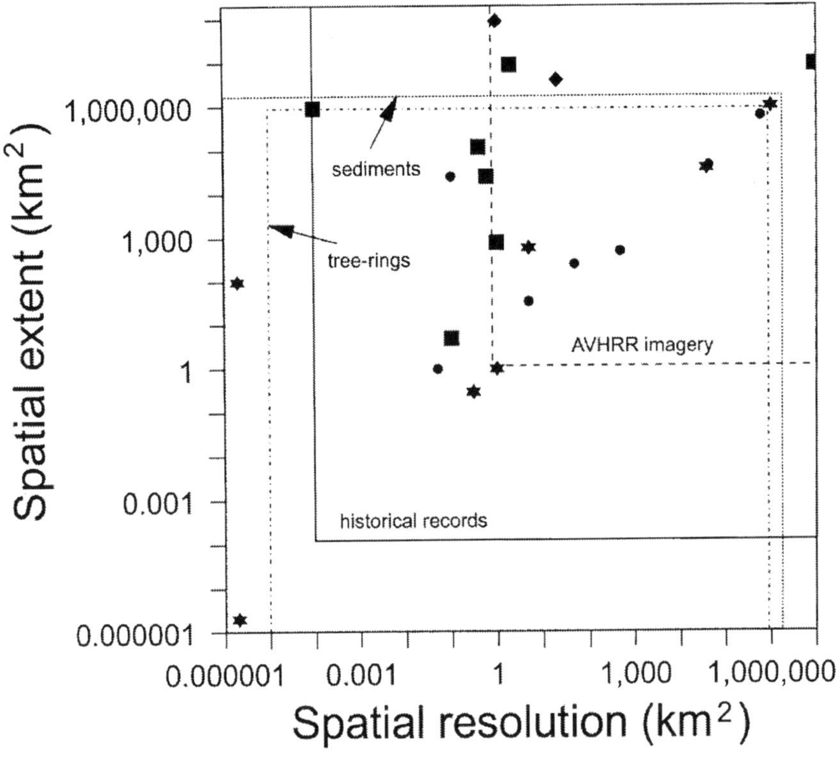

Figure 1. The spatial resolution and extent of the observations made by different fire reconstruction methods. Lines mark the spatial range over which a study might be conducted. Symbols mark the spatial resolution and extent of studies mentioned in the text: historical records (squares), AVHRR imagery (diamonds), sediments (stars) and tree-rings (circles).

Some methods are particularly useful over certain scales of observation. For example, tree-ring and sediment studies are uniquely applicable over small areas and long time periods, while historical records and AVHRR images are more useful over large areas and are only applicable over shorter time periods. This uniqueness allows different methods to address different questions. For example, sediment methods allow the rate of biomass burning to be reconstructed for times prior to the Industrial Revolution, while AVHRR images allow the whole globe to be studied, a task that is impossible using tree-rings.

There are significant ranges of space and time over which all four methods are applicable: between the spatial resolution and extent of 1 to 1,000,000 km^2, and between the temporal resolution and extent of 1 to 20 years. This zone of overlap offers the potential for a comparison of the measures of area burned obtained by the different methods. A calibration of the different methods would then enable the measures made by one method at one scale to be translated to a different scale. However, before such comparisons can be made, some understanding must be reached regarding the different measures of area burned that are employed by the different methods of reconstruction.

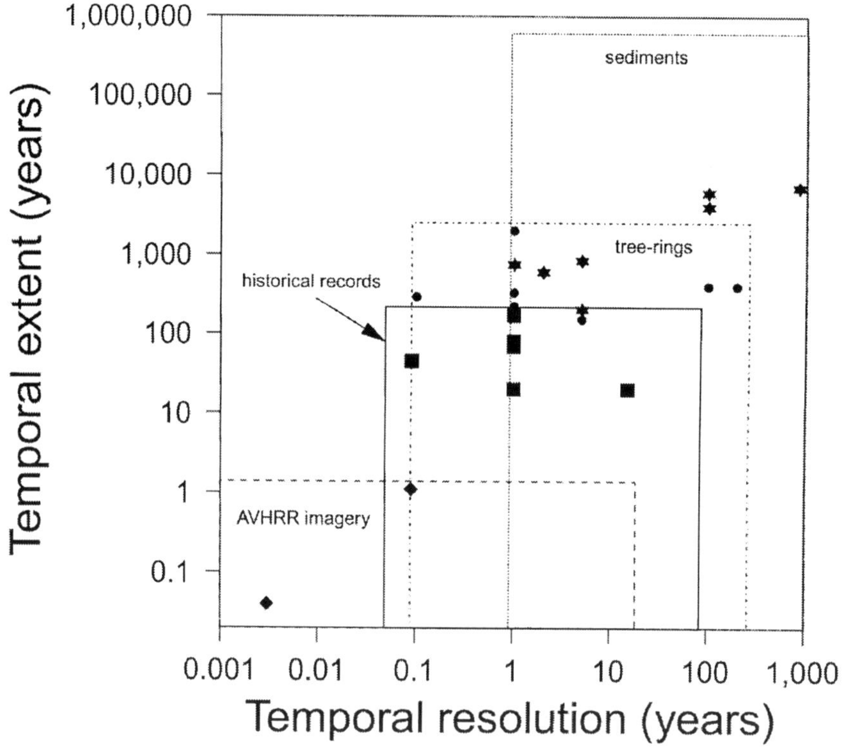

Figure 2. The temporal resolution and extent of the observations made by the different fire reconstruction methods. Lines mark the spatial range over which a study might be conducted. Symbols mark the temporal resolution and extent of studies mentioned in the text: historical records (squares), AVHRR imagery (diamonds), sediments (stars) and tree-rings (circles).

4. METHODS OF MEASURING AREA BURNED

The different methods of reconstructing area burned are, in addition to being employed over different spatial and temporal scales, measured using different methods. I first describe and then compare the commonly used measures. This should enhance the ability to translate the measures of area burned used in different types of studies.

4.1 Burned / not burned

The most basic method of describing area burned is the binary measure of burned or not burned. This method is implicit in the use of satellite imagery that can only detect whether a pixel did or did not burn. For example, if fire is detected in an AVHRR GAC pixel, it means that some, but not necessarily all, of the ca. 4.8 km^2 pixel-area has burned (Eva and Lambin, 1998). As a result, if fires are commonly detected within spatially isolated pixels, the area of the pixel should not be measured as the area burned, but the pixel should simply be characterized as burned or not. The foregoing assumes that the inference of fire from satellite data is correct although, of course, it is possible that the inference itself is not correct (e.g. Kasischke and French, 1995).

The number of pixels that are inferred to be burned is sometimes converted to an area burned through the use of a calibration equation (e.g. Setzer and Pereira, 1991; Pereira and Setzer, 1996; Scholes *et al.*, 1996). The robustness of the calibration equations is questionable, however, because the size of individual fires varies seasonally and with environmental factors such as climate, topography and vegetation. As a result, it has been suggested that the use of a single calibration equation would result in the area burned in arid regions being underestimated, and the area burned in moist regions being overestimated (Scholes *et al.*, 1996).

An alternative approach that might be useful is the use of the binary measure of fire occurrence to model the probability of fire occurrence. For example, such data have been used in a binomial logistic regression, along with environmental characteristics such as weather, topography and distance to human infrastructure, to predict the probability of fire occurrence in different spatial locations (Chou *et al.*, 1993a) and time periods (Vega Garcia *et al.*, 1995). This method, however, again requires the development of region-specific models.

                                           Chris Larsen

4.2   Area burned

The total area of vegetation burned in a defined time period can be calculated using historical records. It can also be determined using satellite imagery if multiple, adjacent pixels are detected as having burned (e.g. Cahoon *et al.*, 1991; Kasischke and French, 1995).

Tree-ring based records, both fire scars and tree-ages, can also be used to reconstruct area burned (e.g. Heinselman, 1973). The remnant patches of older forest can be speculatively patched together to infer the original area of a fire (e.g. Heinselman, 1973). Alternatively, the average area burned per year over multiple decades can be estimated by applying survival analysis to time-since-last-fire data (e.g. Reed *et al.*, 1998). These methods are fairly accurate for recent fires. The accuracy decreases for older fires because more recent fires will have burned over portions of the patches and fire-scarred trees that were created by the older fires. The absolute accuracy of these methods is not known because the over-burning process that creates the remnant patches of old forest takes place over hundreds of years.

Sedimentary records indicate when a fire occurred at or near the location of the sediment. The large temporal resolution of data with a small spatial resolution such as provided by soil studies, and the large spatial resolution of data with a small temporal resolution such as provided by annually laminated lake sediments, has led to there being no reconstruction of area-burned from sedimentary records.

4.3   Fire rotation

The fire rotation, introduced by Heinselman (1973), is equal to the spatial extent of the study area divided by the mean annual proportion of the study area that burned. This measure has been commonly used by foresters and ecologists as a means of expressing how long it would take to burn a large proportion of the study area (e.g. references in Johnson and Gutsell (1994)). The area-burned data could be obtained using historical records, satellite data or tree-ring records. For example, if satellite data suggested that, within one year, 100 km^2 of a 1000 km^2 area had burned, the fire rotation would be: 1000 km^2 / 100 km^2 yr^{-1} = 10 years. Assuming that the rate of burning would remain the same in subsequent years, this suggests that in ten years, an area equal in size to the whole 1000 km^2 study area would burn.

Further, assume that the satellite study was extended to cover 10 years of data, that 100 km^2 of area was burned every year, and that only 700 km^2 of the landscape had been burned during that time period. (The latter observation would be true because some portions of the land will have been burned two or three times.) If the fire rotation over the 10–year period was calculated using the proportion of the landscape that had been burned, the fire rotation would be underestimated as: 1000 km^2 / 700 km^2 * 10 yr^{-1} = 1000 years / 70 = 14.3 years. It is thus important to calculate the natural fire rotation using the mean annual proportion of the study area that burned, and not the proportion of the landscape that burned at least once over the temporal extent of the study. Indeed, even an annual resolution would underestimate the fire rotation if fires over-burn sections of the study area within a single year.

Although the fire rotation can be calculated using one year of data, as the annual area burned can vary markedly due to inter-annual variations in factors such as climate, the measure will become more meaningful as more years of data are employed.

5.   FIRE CYCLE

The fire cycle is the number of years required to burn an area equal in size to the study area (Van Wagner, 1978). It is equivalent to the fire rotation but, while the fire rotation is typically calculated using historical records, the fire cycle is calculated using tree-age data. The different terms are thus used to indicate that they have been calculated using different types of data. The frequency distribution of time-since-last-fire points or areas, reconstructed using tree-ages, can be analysed using statistical methods adopted from survival analysis (e.g. Reed, 1994; Reed *et al.*, 1998). These methods enable the fire cycle in different time periods and spatial locations to be estimated and tested for significant differences. The value of using tree-age data is that it enables the fire cycle to be calculated for time periods hundreds of years into the past.

6.   MEAN FIRE INTERVAL

A fire interval is the time between two consecutive fires. For example, if a tree contained fire scars in the years 1800 and 1810, the fire interval was 10 years. The mean fire interval (MFI) is equal to the sum of individual intervals divided by the number of intervals. For example, assume that a tree contained fire scars in the years 1800, 1810, 1860 and 1900. The intervals are 10, 50 and 40 years, for a sum of 100 years, and there are 3 intervals; the mean fire interval is thus 33.3 years. More accurate estimates of the MFI are obtained through the use of parametric methods that are more appropriate to the data (e.g. Clark, 1989; Johnson and Gutsell, 1994).

MFI decreases as the area from which the fire evidence is collected increases (Arno and Peterson, 1983). For example, assume that there are two trees: tree one has fire scars in the years 1800, 1810, 1860 and 1900, and tree two has fire scars in the years 1800, 1820, 1840 and 1900. Each tree has an MFI of 33.3 years. If we calculate the MFI using a composite of the fire-scar years (that is, 1800, 1810, 1820, 1840, 1860 and 1900; cf. Arno and Sneck, 1977), then the MFI is 20 years.

As the MFI is spatially dependent, inter-comparisons of the MFI are only valid when they are estimated over similarly sized areas. This is best achieved by restricting the area to individual trees (Clark, 1990; Johnson and Gutsell, 1994). In addition, rather than creating a composite of fire-years, the fire-intervals are pooled. For example, the pool of fire-intervals from the trees described above are, from tree one, 10, 50 and 40 years and, from tree two, 20, 20 and 60 years. The mean of these six intervals is still 33.3 years but, as N is now larger, the standard error of the parameter estimate will be lower.

MFIs have been most commonly calculated using fire scars on trees. They have also been calculated from sedimentary records of local fires (e.g. Clark, 1990; Carcaillet, 1998; Larsen and MacDonald, 1998). However, as soil records of fire can be eroded away and as peat records of fire can be burned away, they both tend to produce lengthened estimates of the MFI. In contrast, lake-sediment data should provide shortened MFI estimates as they may record, for example, two medium-sized local fires that did not burn the same portion of the local landscape.

As the calculation of MFIs requires an interval between at least two fires at one point, historical records are not useful unless fires are very frequent. The MFI has thus not been calculated using historical records. Care should be taken when applying it to satellite pixels unless it is not known whether a complete pixel did or did not burn.

6.1 Fire frequency

The term fire frequency is commonly used as an ordinal measure of how common fire is in an area; for example, fire frequency has increased in the tropics during the 20th century. More specifically, fire frequency is the probability of a fire occurring at a spot (Johnson and Gutsell, 1994). The fire frequency is thus the inverse of the fire rotation, mean fire interval or fire cycle. For example, if the fire rotation was 100 years, then the fire frequency was 0.01. If the fire rotation was 10 years, then the fire frequency was 0.1. Thus, as the fire rotation decreases, the fire frequency increases. While the equivalence of these measures may make them seem redundant, it is important to recognize that since different authors use different terms, an understanding of them all is required to translate their results.

7. A COMPARISON OF THE METHODS OF MEASUREMENT

Whether an area burned or did not burn is the fundamental manner of measuring fire occurrence. Historical records elaborate on this measure by providing the area over which the fires did burn. Additional knowledge of the area over which fires did not burn then allows the fire rotation to be calculated. Similarly, whether a fire did or did not burn is also at the root of fire scar and sedimentary evidence of fire occurrence. The MFI, although not defined this way earlier, can be calculated as the ratio of the years during which fires did not burn to the number of years during which fires did burn.

The fire rotation, MFI and fire cycle, although developed for different data types, are all dimensioned in years, and should provide similar quantitative measures. For example, if the fire cycle is 100 years then, as it takes 100 years to burn an area equal in size to the study area, the MFI reconstructed from lakes within the same area should also approximate 100 years. For a statistical comparison of the methods, see Johnson and Gutsell (1994).

8. DISCUSSION

The goals of this paper were to: review the different methods by which area burned is reconstructed; compare them in terms of the spatial and temporal resolutions and extents of their observations; and describe and compare the common methods by which area burned and fire frequency are measured. While some methods of reconstructing area burned are uniquely applicable over certain scales of observation, all of the methods overlap between the spatial resolution and extent of 1 to 1,000,000 km^2, and between the temporal resolution and extent of 1 to 20 years. The different methods of measuring area burned are inter-convertible.

It thus appears that while different researchers tend to focus on their preferred methods, it should be possible, within those spatial and temporal scales, to combine all of the different methods to provide a synoptic view of area burned. There have, however, been few comparisons between the different methods. The likely reason for this is that researchers have been refining the unique methods required to reconstruct the area burned in different areas of the world. For example, the tree-ring based approaches that have been available for close to 100 years have little to no application in the tropics. There has thus been much focus on the development of remote sensing methods that are applicable to large, remote areas such as the tropics.

The comparisons between different methods that have been done indicate that the different methods provide somewhat similar estimates of area burned. Differences, however, indicate that a comparison of results obtained using different methods will require some method of calibration. For example, in northern Alberta, Larsen (1997) found a 63–year fire cycle using a 130–year record of tree-ages, and a 90–year fire rotation using 42 years of fire-map records. In the white spruce (*Picea glauca*) forests of northern Alberta, Larsen and MacDonald (1998) found a 113–year fire-cycle estimate using a 250–year long record of white spruce tree-ages, and an 89–year MFI using an 840–year long pollen and charcoal record from lake-sediments. Clark (1990) found strong similarities between the MFI estimates obtained using tree-ring fire scars and charcoal layers in lake-sediments in 4 different time periods. Kasischke and French (1996) found that AVHRR detected 78% of the total area burned in Alaska over two summers as suggested by Alaska Fire Service records and fire-scar maps created using light aircraft. They suggest that AVHRR might not under-represent area burned as much as it appears, because the maps created using light-aircraft often fail to distinguish remnant unburned areas within large burn-scars (as also observed in northern Alberta by Larsen).

It appears that the different methods of reconstructing and measuring area burned may be inter-convertible. However, if these methods are to be employed for the estimation of gas emissions, it would valuable to determine how well the different methods of reconstructing fire are able to differentiate fire severity. For example, the fires recorded by fire scars on trees are typically low-intensity ground-fires while the fires recorded by tree-ages are typically high-intensity crown-fires. Ground and crown fires will cause different proportions of the above-ground biomass to be combusted, leading to different productions of greenhouse gases. Thus, while the ability to translate the measures of area burned provided by the different methods of reconstruction is valuable, it is insufficient to obtain good estimates of greenhouse gas production unless accompanied by information regarding fire severity.

9.   ACKNOWLEDGEMENTS

Research support was provided through the National Science Foundation Award SBR 88–10917 to the National Center for Geographic Information and Analysis. I thank the two anonymous reviewers for their comments on this paper.

10. REFERENCES

Agee, J.K., Finney, M. and de Gouvenain, R. 1990. Forest fire history of Desolation Peak, Washington. Canadian Journal of Forest Research 20: 350–356.

Arno, S.F. and Peterson, T.D. 1983. Variations in the estimates of fire intervals: a closer look at fire history on the Bitterroot National Forest. USDA Forest Service Research Paper INT–301.

Arno, S.F. and Sneck, K.M. 1977. A method for determining fire history in coniferous forests of the Mountain West. USDA Forest Service General Technical Report INT–42.

Aronoff, S. 1993. Geographic Information Systems. WDL Publications, Ottawa.

Baisan, C.H. and Swetnam, T.W. 1990. Fire history on a desert mountain range: Rincon Mountain Wilderness, Arizona, USA. Canadian Journal of Forest Research 20: 1559–1569.

Bond, W.J. and van Wilgen, B.W. 1996. Fire and Plants. Chapman & Hall, London.

British Columbia Forest Service. 1996. Historical fire CD. Protection Branch. Victoria.

Cahoon, E.R., Levine, J.S., Cofer, W.R. III, Miller, J.E., Minnis, P., Tennille, G.M., Yip, T.W., Stocks, G.J., and Heck, P.W. 1991. The great Chinese fire of 1987: a view from space. In Global Biomass Burning: Atmospheric, Climatic , and Biospheric Implications, J.S. Levine, editor, MIT Press, Cambridge MA, p. 61–66.

Canadian Council of Forest Ministers. 1995. Compendium of Canadian forestry statistics: 1994 summary. National Resources Canada. Ottawa.

Carcaillet, C. 1998. A spatially precise study of Holocene fire history, climate and human impact within the Maurienne valley, North French Alps. Journal of Ecology 86: 384–396.

Chou, Y.H., Minnich, R.A., and Chase, R.A. 1993a. Mapping probability of fire occurrence in San Jacinto Mountains, California, USA. Environmental Management 17: 129–140.

Chou, Y.H., Minnich, R.A., and Dezzani, R.J. 1993b. Do fire sizes differ between Southern California and Baja California? Forest Science 39: 835–844.

Clark, J.S. 1989. Ecological disturbance as a renewal process: theory and application to fire history. Oikos 56: 17–30.

Clark, J.S. 1990. Fire and climate change during the last 750 yr in northwestern Minnesota. Ecological Monographs 60: 135–159.

Clark, J.S. and Royall, P.D. 1995. Particle-size evidence for source areas of charcoal accumulation in late Holocene sediments of Eastern North American lakes. Quaternary Research 43: 80–89.

Clark, J.S. and Royall, P.D. 1996. Local and regional sediment charcoal evidence for fire regimes in presettlement north-eastern North America. Journal of Ecology 84: 365–382.

Cofer, W.R. III, Koutzenogii, K.P., Kokorin, A, and Ezcurra, A. 1997. Biomass burning emissions and the atmosphere. In Sediment Records of Biomass Burning and Global Change, J.S. Clark, H. Cachier, J.G. Goldammer and B. Stocks, editors, Springer-Verlag, New York, p. 189–206.

Cofer, W.R. III, Winstead, E.L., Stocks, B.J., Goldhammer, J.G. and Cahoon, D.R. 1998. Crown fire emissions of CO_2, CO, H_2, CH_4, and TNMHC from a dense jack pine boreal forest fire. Geophysical Research Letters, 25: 3919–3922.

Crutzen, P.J., Heidt, L.E., Krasnec, J.P., Pollock W.H., and Seiler, W. 1979. Biomass burning as a source of atmospheric gasses CO, H_2, N_2O, NO, CH_3Cl,, and COS. Nature 282: 253–56.

Engelmark, O, Kullman, L., and Bergeron, Y. 1994. Fire and age structure of Scots pine and Norway spruce in north Sweden during the past 700 years. New Phytologist 126: 163–8.

Eva, H. and Lambin, E.F. 1998. Remote sensing of biomass burning in tropical regions: sampling issues and multisensor approach. Remote Sensing of Environment 64: 292–315.

Gutsell, S.L. and Johnson, E.A. 1996. How fire scars are formed: coupling disturbance process to its ecological effect. Canadian Journal of Forest Research 26: 166–174.

Harrington, J.B., Flannigan, M.D., and Van Wagner, C.E. 1983. A study of the relations of components of the Fire Weather Index to monthly provincial area burned by wildfire in Canada, 1953–1980. Environment Canada, Canadian Forestry Service Report PI–X–25.

Heinselman, M.L. 1973. Fire in the virgin forests of the Boundary Waters Canoe Area, Minnesota. Quaternary Research 3: 329–382.

Johnson, E.A. and Gutsell, S.L. 1994. Fire frequency models, methods and interpretations. Advances in Ecological Research 25: 239–287.

Johnson, E.A. and Larsen, C.P.S. 1991. Climatically induced change in fire frequency in the southern Canadian Rockies. Ecology 72: 194–201.

Kasischke, E.S. and French, N.H.F. 1995. Locating and estimating the areal extent of wildfires in Alaskan boreal forests using multiple-season AVHRR NDVI composite data. Remote Sensing of Environment 51: 263–275.

Kuhry, P. 1994. The role of fire in the development of *Sphagnum*-dominated peatlands in western boreal Canada. Journal of Ecology 82: 899–910.

Larsen, C.P.S. 1996. Fire and climate dynamics in the boreal forest of northern Alberta between AD 1850 and 1989. The Holocene 6: 449–456.

Larsen, C.P.S. 1997. Spatial and temporal variations in boreal forest fire frequency in northern Alberta. Journal of Biogeography 24: 663–673.

Larsen, C.P.S. and MacDonald, G.M. 1998. An 840–year record of fire and vegetation in a boreal white spruce forest. Ecology 79: 106–118.

Lavoie, C. and Payette, S. 1996. The long-term stability of the boreal forest limit in subarctic Quebec. Ecology 77: 1226–1233.

Levine, J.S. (editor). 1991. Global biomass burning: atmospheric, climatic, and biospheric implications. MIT Press, Cambridge MA.

Levine, J.S., Cofer, W.R. III, Winstead, E.L., Rhinehart, R.P., Cahoon, D.R., Sebacher, D.I., Sebacher, S. and Stocks, B.J. 1991. Biomass burning: combustion emissions, satellite imagery, and biogenic emissions. In: Global biomass burning: atmospheric, climatic, and biospheric implications, J.S. Levine, editor, MIT Press, Cambridge MA, p. 264–271.

MacDonald, G.M., Larsen, C.P.S., Szeicz, J.M., and Moser, K.A. 1991. The reconstruction of boreal forest fire history from lake sediments: a comparison of charcoal, pollen, sedimentological and geochemical indices. Quaternary Science Reviews 10: 53–71.

Millspaugh, S.H., and Whitlock, C. 1995. A 750–year fire history based on lake sediment records in central Yellowstone National Park, USA. The Holocene 5: 283–292.

Pereira, M.C. and Setzer, A.W. 1996. Comparison of fire detection in savannas using AVHRR's channel 3 and TM images. International Journal of Remote Sensing 17: 1925–1937.

Pyne, S.J. 1995. World Fire: the Culture of Fire on Earth. Henry Holt and Co., New York.

Reed, W.J. 1994. Estimating the historical probability of stand-replacement fire using the age-class distribution of undisturbed forest. Forest Science 40: 104–119.

Reed, W.J., Larsen, C.P.S., Johnson, E.A. and MacDonald, G.M.. 1998. Estimation of temporal variations in historical fire frequency from dendrochronological time-since-fire map data. Forest Science 44: 465–475.

Robinson, J.M. 1991. Problems in global fire evaluation: is remote sensing the solution? In: Global biomass burning: atmospheric, climatic, and biospheric implications, J.S. Levine, editor, MIT Press, Cambridge MA, p. 67–73.

Scholes, R.J., Kendall, J., and Justice, C.O. 1996. The quantity of biomass burned in Southern Africa. Journal of Geophysical Research 101(D19): 23,667–23,276.

Seiler, W. and Crutzen, P.J. 1980. Estimates of gross and net fluxes of carbon between the biosphere and the atmosphere from biomass burning. Climatic Change 2: 207–247.

Seischab, F.K. and Orwig, D. 1991. Catastrophic disturbance in the presettlement forests of Western New York. Bulletin of the Torrey Botanical Club 118: 117–122.

Setzer, A. and Pereira, M. 1991. Amazonia biomass burnings in 1987 and an estimate of their tropospheric emissions. Ambio 20: 19–22.

Sugita, S., MacDonald, G.M., and Larsen, C.P.S. 1997. Reconstruction of fire disturbance and forest succession from fossil pollen in lake sediments: potentials and limitations. In Sediment Records of Biomass Burning and Global Change, J.S. Clark, H. Cachier, J.G. Goldammer and B. Stocks, editors, Springer-Verlag, New York, p. 387–412.

Swetnam, T.W. 1993. Fire history and climatic change in giant sequoia groves. Science 262: 885–889.

Swetnam, T.W. and Betancourt, J.L. 1990. Fire–Southern Oscillation relations in the southwestern United States. Science 249: 1017–1020.

Taylor, K.C., Mayewski, P.A., Twickler, M.S. and Whitlow, S.I. 1996. Biomass burning recorded in the GISP2 ice core: a record from eastern Canada? The Holocene 6: 1–6.

Van Wagner, C.E. 1978. Age-class distribution and the forest fire-cycle. Canadian Journal of Forest Research 8: 220–227.

Van Wagner, C.E. 1988. The historical pattern of annual area burned in Canada. The Forestry Chronicle 64: 182–185.

Vega Garcia, C., Woodard, P.M., Titus, S.J., Adamowicz, W.L., and Lee, B.S. 1995. A logit model for predicting the daily occurrence of human caused forest fires. International Journal of Wildland Fire 5: 101–111.

Whitney, G.G. 1994. From Coastal Wilderness to Fruited Plain. Cambridge University Press, Cambridge.

Interactions Between Biomass Burning and Climate: Conclusions Drawn from the Workshop

JOHN L. INNES[1], MARTIN BENISTON[2] and MICHEL VERSTRAETE[3]
[1]*University of British Columbia, Vancouver, Canada*
[2]*University of Fribourg, Switzerland*
[3]*Joint Research Center, CEC, Ispra, Italy*

Abstract: An international workshop with the title: "Biomass burning and its inter-relationships with the climate system" was held in Wengen, Switzerland, from 28 September to 2 October 1998. The workshop was attended by some 50 scientists from 12 countries. It was co-sponsored by the University of Fribourg, ENAMORS (the European Network for the development of Advanced Models to interpret Optical Remote Sensing data), the Swiss National Science Foundation and the Swiss Federal Institute for Forest, Snow and Landscape Research.

A wealth of new material was presented on the inter-relationships between climate and biomass burning. Forest fires were the subject of particular attention as a result of their world-wide prominence during the past 12 months. The period 1997–98 was characterised by a strong El Niño event, resulting in much drier conditions than normal in areas such as Indonesia and Brazil. This, in turn, resulted in many of the small-scale fires lit to clear land for agricultural activities getting out of control and spreading to primary forest areas. Devastating forest fires also occurred elsewhere, and large areas of forest were burned in 1997–98 in areas as disparate as Canada, Russia, Mexico and Greece.

While there has been enormous progress in our understanding of how biomass burning affects climate, there are still many shortcomings. Even with the available technology and resources, it is still difficult to assess exactly the area of the Earth's surface burned in the last 12 months and how much of carbon dioxide and other radiatively-active greenhouse gases were released into the atmosphere. There is little doubt in the scientific community that satellite-based remote sensing systems provides the key to the global monitoring of biomass burning. A significant amount of research has been directed towards developing new algorithms for fire detection and monitoring, which in turn, has improved estimates of the global extent of biomass burning considerably. It is expected that these capabilities will continue to improve as new satellite systems are deployed over the next several years, especially new geostationary meteorological satellites and polar-orbiting platforms such as EOS–AM (NASA), ENVISAT (ESA) and ADEOS–II (NASDA).

1.   SCIENTIFIC HIGHLIGHTS OF THE MEETING

1.1   New information and progress

The 1997–98 Indonesian fires released much more carbon into the atmosphere than the oil-well fires in Kuwait following the Gulf War. The role of biomass burning as a source of precursors for photo-oxidant formation is now recognized. It is particularly important in the Tropics, and ozone derived from precursors emitted during burning has been shown to reach peak hourly concentrations of 120–150 ppb in west Africa. The significance of these phytotoxic concentrations for tropical forests requires investigation.

The detection of active fires on the basis of their thermal emissions is relatively well understood and mastered, even though the monitoring of these events with current space technologies faces definite difficulties: the polar-orbiting platforms which provide the highest and most appropriate spatial resolutions are often not observing the areas most likely to burn at the right time. On the other hand, geostationary satellites capable of monitoring the Earth typically every half an hour do so at much lower spatial resolution and can only spot the largest fire events. Significant progress has recently been achieved in the monitoring of active fires from space, and products are being made available operationally on the Internet by various agencies and institutions.

The assessment of the extent of burned areas, however, remains the most elusive challenge, inter alia because (1) only a fraction of the above-ground vegetation burns in uncontrolled fires, (2) the spectral response of partially burned areas is not well known, and (3) the vegetation tends to regrow quickly, at least in some ecosystems such as grasslands. The next generation of advanced space sensors will help in this endeavour, provided substantial research and development efforts are also undertaken to define, test, and implement high performance algorithms to better extract the desired information from the measured signals.

These emerging capabilities will offer new opportunities to develop, improve and implement real-time or near-real-time applications such as fire early warning systems, and support to fire damage control activities. The Centre de Suivi Ecologique in Dakar, Senegal has already experimented such a prototype system and is capable of warning local forest managers of the presence of fires in their areas of responsibility. Similar systems elsewhere reveal the value of the approach.

The estimation of the amount and type of smoke and particulate emissions resulting from fire activities, and the impact of these emissions on the regional climate, environment and health of the populations downwind from the burning regions will constitute the next challenge. Addressing these issues will require interdisciplinary models capable of integrating data streams from a multiplicity of sources, at various scales and resolutions."

1.2 Future occurrence of fires

Some model predictions suggest that the frequency and intensity of fires may increase as a consequence of climate change in some areas in the future, particularly in the boreal region. This is important as about 37% of the total terrestrial global carbon pool (plant biomass and soil carbon) is contained within the boreal zone. Increases in fire frequency may be enhanced by reductions in local fire-fighting capacities (because of budget restrictions) and increased fuelwood (because of reduced prescribed burning as a result of air quality legislation). Changes in the frequency–intensity relationships of fires have considerable implications for carbon budgets, as the amount of carbon released is dependent on the nature of the fires. The changes may also bring about changes in forest species composition. Changes in the occurrence of fires in some parts of the world are believed to have already taken place as a result of climatic changes over the last 100 years.

1.3 Reliable estimates of GHG and aerosol emissions for the IPCC

Airborne measurements in smoke from biomass burning in Brazil (obtained as part of the SCAR–B Project) show that the compositions and properties of smoke particles change rapidly as smoke ages (due to condensation of gases and particle coagulation). Consequently, emission factor measurements made close to fires may not be appropriate in assessing regional and global effects of smoke on atmospheric chemical composition and the earth's radiation balance.

Although the globally-averaged direct radiative forcing due to smoke from biomass burning is probably small (-0.3 watts per square meter (cooling)), local and regional effects can be large.

1.4 Policy implications of fire and biomass burning

Biomass burning is an important factor contributing to reduced visibility along the east coast of the USA, and has significant implications for legislation protecting visibility. Biomass burning is also likely to be important when calculating national carbon budgets. Because of increased fire activity, the strength of the North American boreal forest as a carbon sink has been reduced on the order 0.03 Gt per year over the past two decades. This has been caused by a steady increase since 1970 in the area of Canadian forests burned annually.

2.   SEVERAL AREAS WERE IDENTIFIED AS DESERVING FURTHER ATTENTION:

Mechanisms which encourage interdisciplinarity should be encouraged. This would include the organisation of further workshops along the lines of the Wengen model, with a specific goal of embracing the socio-economic research community. There is also a need to foster mechanisms for direct and regular interactions between scientists and decision makers from the local community to the government levels.

Greater exchange of scientists should occur, particularly from the countries most affected by biomass burning. This should include encouraging more scientists from less-developed countries to attend specialized workshops. There is a need to develop a harmonised terminology. This would help the exchange of information between different cultures. As in most scientific areas, capacity in developing countries should be increased, and existing mechanisms, such as the START (Global Change System for Analysis, Research and Training) programme should be strengthened.

Human resources need to keep up with remote-sensing technology. Currently, many interpretation problems are more the result of the lack of suitably qualified personnel than due to shortcomings in the available technology. Research results need to made available as quickly as possible on the Internet, and should also be made accessible to developing countries.

There is a need to link fire studies to the assessment of the sustainability of ecosystems. Fire is a highly disturbing factor for many ecosystems, with burning breaking the existing equilibrium between climate/soil conditions/landuse/land cover. Examples include dry dense forests (e.g. primary dry forests on the West coast of Madagascar), seasonal semi-deciduous forests (e.g. the northern and southern edges of the Congo and Amazon Basins), coastal forests (such as mangroves), some coniferous forests (e.g. those in southern China) and others. There is therefore a real need to link the studies of fire distribution/occurrence with the sustainability of ecosystems in which fires are occurring. Within given ecosystems, burning will act as a maintenance factor; within others, it will act as a disturbing one with strong changes of the surface characteristics.

3. SEVERAL RESEARCH PRIORITIES WERE ALSO IDENTIFIED

1. Priorities for better or more observations:

A new generation of Earth Observing instruments is being launched by the major national and international space agencies. These new sensors will provide the scientific community with instruments offering much enhanced performances. However, significant R&D support will be needed in the short term to ensure that appropriate algorithms and methods of data interpretation are designed, implemented and tested, so that much better and more reliable products and services can be delivered to the user community in a timely manner. Specifically, the spectral and directional signatures of burned areas should be fully documented in a number of ecosystems with high priority, and made widely available to support the effective development of advanced optimized methods to analyze remote sensing data for the purpose of monitoring the extent and properties of these areas. Priorities include:

1.1 Reliable and accurate measurement of the extent of burned areas, as well as of their radiative and structural properties

1.2 Estimation of the amount and state of above-ground biomass and fuel load, but also of the quantity of peat and other underground stocks of carbon

1.3 Assessment of the burning efficiency factors and emission rates at appropriate scales and resolutions, including further emphasis on the retrieval of atmospheric chemistry data from biomass burning

1.4 Development of complete carbon accounting and its anthropogenic implications in order to quantify the terrestrial sinks?

1.5 Better quantification of several sources of greenhouse gases associated with biomass burning (e.g., agricultural residues, land-fills).

Long series of data, which are also accessible to scientists and others, are invaluable. Where such series exist, they need to be safe-guarded. In many areas, new investigations need to be started and mechanisms introduced to ensure their continuity. Such series are important as (1) they provide the effective definition of a reference condition against which any subsequent changes can be assessed, (2) they provide documentation of the natural and human-induced variability, both seasonally and from year to year, over a significant period of time, thereby providing background information on the natural or expected changes on these time scales, and (3) they answer the need for continuing monitoring during and after environmental or resource management policies have been approved and implemented, thereby enabling the evaluation of their effectiveness.

2. Priorities for better or more accurate models:

Establishment, evaluation and operational exploitation of comprehensive dynamic models of vegetation capable of predicting the state and evolution of plant canopies, the amount of fuel, the risk of fire and the rates of gas and particulate emissions (when burning), as a function of soils and nutrients, climate, vegetation type, past fire history, etc.

3. Priorities for validation and testing of the results:

Identification and strengthening of a global network of long-term field stations where comprehensive measurements are being made, and establishment of an agreed upon list of key variables to monitor

4. Other critical areas of research:

4.1   Addressing scaling issues to help match local field observations with areal-averaged measurements from space, through appropriate models

4.2   Delivery of fire and biomass burning information in near-real time to the global research community, but also to managers and policy makers

4.3   Assessment of feedback of changing climate on natural forests/ecosystems/biodiversity (e.g., how would a changed climate influence the frequency / intensity / areal extent of biomass burning)

4.4   Quantification of the loss of species in sensitive ecosystem zones due to biomass burning (relevant to biodiversity) ?

4.5   Implications of biomass burning for soil quality and watersheds etc. ?

4.6   Relationships of deforestation, reforestation and afforestation activities to poverty, employment, resettlement, agriculture etc. ?

Abbreviations and Acronyms

AATSR	Advanced along track scanning radiometer
ACT	Australian Capital Territory
ADEOS	(NASDA)
AFWA	Air Force Weather Agency (USA)
AFWIN	Air Force Weather Information Network (AFWA)
AMIP	Atmospheric Model Intercomparison Program
AMS	American Meteorological Society
ATSR	Along-track scanning radiometer (ERS)
AVHRR	Advanced Very High Resolution Radiometer (NOAA)
CART	Classification and regression tree
CCM	Community climate model (NCAR)
CIRA	Cooperative Institute for Research on the Atmosphere (CSU)
CIRES	Cooperative Institute for Research in the Environmental Sciences
CNES	Centre National d'Études Spatiales
CNRS	Centre National de la Recherche Scientifique
CPT	Canadian Polar Trough
CSU	Colorado State University
CTM	Chemical transport model
DIS	Data and Information System (IGBP)
DMSP	Defense meteorological satellite program
DU	Dobson units
ECMWF	European Center for Medium-range Weather Forecasts
ENAMORS	European network for the development of advanced models to interpret optical remote sensing data
EOS	Earth Observing System (NASA)
EPA	Environmental Protection Agency (USA)
EPM	Emission production model

ERA	ECMWF re-analysis
ERBE	Earth radiation budget experiment
ERS	European remote sensing satellite
ESA	European Space Agency
FARSITE	Fire Area Simulator
FDF	Florida Division of Forestry
FERA	Fire and Environmental Research Applications Team (USDA Forest Service)
FOFEM	First order fire effects model
FSL	Forecast Systems Laboratory (NCEP)
GAC	Global area coverage (AVHRR)
GCM	Global climate model
GEBA	Global energy balance archive
GEIA	Global emission inventory analysis (IGAC)
GEMI	Global Environmental Monitoring Index
GFP	Global fire product
GOES	Geostationary operational environmental satellite
GW	GigaWatt
HRPT	High resolution picture transmission format (AVHRR)
HIRETYCS	High resolution ten years climate simulation
IFOV	Instantaneous-field-of-view
IGAC	International Global Atmospheric Chemistry programme
IGBP	International Geosphere–Biosphere Programme
IIASA	International Institute for Applied Systems Analysis
IWAQM	Interagency Working Group on Air Quality Modelling
JERS	Japanese earth resources satellite
LAC	Local area coverage (AVHRR)
Landsat	Land satellite
MAS	ER–2 MODIS airborne simulator (NASA)
MFI	Mean fire interval
MM5	Mesoscale model version 5 (NCEP)
MODIS	Moderate resolution imaging spectrometer
MOZART	Model for ozone and related chemical tracers
MSS	Multispectral scanner system (Landsat)
Mtoe	Million tonnes of oil equivalent
MTF	Modulation transfer function

NAAQS	National ambient air quality standards
NASA	National Aeronautics and Space Administration (USA)
NASDA	National Space Development Agency of Japan
NBP	Net biome production
NCAR	National Center for Atmospheric Research (USA)
NCEP	National Center for Environmental Protection (NOAA–NWS)
NDVI	Normalized difference vegetation index
NEP	Net ecosystem production
NMHC	Non-methane hydrocarbons
NOAA	National Oceanic and Atmospheric Administration
NPP	Net primary production
NWS	National Weather Service (NOAA)
OAQPS	Office of Air Quality Planning and Standards (EPA)
OLS	Operational linescan system (DMSP)
P	Precipitation
PBL	Planetary boundary layer
PDT	Pacific daylight time
PE	Potential evapotranspiration
$PM_{2.5}$	Particulate matter (diameter $\leq$2.5 microns)
PM_{10}	Particulate matter (diameter $\leq$10 microns)
PNA	Pacific-North American
P-PE	Annual effective precipitation
PRF	Point response function
PSF	Point spread function
RAMS	(CSU)
RCM	Regional climate model
RCZ	Regional charcoal zone
SAGE	Stratospheric aerosol and gas experiment
SAR	Synthetic aperture radar (ERS/JERS)
SASEM	Simple approach smoke estimation model
SAVI	Soil adjusted vegetation index
SCAR–C	Smoke, Clouds And Radiation programme
SPOT	Systeme pour l'observation de la terre
SRB	Surface radiation budget project
SST	Sea surface temperature

START	Global Change System for Analysis, Research and Training
S/TIPS	State/Tribal implementation plans
T_b	Bio-temperature
THFM	Thousand hour time lag fuel moisture
TM	Thematic mapper (Landsat)
TOA	Top-of-atmosphere
TOMS	Total ozone mapping spectrometer
UNEP	United Nations Environment Programme
USDA	United States (of America) Department of Agriculture
USDA/FS	United States (of America) Department of Agriculture Forest Service
USDOI	United States (of America) Department of Interior
USDOI/FWS	United States (of America) Department of Interior Fish and Wildlife Service
USDOI	United States (of America) Department of Interior National Park Service
USGS	United States (of America) Geological Survey
UW	University of Washington
WHO	World Health Organisation
WMO	World Meteorological Organisation

Index

Advances in Global Change Research

1. P. Martens and J. Rotmans (eds.): *Climate Change: An Integrated Perspective.* 1999
ISBN 0-7923-5996-8
2. A. Gillespie and W.C.G. Burns (eds.): *Climate Change in the South Pacific: Impacts and Responses in Australia, New Zealand, and Small Island States.* 2000
ISBN 0-7923-6077-X
3. J.L. Innes, M. Beniston and M.M. Verstraete (eds.): *Biomass Burning and Its Inter-Relationships with the Climate Systems.* 2000   ISBN 0-7923-6107-5

KLUWER ACADEMIC PUBLISHERS – DORDRECHT / BOSTON / LONDON

DATE DUE

DEMCO, INC. 38-2931